AUTOMATIC TRANSMISSION SERVICE

A TEXT/WORKBOOK

William L. Husselbee

RESTON PUBLISHING COMPANY, INC.

A Prentice-Hall Comapny

Reston, Virginia

Library of Congress Cataloging in Publication Data

Husselbee, William L.
   Automatic transmission service.

   Includes index.
   1. Automobiles--Transmission devices, Automatic--
Maintenance and repair. I. Title.
TL263.H88   629.2'446   81-1985
ISBN 0-8359-0266-8   AACR2

© 1981 by
RESTON PUBLISHING COMPANY, INC.
A Prentice-Hall Company
Reston, Virginia 22090

All rights reserved. No part of this
book may be reproduced in any way,
or by any means, without permission
in writing from the publisher.

10 9 8 7 6 5 4 3 2 1

Printed in the United States of America

AUTOMATIC TRANSMISSION SERVICE

CONTENTS

PREFACE   ix

SECTION 1--SHOP EQUIPMENT AND TOOLS   1
    Torque Converter Flushers, 1
        Function, 1; Design, 2; Operating Characteristics, 4
    Converter End-Play Gauges, 5
        Universal Gauge Design, 6
    Torque Converter Drain Plug Installing Sets, 7
    Automatic Transmission Dynamometers, 9
        Various Problems Uncovered by the Dynamometer, 9; Design, 10
    Portable Diagnostic Test Equipment, 17
        Hydraulic Gauge Sets, 17; Vacuum Pump and Gauge Assembly, 18
    Transmission Cleaning Equipment, 20
        Steam Cleaners, 21; Jet Cleaners, 22; Parts Washers, 23; Abrasive
        Blasters, 26
    Clutch Spring Compressors, 29
    Bushing Equipment, 35
    Lifting Equipment, 38
        Hydraulic Lifts, 38; Hydraulic Jacks, 40; Safety Stands, 42;
        Transmission Jacks, 44
    Transmission Work Benches, 45
    Transmission Fixtures or Stands, 46
    Thread Repair Tool Sets, 48
    Power Tools, 49
    Pullers, 53
        Gear Type Pullers, 53; Seal Pullers, 56; Pump Pullers, 56

Snap Ring Pliers, 57
Other Hand Tools and Special Equipment, 58
Check-Up Questions, 60

SECTION 2—MEASURING DEVICES AND FASTENERS 63

Measuring Tools, 63
   Torque Wrenches, 63; Micrometers, 68; Feeler Gauges, 71; Dial Indicators, 73
Fastening Devices, 75
   Machine Screws, 75; Bolts, 78; Nuts, 79; Washers and Cotter Pins, 80; Drive Keys, Balls, and Pins, 81; Splines, 82; Snap Rings, 82
Check-Up Questions, 84

SECTION 3—AUTOMATIC TRANSMISSION PROBLEM DIAGNOSIS 86

Preliminary Fluid Level and Condition Checks, 87
   Fluid Level, 87; Fluid Condition, 90
Fluid Leakage Checks, 91
   Typical Checking Procedure, 93; Converter Housing Visual-Leak Inspections, 95; Black-Light Tests, 97; Cooler Leakage Tests, 98
Other Preliminary Checks, 98
   Idle-Speed Checks, 98; Linkage Checks, 99; Vacuum System Checks, 100; Electrical Transmission Circuits, 103
Operational Tests, 103
   Stall Tests, 103; Road Tests, 106; Pressure Tests, 110
Diagnosis Guides, 116
Additional Diagnostic Tests, 118
   Air Checks, 118; Noise Detection, 121; Hoist Test, 121; Dynamometer Tests, 122
Check-Up Questions, 131

SECTION 4—CHANGING THE TRANSMISSION FLUID AND FILTER 133

Draining the Fluid from the Transmission, 134
Inspecting and Servicing the Pan, 136
Filter or Screen Service, 139
Pan Installation, 141
Draining the Converter, 143
Pumping the Fluid from a Converter, 144
Filling the Transmission with New Fluid, 147
Check-Up Questions, 148

SECTION 5—BAND AND LINKAGE ADJUSTMENTS 150

Band Adjustments, 150
   Effects of Improper Band Adjustments, 150; Frequency of Band Adjustments, 151; Types of Band Adjustments, 152; Common Band Adjustment Service Operations, 154; Equipment and Tools Necessary To Adjust Bands, 154; Band Adjustment Specifications, 154; Typical Internal Band Adjustment Procedures, 156; Typical External Band Adjustment Procedures—C-4 Transmissions, 158
Linkage Adjustments, 161
   Gearshift (Manual Valve) Linkage, 161; Accelerator, Throttle, and Kickdown Valve Linkages, 164
Check-Up Questions, 169

CONTENTS   vii

SECTION 6—TRANSMISSION AND SEAL REMOVAL AND INSTALLATION   171
    Transmission Removal, 172
        General Instructions, 172; Procedures, 172
    Front Seal Replacement, 181
    Extension Housing Seal Replacement, 184
    Transmission Installation, 185
    Check-Up Questions, 193

SECTION 7—TORQUE CONVERTER AND HYDRAULIC PUMP INSPECTION, TESTING, AND   195
    SERVICE
    Converter Inspection and Service, 196
        Serviceability Checks, 196; Converter Flushing, 205; Other Converter
        Service, 208
    Hydraulic Pump Inspection and Service, 212
        Teardown and Inspection, 212; Service and Reassembly, 217
    Check-Up Questions, 220

SECTION 8—SUBASSEMBLY CLEANING, INSPECTION, AND SERVICE   222
    Transmission Hard Parts, 223
        Transmission Case, 223; Extension Housing, 230; Drums, 233; Brake
        Drum Cleaning and Inspection, 238; One-Way (Overrunning) Clutches, 238;
        Servos and Accumulators, 241; Governors, 243; Planetary Gear Trains,
        245; Transmission Shafts, 247; Thrust Washers, 248; Bearings, 250;
        Bushings, 251; Valve Bodies, 254
    Soft Parts, 257
        Clutch Plates, 257; Bands, 258; Metal and Teflon Sealing Rings, 259;
        Rubber Sealing Rings, 260; Metal-Clad Seals, 261; Gaskets, 262
    Transmission Cooling System Inspection and Service, 262
        Inspection, 263; Service, 265
    Check-Up Questions, 269

SECTION 9—TRANSMISSION OVERHAUL   272
    General Instructions, 272
    Disassembly, 273
        Transmission End-Play Checks, 274; Removal of Case and Extension
        Housing Components, 276
    Subassembly Overhaul, 288
        Case, 288; Case Manual Lever Seal, 289; Intermediate Servo, 291; Low
        Reverse Servo, 293; Front Pump, 293; Reverse-High Clutch, 297; Forward
        Clutch, 301; Forward Planetary Gear Train, 305; Reverse Planetary Gear
        Train, 309; Low Reverse Brake Drum, 310; Output Shaft and Distributor
        Sleeve, 312; Governor Assembly, 312; One-Way Clutch, 315; Low Reverse
        Band, 316; Extension Housing, 316; Valve Body, 318
    Transmission Reassembly, 322
        General Instructions, 322; Assembly Procedure, 323
    Check-Up Questions, 339

APPENDIX   342

INDEX   347

PREFACE

Automatic Transmission Service: A Text/Workbook is basically a shop/lab manual. Its function is to provide the reader with the "how to" information necessary to troubleshoot, service, inspect, and repair the modern automatic transmission.

Automatic Transmission Service: A Text/Workbook covers every aspect of shop training necessary to enter the field of automatic transmission repair. The manual has nine well-thought-out and complete sections covering shop equipment and tools; measuring devices and fasteners; automatic transmission problem diagnosis; changing the transmission fluid and filter; band linkage adjustments; transmission and seal removal and installation; torque converter and hydraulic pump inspection, testing, and service; subassembly cleaning, inspection, and service; and transmission overhaul.

Each section also contains various types of training aids. For example, each section contains many illustrations and photos which assist the reader in understanding the material or procedures contained in the section. Important equipment necessary to repair automatic transmissions. The remaining sections provide step-by-step procedures for troubleshooting, servicing, inspecting, and repairing the automatic transmission.

Each section also contains various types of training aids. For example, each section contains many illustrations and photos which assist the reader in understanding the material or procedures contained in the section. Important steps or precautions are emphasized. Finally, at the end of each section are check-up questions designed to test the reader's understanding of the subject matter found in the section, with the answers to the questions provided in the Appendix.

Because this is a lab or shop training manual, it contains very little information on how an automatic transmission functions. However, when this manual is used in conjunction with the texbook <u>Automatic Transmission Fundamentals</u>, the two-volume package forms a valuable training aid and reference for the prospective mechanic or for any vocationally oriented program in automatic transmissions. <u>Automatic Transmission Fundamentals</u> acts as the classroom text and reference book, covering automatic transmission construction and theory of operation, and <u>Automatic Transmission Service: A Text/Workbook</u> deals with the "hands-on" portion of the training. It is a well-known fact that a good balance between theory and hands-on training is necessary to train competent automatic transmission technicians.

The material contained in this manual represents an overview of the many types of tools and equipment along with the diagnosis, service, inspection, and repair procedures commonly used by the industry. It does not cover the overhaul procedures for every automatic transmission on the market because this information would fill many volumes. Instead, this manual presents those concepts that apply to every unit or that can be applied to the majority of transmissions the mechanic will encounter in the field today.

William L. Husselbee

AUTOMATIC TRANSMISSION SERVICE

SECTION

1

Shop Equipment and Tools

REFERENCES: Manufacturers' instructions for the operation of each piece of equipment.
Automatic Transmission Fundamentals, Chapters 2, 3, 4, 5, and 13.

To properly maintain and repair an automatic transmission, a shop must have certain equipment and service tools. Although some of this equipment and tooling may have some use in other areas of automotive repair, the majority of the items, discussed in this section, you will find only in an automatic transmission repair shop. It is not the intent of this section to cover all of the equipment and special tools necessary to rebuild every kind of automatic transmission, but it will present an overview of those tools commonly used to repair the majority of units.

TORQUE CONVERTER FLUSHERS

Function

In any automatic transmission repair facility, a torque converter flushing machine (Fig. 1-1) is a valuable piece of equipment. This machine's function is to recycle used converters and eliminate costly comebacks. To accomplish these tasks, the machine provides a special flushing process necessary to thoroughly clean a sealed converter.

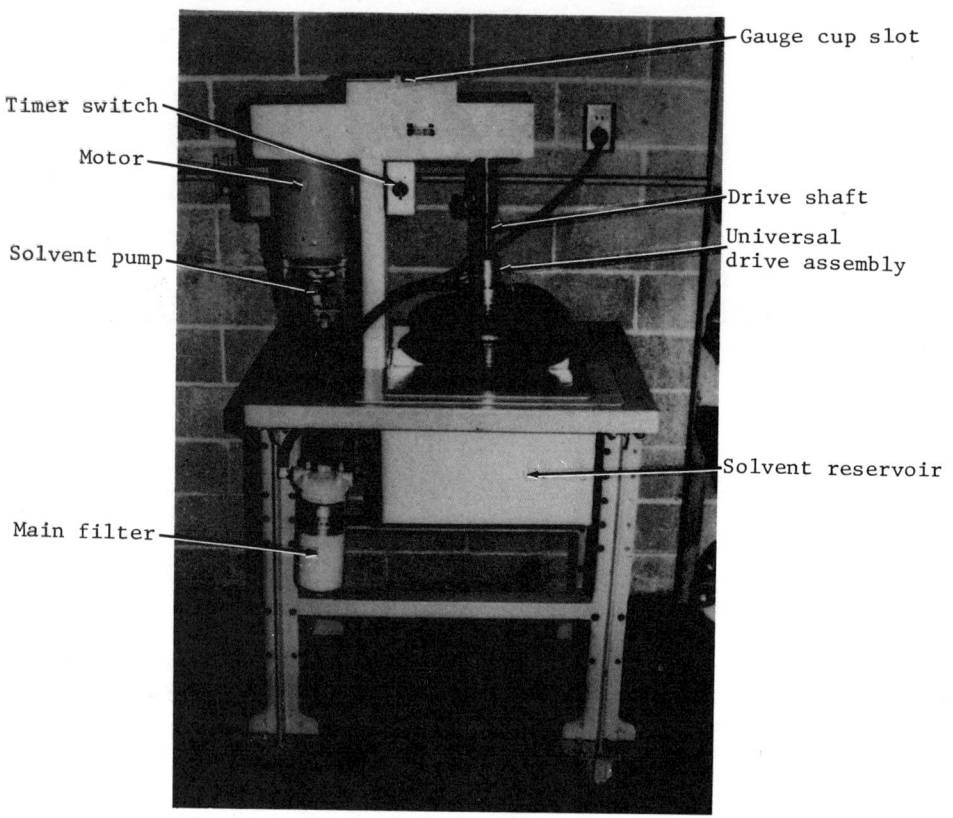

FIG. 1-1   Typical torque converter flushing machine.

The flushing process does two things. First, it removes any accumulated wear particles of gears, bushings, bands, and clutch plates held in suspension within the converter. These particles would otherwise recirculate back through a newly rebuilt transmission and damage the unit. Second, the special solvent, used in the flushing process, tends to dissolve and carry away any built-up varnish from within the torque converter. Varnish, remember, is a by-product of oxidized fluid, and if allowed to remain in the converter, it will contaminate the new fluid.

Design

The flushing machine shown in Fig. 1-1 has an electric motor and timer, universal drive assembly, solvent reservoir, two filters, and a solvent pump. The electric motor turns an adjustable drive shaft that engages into the universal drive assembly and also drives the solvent pump. The timer switch is not only the on-off control for the electric motor and its driven components, but it permits the operator to time the flushing operation.

Shop Equipment and Tools

In other words, the technician can set the timer for up to 30 minutes of cleaning, and at the end of this time period, the timer will shut the machine off.

The universal drive assembly (Fig. 1-2) has two functions. First of all, its drive tip engages into the splines of the turbine, and with the drive assembly inserted into the drive shaft of the machine, the electric motor can spin the turbine inside the converter. This spinning action causes the turbine to produce a vortex flow in the cleaning solvent, contained inside the converter.

Second, the universal drive assembly has two quick-disconnect fittings which connect hoses to the assembly. A hose from the solvent pump attaches to the upper fitting; and when the machine is on, solvent moves into the fitting, through the inlet chamber and drive shaft, and finally passes into the area under the turbine. The lower fitting connects a hose to the drive assembly from the machine's sump. When the machine is in operation, solvent passes into the lower chamber from the impeller side of the converter, moves out of the fitting, and returns via the hose to the solvent sump.

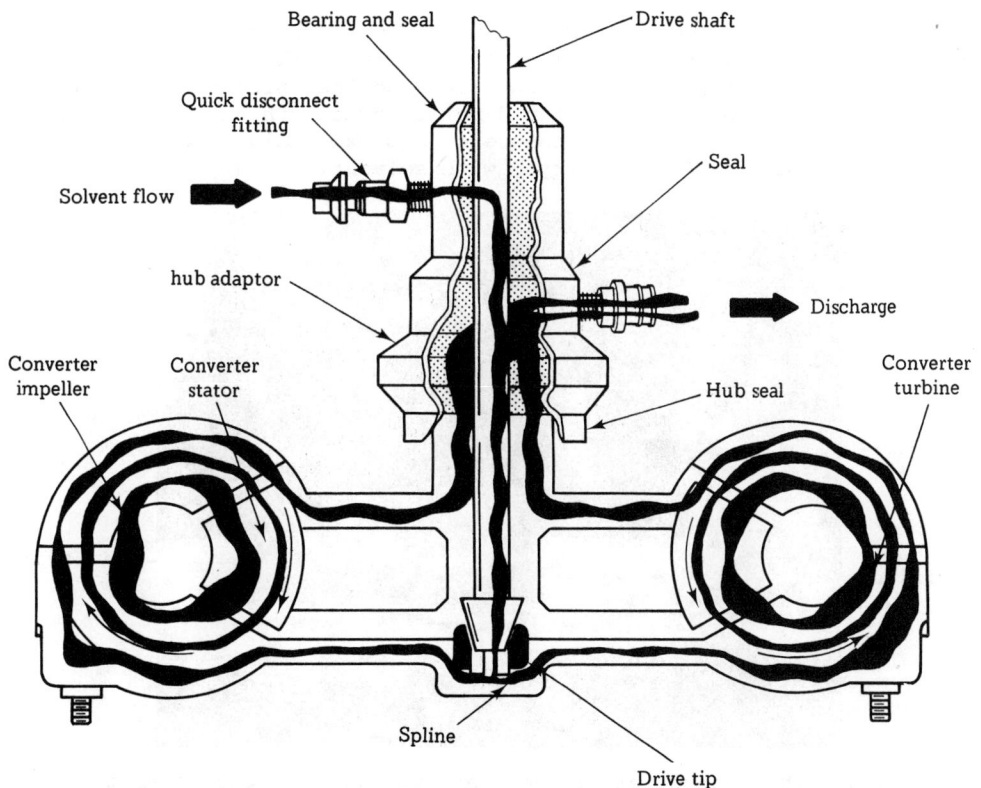

FIG. 1-2 Universal drive assembly of the flushing machine.

The sump is the solvent reservoir for the machine. It contains a quantity of a cleaning agent such as mineral spirits, Varsol, or safety solvent. Also, a line attached to the sump carries the cleaning agent to the pump whenever the operator activates the timer switch.

The machine also has two filters (Fig. 1-3). A magnet holds the first filter to the bottom of the solvent tank where it screens out any large particles from the cleaning agent. The location of the second filter is in the line between the tank and the pump. This filter removes and suspends any smaller particles so that the cleaning agent will not carry them into the pump and eventually back into the converter.

Finally, the solvent pump has the responsibility of supplying the converter with a pressurized supply of cleaning agent. When the electric motor is on, it drives the pump. The pump, in turn, moves the fluid first from the solvent tank, then through the filters, and finally into the converter via the universal drive assembly.

Operating Characteristics

This type of flushing machine utilizes the vortex method to circulate the agent required to clean the inside of the converter. With the machine in operation, the pump supplies cleaning fluid to the inlet chamber, via the quick-disconnect fitting. The fluid moves through the hollowed drive shaft and under the turbine at a pressure of approximately 30 psi (Fig. 1-4). This action keeps the converter charged with solvent.

FIG. 1-3   This converter flusher has two filters, one located at the bottom of the reservoir, and the other situated between the reservoir and the pump.

Shop Equipment and Tools

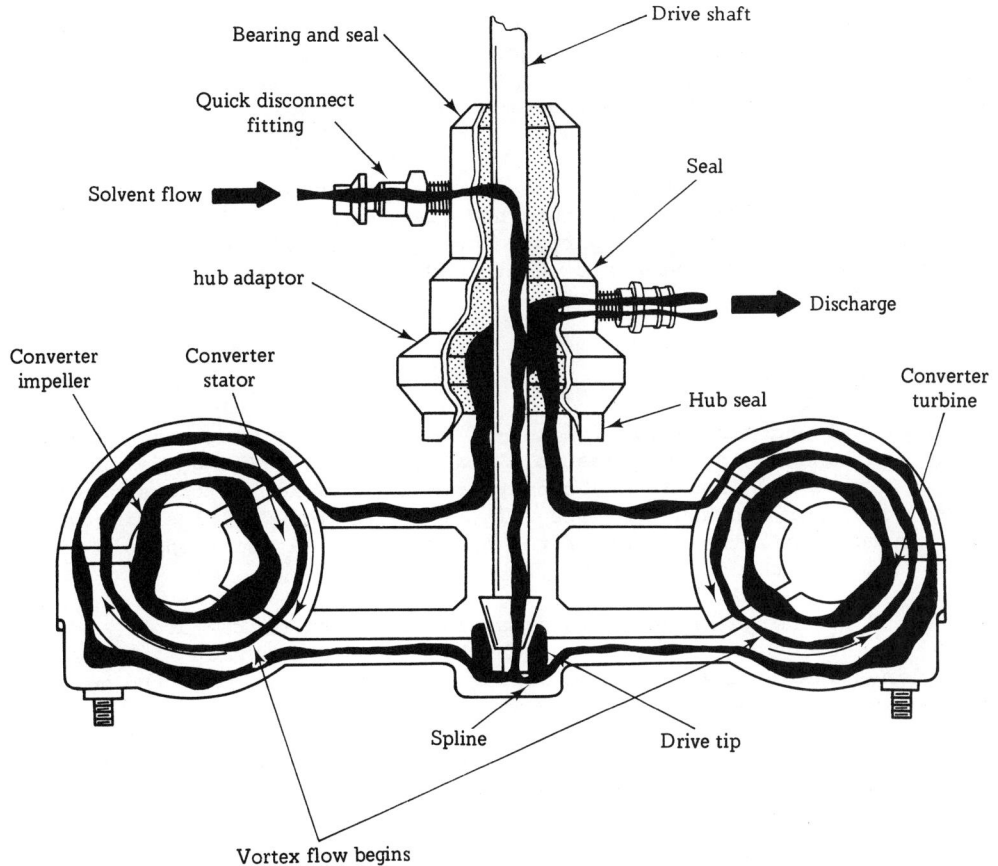

FIG. 1-4 Flusher operation.

At the same time, the drive tip has engaged the turbine splines, and it is turning the turbine at about 250 rpms. The turbine, when spinning, is a large centrifugal pump due to its physical structure. As a result, the spinning turbine creates a vortex flow or flow due to centrifugal force.

This vortex flow begins at the inside center of the turbine and terminates at the curved edge where the cleaning agent moves outward from all the vaned passages at a tremendous velocity. This vortex flow then moves through the impeller and stator, and returns to the turbine. This process continues over and over as long as the timer is on. As a result of this flushing action, the cleaning agent lifts up the particles and varnish and carries them out of the discharge port and into the sump and filters.

CONVERTER END-PLAY GAUGES

Because all modern automotive torque converters are sealed units, it is impossible for the mechanic to visually inspect their internal components for wear. Consequently, special equipment is necessary. Figure 1-5 shows two pieces of such equipment used to check for internal wear in a sealed converter. The device on the left is a factory device and can test only one converter type. The piece of equipment on the right, however, is a universal unit that can check the majority of automotive converters.

Both of these devices check the end-play of the turbine. Now, since bearings or thrust washers control this end-play, any excessive movement indicates abnormal wear of these parts. Consequently, if the end play is beyond specifications, the mechanic should not reuse the converter without first having it rebuilt.

Universal Gauge Design

The universal end-play gauge (Fig. 1-6) consists of a gauge cup, gauge shaft, tip and handle, and an adjustable dial indicator. The gauge cup slides into a slot machined into a converter mount bracket on the side of the torque converter flushing machine (Fig. 1-1) mentioned earlier in this section. In addition, the center of the gauge cup has a machined hole that acts as a guide for the gauge shaft.

On one end of the guide shaft is a tapered tip, and on the other end is a handle. The design of this tip is such that it makes contact with the turbine splines inside the sealed converter. The handle permits the mechanic to easily move the guide shaft up or down within the gauge cup.

The dial indicator is an adjustable, precision instrument that actually measures the amount of turbine end play in thousandths of an inch. The instrument is adjustable in that the mechanic can move it to any given

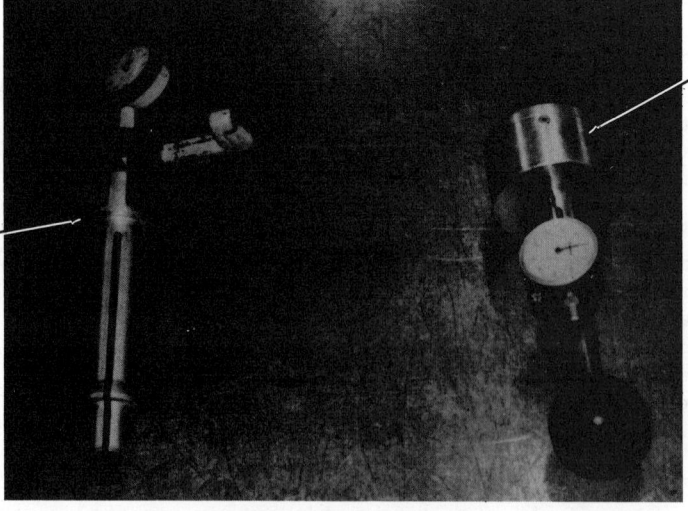

FIG. 1-5   Several typical torque converter end-play gauges.

Shop Equipment and Tools

position on the guide shaft and then lock it in place. This device also has a dial face that the technician can rotate, to zero the gauge, by simply loosening a thumbscrew.

The dial indicator itself has a needle and a dial with a scale that reads from 0 to .100 inch. An indicator stem that senses turbine end-play contacts the bottom of the gauge cup and moves the needle in proportion to the amount of end-play. With this arrangement, the mechanic can easily determine the total end-play by the amount of needle deflection across the scale.

TORQUE CONVERTER DRAIN PLUG INSTALLING SETS

Many sealed torque converters come from the factory without a drain plug; this makes servicing of the converter much more difficult. For instance, when it is time to change the fluid in the transmission and converter, the mechanic cannot drain the dirty fluid from the converter. Instead, the technician will have to pump the fluid out, which is a time consuming process. In addition, when a mechanic wants to clean the converter and check its end-play, he will have a hard time draining the old fluid out of a converter that has no drain plug. To offset these service problems, special equipment is now available to drill and tap converters for installation of a drain plug.

Figure 1-7 shows a drill, guide, and tap set used for the purpose of installing drain plugs in General Motors torque converters. This set consists of a mount bracket and attaching hardware, drill guide, and tap guide, in addition to a drill and tap. The mount bracket has two slotted openings, one at each end of the unit, which permit the mechanic to attach the bracket to different-size converters using the hardware provided with the set. The bracket also has a leg or extension welded at an angle to the main bracket.

This extension serves several functions. First, it houses a locking receptacle for the drill or tap guides. Second, it is the portion of the

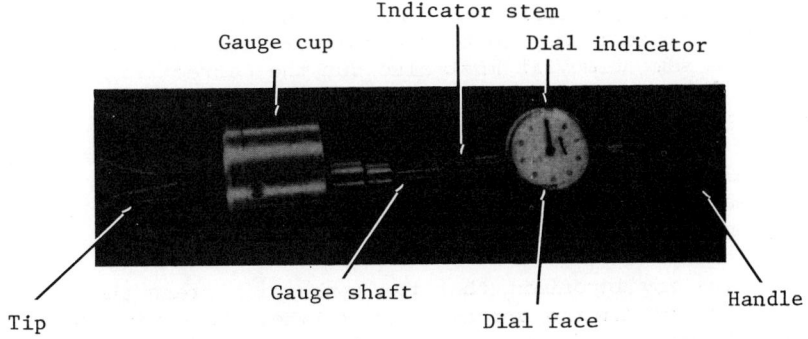

FIG. 1-6   Design of the universal end-play gauge.

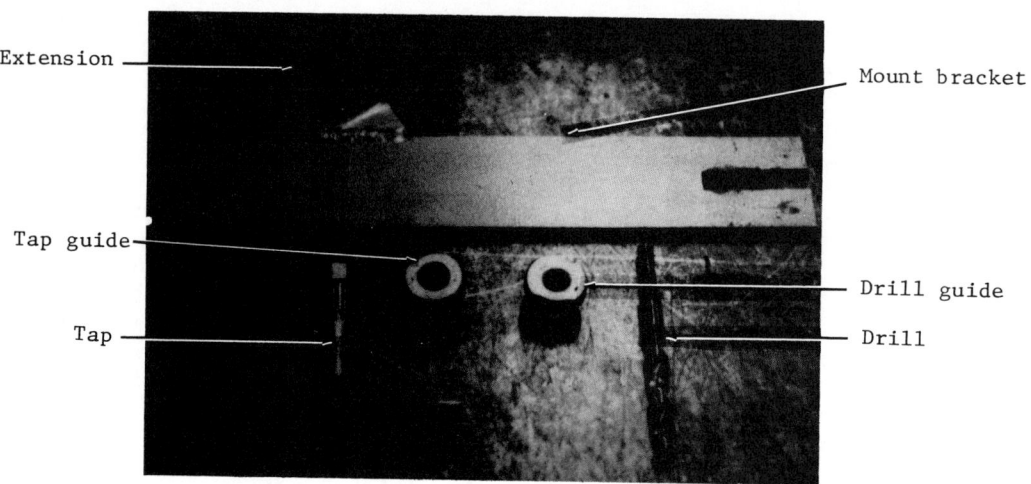

FIG. 1-7 Drill and guide set used to install drain plugs in torque converters.

bracket that actually determines the general area of the converter where the mechanic will drill and tap to install the plug.

The drill and tap guides serve about the same function but for different tools. For example, the drill guide, when inserted into the receptacle, acts as an alignment tool for the drill so that it will maintain the proper drilling angle and not wobble during the drilling process. As a result, the mechanic can machine a nearly perfectly round hole in the converter at the correct angle and to the proper size.

On the other hand, the tap guide, when installed into the receptacle, serves as a guide for the tap during the threading process. In other words, this guide keeps the tap in alignment so that the threads are cut into the converter housing at the correct angle. The guide also prevents the tap from wobbling, which can cause the tap to excessively enlarge the hole.

The drill, of course, machines the hole in the converter housing. This drill is special in that it has to be the correct size to properly fit into the guide. At the same time, it must also cut the correct size hole for the tap.

The tap used with this set is 1/8-inch pipe tap. This device cuts a tapered, internal pipe thread into the converter housing. This thread design is necessary to prevent leakage around the threads after the mechanic installs the plug.

Finally, it is very important that the technician clean the torque converter after this machining process is complete. Otherwise, metal chips trapped in the converter will circulate in the fluid and severely damage the transmission. Therefore, a mechanic should never drill and tap a converter

Shop Equipment and Tools

that is still in the vehicle, because there is no proper way to clean out the unit.

AUTOMATIC TRANSMISSION DYNAMOMETERS

Many repair shops that specialize in rebuilding automatic transmissions now use a special transmission dynamometer to test remanufactured and malfunctioning units. This machine (Fig. 1-8) simulates the operating conditions the transmission encounters while in service in a motor vehicle without the need of the unit being in the vehicle itself. In other words, the machine bench checks the rebuilt or malfunctioning transmission for various problems with the unit operating under various load conditions, and the technician can spot and repair any defects in the transmission before installing it in the vehicle.

Various Problems Uncovered by the Dynamometer

The dynamometer can reveal many types of defects in an automatic transmission. The types of malfunctions detected during a test and the frequency of their occurrence will depend on two things: (1) the skill of the machine operator; and (2) the type of transmission tested at any given time. With these facts in mind, let's look at some of the problems most commonly discovered by this piece of equipment and some of their causes.

1. Low or no hydraulic pressure: This problem can result from such things as a defective hydraulic pump, missing or stuck valves in the valve body, loose-attaching bolts, and internal leakage due to defective gaskets or sealing rings.

2. Improper or no upshift pattern: A common defect brought on by such things as a malfunctioning governor, throttle valve, or stuck shift valves within the valve body.

3. No downshift: In this situation, the transmission remains in high or intermediate when it should automatically downshift to low. This condition occurs when there is high governor pressure caused by a stuck governor valve or a shift valve (or valves) stuck in the open position.

4. Bushing seizure or failure: A problem usually caused by the lack of proper lubrication or by foreign materials trapped between the bushing and the component it supports.

5. No drive forward or reverse: A condition, if the transmission has normal hydraulic pressure, commonly caused by a defective band, defective clutch assembly, or by a manual valve that is out of adjustment or not properly connected to its actuating linkage.

6. External leaks: A problem that is easily seen and usually resulting from a defective gasket or seal.

7. Unacceptable noise levels: This situation can be the result of worn planetary gear train components, bushings, and thrust washers. In addition, the noise may be hydraulic pump whine brought on by high hydraulic pressure, worn pump components, or air in the system.

8. Internal unbalance that sets up a vibration pattern: This condition may be due to such problems as an unbalanced torque converter, worn bushings, or transmission shafts that are bent or out of alignment.

9. Servo lugs (ends) broken off a band: This particular condition will cause a transmission not to operate at all in at least one gear ratio. The lug itself snaps off the band as a result of metal fatigue, compounded usually by high hydraulic pressure on the band as it activates the servo.

10. Wrong valve body or a valve body that is contaminated with gum or varnish: These particular conditions will cause a wide variety of transmission malfunctions including high or low hydraulic pressures, erratic or no upshifting or downshifting, or a possible no drive condition in forward or reverse.

Finally, a bonus feature of pretesting a rebuilt transmission is that it thoroughly lubricates the transmission internally. This action is desirable under any condition, but it is most important when the unit may be in storage from 1 week to 6 months.

Design

The dynamometer shown in Fig. 1-8 has a drive head section, cradle mount assembly, instrument panel, utility panel, main-control panel, and a brake load section. The drive head section (Fig. 1-9) contains an electrically powered, main-drive motor; input-shaft assembly; a pair of vari-pitch, vee-belt pulleys; a hydraulic, speed-control cylinder; and an input tachometer drive assembly. The main-drive motor supplies the driving torque necessary to rotate the input-shaft assembly. A start-stop switch on the control panel activates this motor which operates at one predetermined speed.

The input-shaft assembly drives the torque converter of the transmission or its input shaft. The converter drive (Fig. 1-10) or direct drive keys into one end of the input-shaft assembly. When the operator turns the start-stop switch to the "start" position, the main-drive motor rotates the input shaft, torque converter, or directly the input shaft of the transmission.

The pair of vari-pitch, vee-pulleys alters the speed (rpm) of the input-shaft assembly. One of these pulleys attaches to one end of the main-drive motor; the other fastens to the free end of the input-shaft assembly. A

FIG. 1-8 Typical automatic transmission dynamometer.

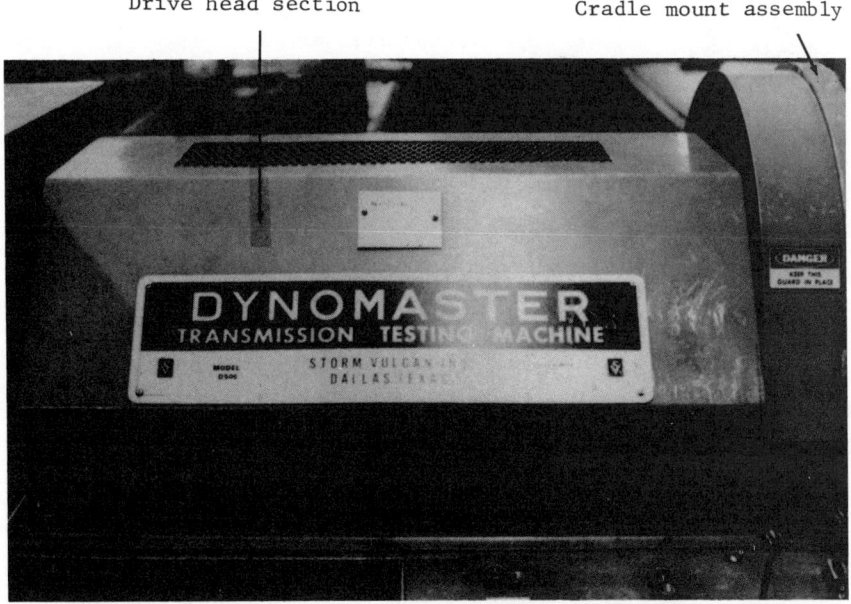

FIG. 1-9 Drive head section of the dynamometer.

wide vee-belt rides between these two pulleys and transmits the driving torque from the motor to the input shaft.

The speed control cylinder has the responsibility of actually changing the width of the pulleys. By altering the width of these pulleys, the belt moves either up or down in the vee-pulley slot. As a result, the working size of the pulleys changes and so does the speed ratio between the motor and the input shaft. In other words, the drive motor runs at a constant rpm; but when the operator moves the input-speed control handle, the hydraulic speed-control cylinder alters the pulley width to raise or lower the input shaft speed.

The input tachometer assembly directs a signal to the input tachometer mounted on the instrument panel, indicating the rpm of the input shaft. A small vee-belt that operates between a pulley on the input shaft and one on the tachometer assembly drives a small generator within the tachometer assembly. This generator creates an electric signal in proportion to input shaft speed; this signal, in turn, causes the tachometer gauge needle to move in proportion to the speed of the input shaft.

The cradle mount assembly (Figs. 1-9 and 1-10) supports and aligns the front section of the automatic transmission in the machine. The transmission itself bolts to a specially machined mount plate; the mount plate, in turn, fits over dowel pins on the cradle mount assembly. Finally, attaching bolts secure the mount plate to the cradle assembly, which also can move back and forth on a slide built onto the bed of the machine. This cradle movement,

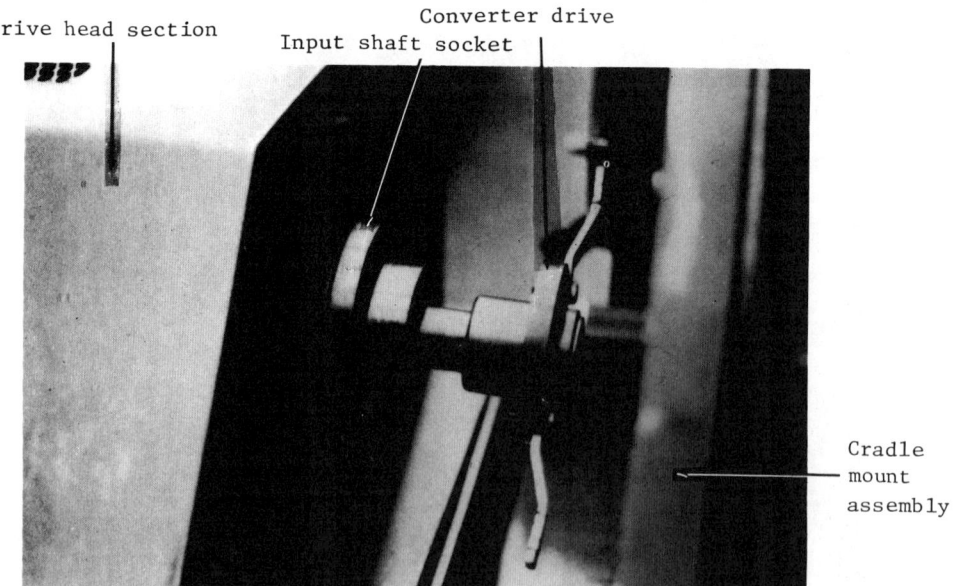

FIG. 1-10 Input shaft, converter drive, and cradle mount assembly of the dynamometer.

Shop Equipment and Tools

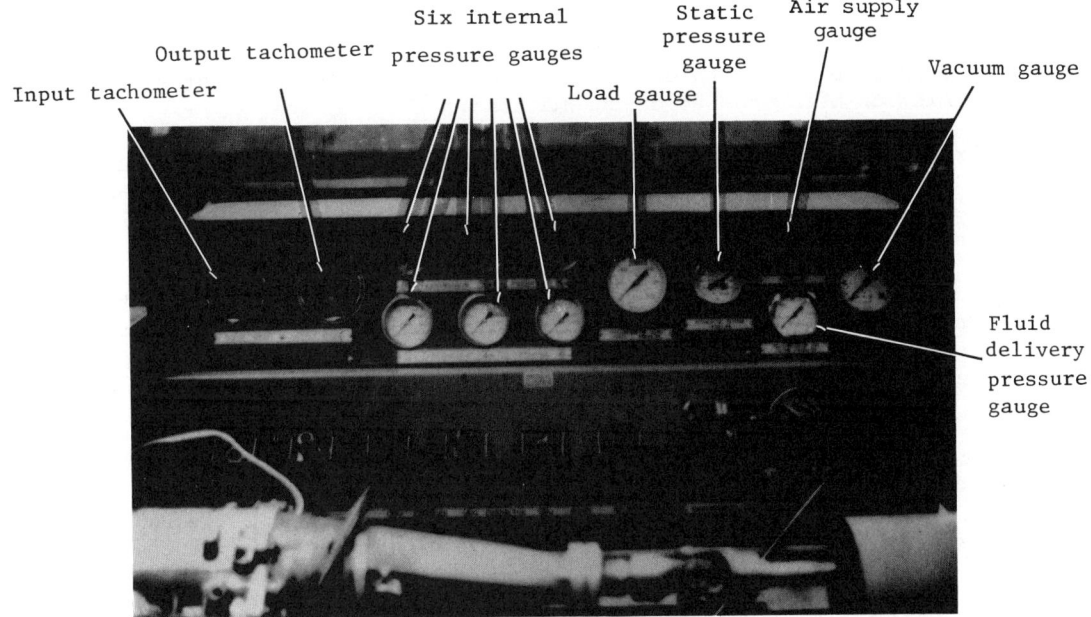

FIG. 1-11 Instrument panel of the transmission tester.

along with a variety of mount plates, permits the operator to adapt the machine so that it can test different types of transmissions.

The instrument panel (Fig. 1-11) houses two tachometer gauges; a set of transmission internal fluid gauges; a dynamometer load application gauge; main air supply gauge; vacuum gauge; transmission fluid delivery pressure gauge; and an accessory-system, static hydraulic pressure gauge. The first tachometer measures the speed (rpm) of the input shaft, and the second tachometer records the relative speed of the output shaft in rpms.

The six transmission fluid-pressure gauges, when connected to the transmission via pressure hoses and fittings, measure the various internal hydraulic pressures developed within the transmission. The actual number of the hydraulic circuits measured by these gauges, of course, depends on the transmission design. For example, the Ford C-6 automatic transmission has two test points: one is for control pressure and the other is for throttle pressure. On the other hand, a Chrysler Torqueflite transmission has five pressure test points: line, front-servo release, lubrication, rear-servo apply, and governor.

The dynamometer load application gauge shows the amount of pressure applied to the machine's load section. The normal test load measured by this gauge, is between 1200-1600 psi. The maximum is 3000 psi.

The main air-supply gauge measures the amount of shop air supplied to the machine. The gauge itself has a dial that reads from 0 to 160 psi, and a reading of 75 psi is adequate to operate the machine's air/hydraulic pump.

The vacuum gauge measures the amount of vacuum the machine's vacuum pump system produces. The gauge reads between 0 and 30 inches Hg (mercury), and it will indicate to the operator any sizable leak in the modulator system of a transmission being tested on the machine.

The transmission fluid delivery pressure gauge indicates the pressure of the hydraulic fluid used to fill a test transmission from the delivery hose. This gauge has a range of readings between 0 and 60 psi, but normal delivery gauge pressure is less than 45 psi.

The accessory system hydraulic gauge measures the static pressure that the machine's air/hydraulic pump produces to operate the various components of the load section. This gauge reads between 0 and 1000 psi, and the operator can adjust static pressure on this gauge by turning the input air-regulator valve located on the control panel.

The utility panel below the instrument panel supports the fluid delivery volume gauge, fluid delivery hose, vacuum hose, and six hydraulic pressure hoses with quick-disconnect fittings (Fig. 1-12). The fluid delivery volume gauge measures the amount of fluid in quarts pumped from the machine's reservoir into the transmission being tested.

The fluid-delivery hose, to the left of this gauge, connects to the machine's delivery pump and carries pressurized fluid to the transmission. The hose itself pulls out from the machine, and it is long enough in length to reach the dip-stick tube near the front of most transmissions.

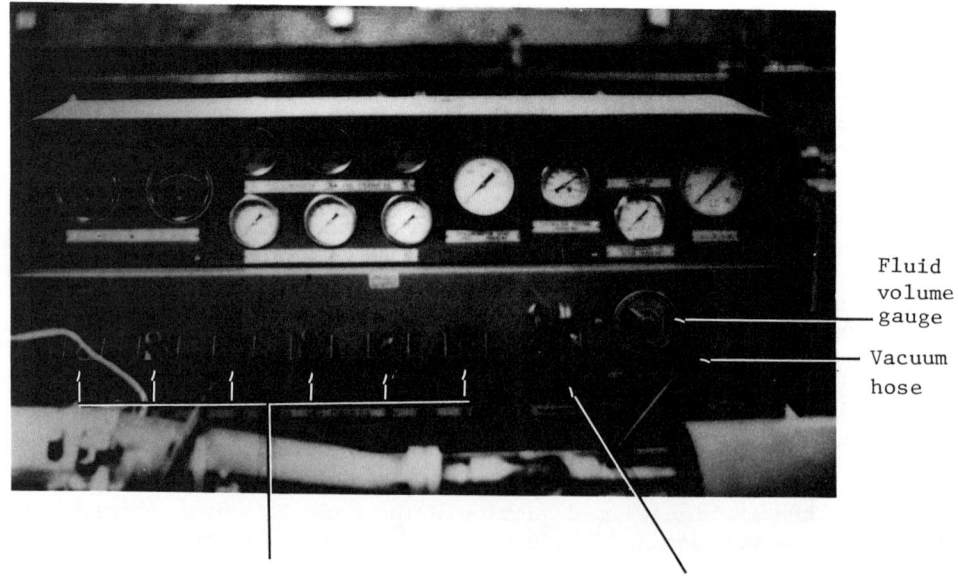

FIG. 1-12 Dynamometer's utility panel.

Shop Equipment and Tools

The operator connects the vacuum hose to the transmission before testing it on the machine. The open end of this hose fits over the vacuum modulator fitting. If the transmission design is such that it has no modulator, the operator will not connect this hose and will leave the vacuum pump switch in the off position.

The six hydraulic pressure hoses attach, via their quick-disconnect fittings, to the hydraulic test points of the transmission. As previously mentioned, these hoses also connect to the pressure gauges on the instrument panel; therefore, the hoses carry the various test pressures to these gauges.

Mounted on the main control panel (Fig. 1-13) are the input air-regulator valve, 12-volt DC switch, vacuum pump switch, transmission fluid supply switch, dynamometer load apply switch, disc-brake hydraulic valve, input-speed control valve, and the input-motor start-stop switch. The input air-pressure regulator valve, as previously stated, controls the amount of input air pressure delivered to the air/hydraulic pump. By rotating this valve, the operator can adjust the machine's static hydraulic pressure.

The 12-volt DC switch is an on-off switch for the machine's 12-volt system. This system tests the electrical kickdown circuit on such transmissions as the T-300 and 400. The switch, when on, allows a small electrical current to pass through two test leads. One lead connects to the kickdown solenoid terminal on the side of the transmission; the other attaches directly to the transmission case--the grounded side of the circuit. The current flow, in turn, activates the kickdown circuit.

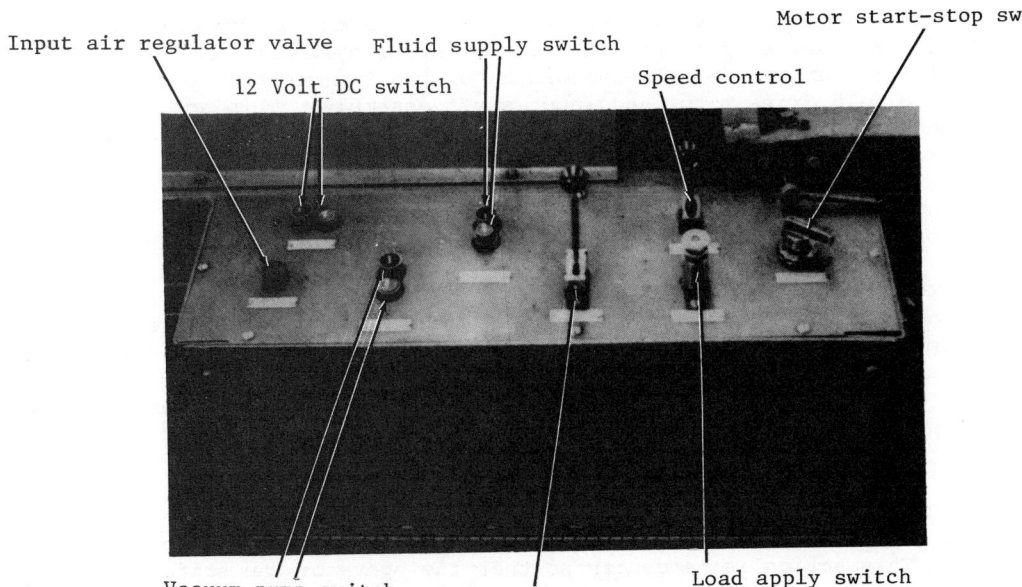

FIG. 1-13 Main control panel of the dynamometer.

The vacuum pump switch controls the operation of the machine's vacuum pump system. It is only an on-off type switch; consequently, when the switch is on, the pump produces a given amount of vacuum up to its capacity. This vacuum, as previously stated, is necessary to test a transmission's vacuum modulator system.

The transmission fluid supply switch activates the pump system used to transfer the fluid from the machine's reservoir to the supply hose on the utility panel. This switch is also an on-off switch and does not control the pressure of the pump. Finally, except when the operator services the transmission with fluid, the switch is usually left off to prevent the pump circuit from overheating.

The dynamometer load apply switch is a regulatory type switch. It controls the amount of accessory-system hydraulic pressure directed to the hydraulic dynamometer. And as previously mentioned, the operator, by adjusting this switch, applies a hydraulic pressure of between 1200 and 1600 psi to the dynamometer during an average test sequence on a transmission.

The disc brake control valve, as its name implies, controls the operation of the disc brake, located in the brake load section. By activating this control valve, the operator can stop the rotation of the output shaft. Furthermore, this valve design is such that it will stay in the "brake apply" position until the operator manually moves it to the off position. This valve design permits the technician to work on the transmission while the machine is running without worrying about the possibility of getting hurt by a rotating output shaft.

The input-speed control valve increases or decreases the rpms of the input shaft. This valve activates the speed control cylinder, and the speed control cylinder changes the relative widths of the two vee-belt pulleys, located on the main drive motor and input shaft assembly. This action either increases or decreases input shaft speed relative to the position of the control valve.

The input-motor start-stop switch activates the main drive motor. It is only an on-off switch. Therefore, the switch cannot regulate motor speed.

The brake load section of this machine (Fig. 1-14) houses the hydraulically operated dynamometer, the disc-brake assembly, the dynamometer oil reserve tank, the splined driveshaft element, and the output tachometer drive assembly. The hydraulically operated dynamometer, which connects to the driveshaft via pulleys and a notched belt, applies a load to the driveshaft. And as previously stated, the amount of this load depends on the position of the load switch on the control panel.

The disc-brake assembly is also a hydraulically operated unit. Its function is to actually stop or hold the driveshaft during the various test procedures. This action is necessary so that the operator can shift the transmission from a forward to reverse ratio without damaging the unit and can stall test it. The brake also acts as a safety device to protect the operator against personal injury.

Shop Equipment and Tools 17

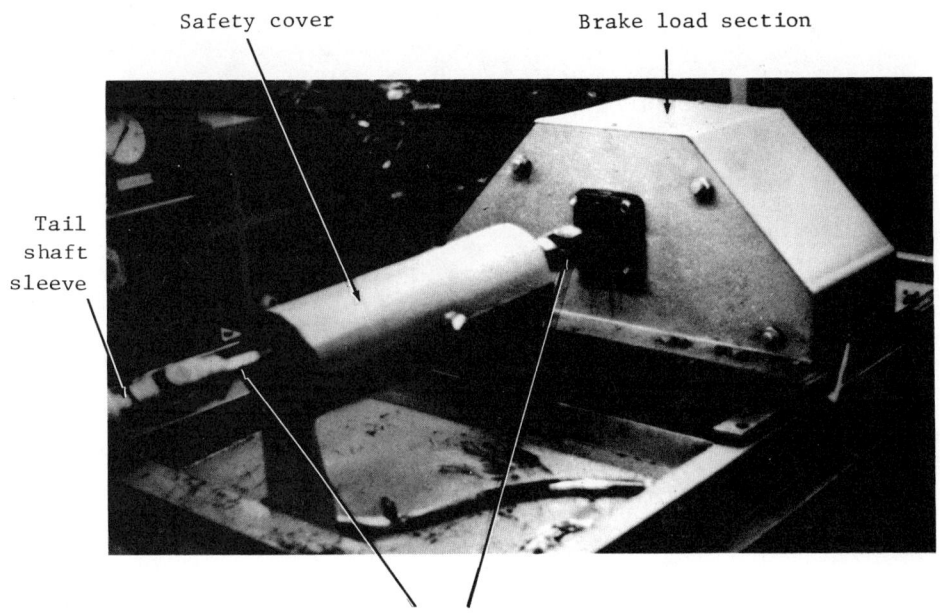

FIG. 1-14 Brake load section and output shaft element of the dynamometer.

The oil reserve tank is the reservoir for the closed dynamometer hydraulic system. The capacity of this tank is 32 gallons of hydraulic fluid.

The driveshaft attaches to the output shaft of the transmission. The driveshaft itself on one end connects into the machine's driveshaft element through a U-joint assembly. The other end of the driveshaft has a splined section that keys to another U-joint assembly (Fig. 1-14). This assembly keys into the tailshaft sleeve which, in turn, splines into the transmission's output shaft.

The output tachometer drive assembly connects directly to the output shaft element. This unit sends an electrical signal to the output tachometer on the instrument panel. In other words, this assembly is also a small generator that develops an electrical signal in proportion to output shaft speed, and this signal operates the tachometer gauge.

PORTABLE DIAGNOSTIC TEST EQUIPMENT

Hydraulic Gauge Sets

Figure 1-15 shows a hydraulic gauge set used to test the internal pressures of an automatic transmission that is still in a motor vehicle.

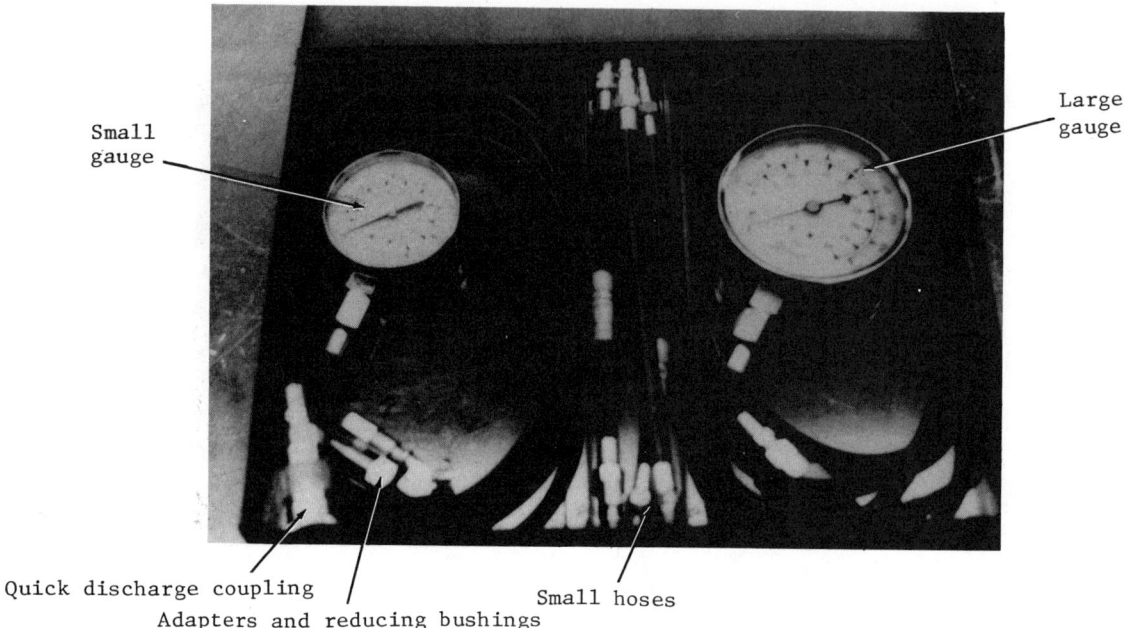

FIG. 1-15 Typical hydraulic test gauge set.

This set consists of two gauges with hoses and a set of adapters. The small gauge tests such pressures as throttle and governor. It has a 2½-inch dial that measures pressure from 0-100 psi (100-700 kPa).

The larger gauge measures the amount of line (control) pressure within an operating transmission. This unit has a 3½-inch dial. The face of the dial has pressure divisions from 0-300 psi (200-2000 kPa).

Both gauges connect to hoses that are 6 feet in length. With this hose length, the technician can temporarily install the gauge inside the vehicle. He can then observe its reading while performing operational tests on the vehicle and transmission.

This set also includes several adapters, small hoses, and reducing bushings. The mechanic will use these components, in conjunction with either or both gauges, for several reasons: (1) to connect the gauges to different types of automatic transmissions; (2) to connect the gauges to a transmission's test points, obstructed by the vehicle's frame, body, or exhaust pipes.

Vacuum Pump and Gauge Assembly

The hand operated vacuum pump and gauge assembly (Fig. 1-16) serves several useful purposes. First, with this device a mechanic can test a

Shop Equipment and Tools

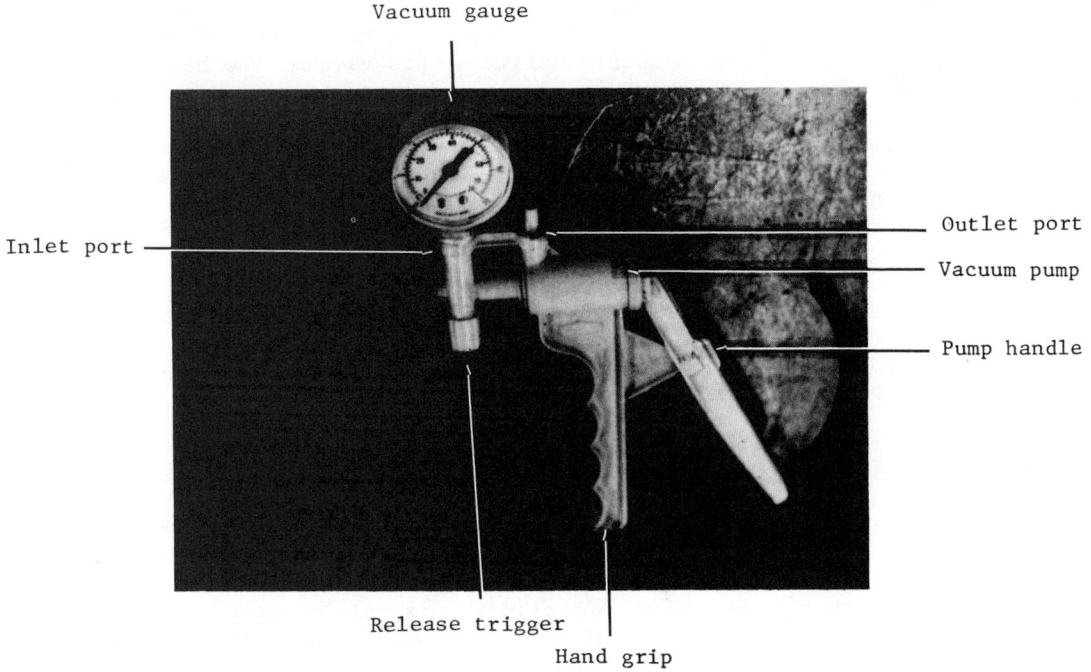

FIG. 1-16 Hand operated vacuum pump and gauge assembly.

modulator diaphragm assembly for vacuum leaks with the modulator on or off the transmission. Second, with this tool and the large hydraulic gauge previously mentioned, the technician can test and adjust the control (line) pressure of a transmission, equipped with a vacuum modulator assembly, with the unit still in the vehicle.

The single piece of equipment shown in Fig. 1-16 has a vacuum gauge, vacuum pump, inlet and outlet ports, release trigger, and a hand grip and pump handle. The vacuum gauge connects externally, via a hose, to the component being tested and internally to the vacuum pump. The dial scale of the gauge has reading ranging from 0 to 30 inches Hg and 0 to 76 mm Hg.

The vacuum pump itself is inside the main housing. The pump handle mechanically activates this pumping mechanism, and the pumping mechanism, in turn, evacuates the air from any component being tested.

This pump requires two valve controlled ports--an inlet and outlet. The inlet port connects, via a hose, to the unit being tested. The outlet port, on the other hand, provides an opening for the pump directly into the atmosphere.

The release trigger has one important function: it vents all or part of any vacuum built up by the pump assembly to the atmosphere. In other words, by coordinating movements of the pump handle and trigger, the operator can easily pump up a _specific_ amount of vacuum with this piece of equipment. As previously mentioned, the mechanic utilizes this vacuum, whether it be 5 inches or 20 inches Hg, to check or adjust modulated control pressure.

The hand grip and handle merely provide a means by which the mechanic can hold and activate the pump assembly. The palm of the mechanic's hand fits over the pump handle with the fingers encircling the hand grip. This construction permits ease of control with one hand, leaving the other free to perform other tasks.

TRANSMISSION CLEANING EQUIPMENT

Automatic transmission repair facilities primarily use four types of cleaning equipment to remove dirt, grease, oxidized fluid, and corrosion from units being rebuilt. These pieces include the steam cleaner, the jet cleaner, the safety-type parts washer, and the abrasive blaster. Most repair shops, for example, use the steam cleaner to remove dirt and grease from a transmission after the mechanic removes it from the vehicle, and just prior to tearing it down. In operation, this unit converts a mixture of water and soap into steam that quickly dislodges the dirt and grease from the outside surfaces of the transmission case.

The jet cleaner, on the other hand, will clean both the outside and inside surfaces of large transmission components. When in service, this machine uses a preheated chemical solution and water under pump pressure to perform the cleaning task inside the machine itself. Simply speaking, the operation involves the spraying of the hot solution onto the parts, through a set of nozzles, as the parts revolve on a turntable assembly. This combined action removes most remaining dirt, grease, and oxidized fluid in 6 to 10 minutes.

The one piece of cleaning equipment most often found in all repair shops is the safety type, solvent filled, parts washer. This unit usually serves the function of cleaning smaller components that are too fragile to be cleaned in or that can get lost in a jet cleaner. The parts washer, depending on the solvent used and the type of machine, removes dirt, grease, and varnish by either washing the parts by hand under a stream of clean solvent or by soaking the parts in the solvent, agitated by air.

Many shops now also use an abrasive type blasting machine to clean certain transmission components. The machine's blasting process serves two functions: (1) certain abrasive blasting will remove paint, rust, and corrosion from metal surfaces without contaminating the surface, changing its critical dimensions, or removing any of its base metal; (2) abrasive blasting removes the glaze built up on clutch drums and steel clutch plates. This action provides an excellent friction characteristic to these components and therefore reduces band or clutch slippage.

Shop Equipment and Tools   21

Steam Cleaners

Figure 1-17 shows a typical steam cleaner used in automatic transmission repair shops. This unit consists basically of a water pump with control switch, heating coil and burner assembly, soap tank and valve, plus a gas valve and pressure hose. An electric motor drives the water pump by means of a vee-belt and pulleys. The pump itself pressurizes tap water sufficiently to push it through the heating coil. In addition, the pump switch--an on-off toggle switch--controls the operation of the pump by either connecting or disconnecting electrical power to its motor.

   The heating coil and burner turn the tap water from the pump into steam. The water passes through the coil under pump pressure. The coil itself sits above a gas fired burner assembly that heats the coil. In other words, cold water enters the coil's inlet, and steam under very high pressure exits from the coil's outlet.

   The soap tank and valve supply soap or cleaning agent to the tap water before it enters the pump. The tank itself is just a reservoir for a mixture of water and a cleaning agent. The valve just regulates the amount of this mixture that leaves the tank on its way to the pump.

   The gas valve controls the combustible gas flow to the burner underneath the coil. This valve is an on-off type that the operator turns on after tap water is flowing through the coil. A safety-type shut-off valve activated by the burner's pilot light prevents gas flow through this main gas valve if the pilot should go out.

FIG. 1-17 Design of a typical steam cleaner.

Attached to the coil's outlet port is a pressure delivery hose. This hose carries the pressurized steam to a combination handle and nozzle assembly that connects to the opposite end of the hose. The operator uses this assembly to direct the flow of steam onto the component he is cleaning.

Jet Cleaners

The jet cleaner assembly shown in Fig. 1-18 consists mainly of an insulated reservoir, a gas burner, a turntable assembly, a set of oscillating jets, and a filter assembly. The insulated reservoir of this machine has a fluid capacity of 300 gallons of cleaning agent. The cleaning agent, in this case, is water mixed in the correct proportion with a chemical that will clean either iron, steel, aluminum, or a combination of these metals.

This mixture must be very hot in order to thoroughly clean the parts; the gas burner assembly performs this function inside of the reservoir itself. This assembly consists of piping, pilot light, and thermostat. The burner piping fits into the lower section of the reservoir, and when the safety-type pilot light ignites the gas in the burner, the heat generated in its piping heats the cleaning agent contained in the reservoir. Finally, a thermostat controls the operation of the burner assembly so that the cleaning agent never drops below a predetermined temperature.

The turntable (Fig. 1-19) fits inside the cleaning chamber. This device serves as the cleaning platform for parts that the machine will clean.

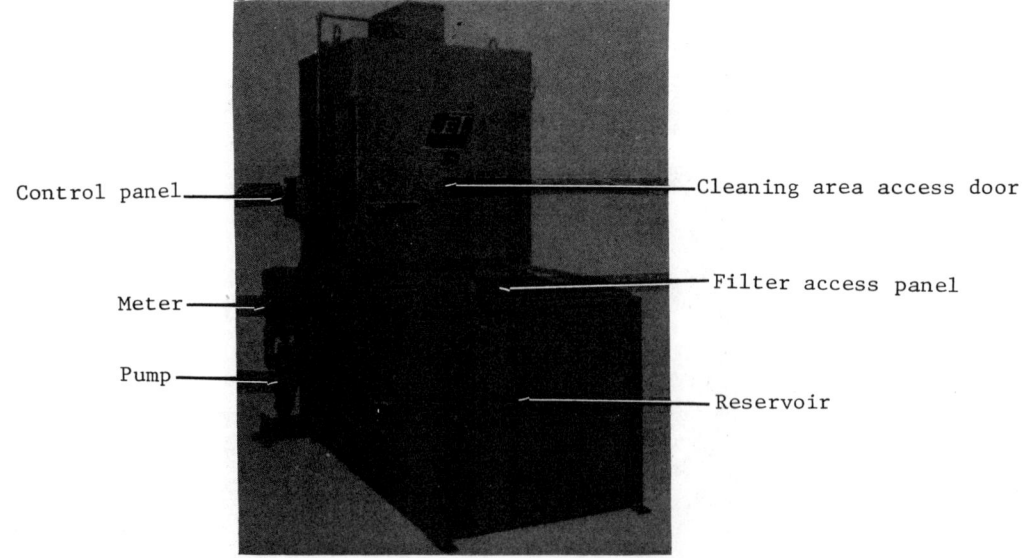

FIG. 1-18 Typical jet-type cleaner.

Shop Equipment and Tools

FIG. 1-19 Motor driven turntable of the jet cleaner.

Larger components sit on the turntable directly; smaller parts fit into a basket that the turntable supports.

So that the machine thoroughly cleans the parts contained on the turntable, an electric motor revolves the turntable at 1/8 rpm. This action permits the cleaning agent from the nozzles sufficient time to reach and clean all areas of the parts in a short period of time. Finally, a turntable switch located on the control panel permits the operator to stop the turntable's rotation in order to load or unload parts from the platform.

A set of oscillating jet assemblies (Fig. 1-20) are also inside the cleaning chamber. The jet nozzles themselves fit onto two hollow tubes that oscillate back and forth. This oscillation results from the action of an electric motor and mechanical linkage. Finally, another switch on the control panel activates a motor-driven pump that supplies pressurized hot cleaning solution to the nozzle assemblies as they oscillate.

The filter assembly (Fig. 1-21) sits below the level of the cleaning chamber but above the reservoir. The cleaning solution, draining from the cleaning chamber, passes through this stainless steel filtering system. The filter and waste container remove and store dirt and particles so that they will not recirculate back through the pump and damage it or contaminate the newly cleaned parts.

Parts Washers

Figure 1-22 illustrates one type of safety-type parts washer. This unit has a reservoir (soaking tank) and pump, twin filtering system, air

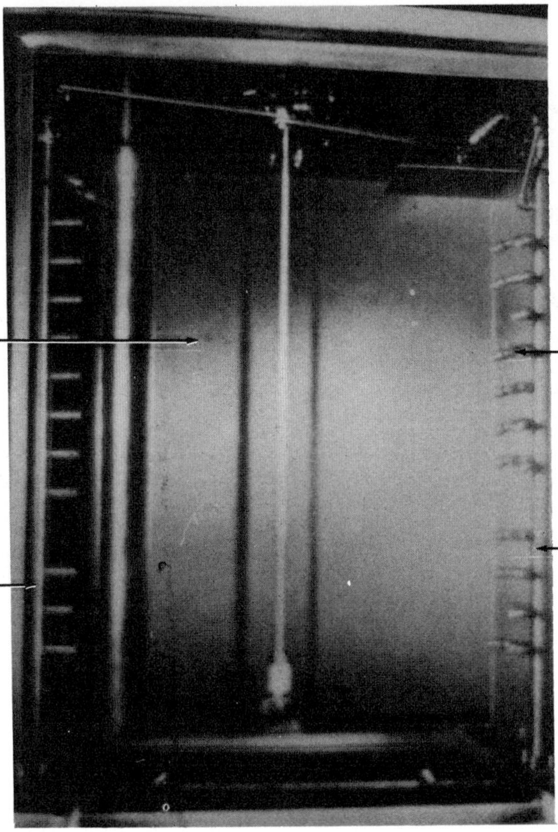

FIG. 1-20 Set of oscillating jet assemblies.

FIG. 1-21 Filter assembly of the jet cleaner.

Shop Equipment and Tools

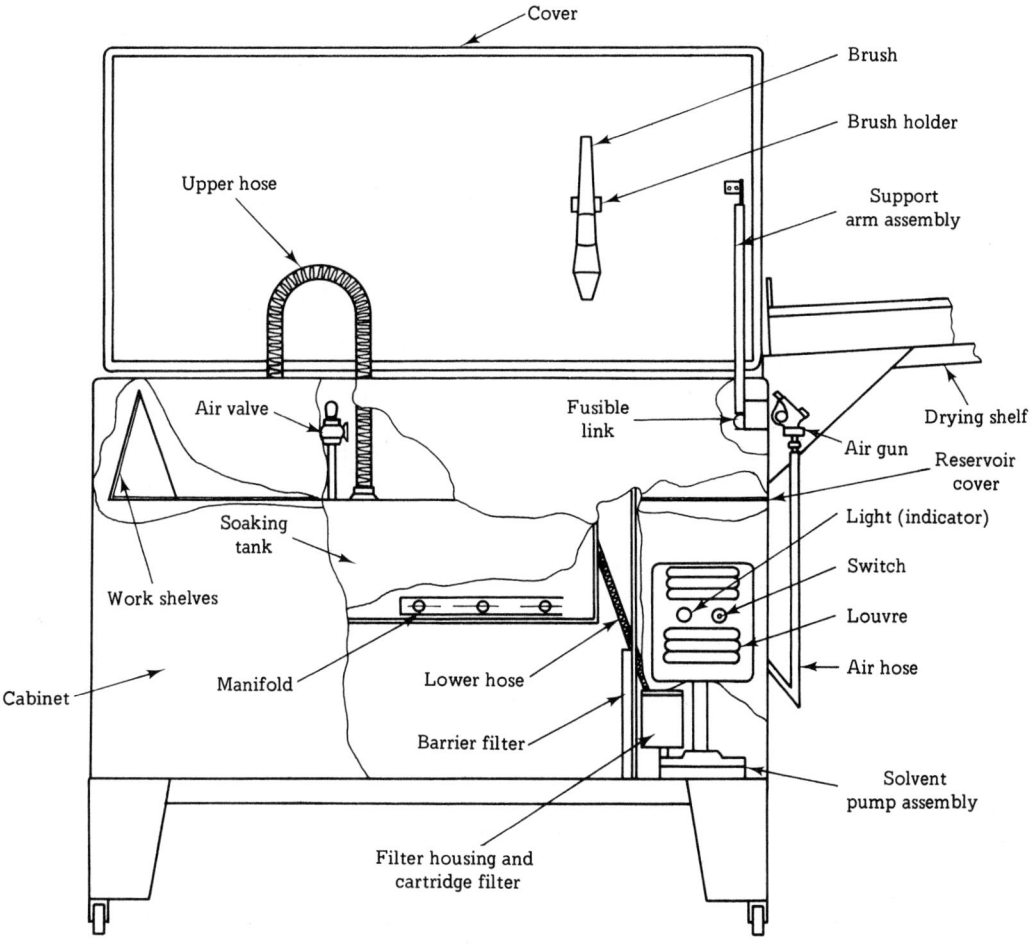

FIG. 1-22 Design of a typical safety-type parts washer.

agitation system, drying shelf, and a cover controlled by a fusible link. The soaking tank is part of this machine's cabinet. The tank itself holds a quantity of safety cleaning solvent, and the recommended type is one that has a high flash point and will not harm a mechanic's skin, hands, or clothing.

Located in the lower right side of the soaking tank is the solvent pump. The function of this pump is to transfer solvent from the tank, via a lower hose, to a braided metal upper hose situated above the level of the solvent in the main tank. The mechanic will use this flowing solvent to wash and rinse off parts he is cleaning.

An on-off electrical switch and light assembly, located in front of the pump but on the outside of the cabinet, activates the pump. When the operator turns the switch on, the pump begins to function and the indicator light glows. This indicator light helps to remind the operator to turn the pump switch off when pump operation is no longer necessary to the cleaning process,

The design of the twin filtering system is such that it keeps all contamination from entering the pump's reservoir so that the pump transfers only clean solvent; consequently, the system protects the pump from particle damage. The system consists of a barrier filter and cartridge filter. The barrier filter separates the main tank reservoir from the pump reservoir. This device stops all larger particles from entering the pump's reservoir; as a result, they sink to the bottom of the main tank.

On top of the pump body is the cartridge-type filter. This filter design is such that it traps any smaller particles of contamination that might get by the barrier filter and flow into the pump's inlet port. Finally, because the cartridge filter traps these particles, the operator will have to replace it on a periodic basis, or the filter will clog and cause pump damage.

The air agitation system consists of an air valve and manifold. The air valve connects to an external air supply via a pressure hose. The valve itself regulates the flow of shop air going through a pipe to the air manifold.

The air manifold rests about at the midpoint in the main tank, beneath the level of the solvent. The function of this manifold is to evenly distribute the air, supplied to the manifold by the air valve, to the surrounding solvent. In other words, if the operator opens the air valve, air passes through the holes in the manifold, causing the solvent to agitate.

The sheet-metal drying shelf mounts to the side of the cabinet. Its construction is such that the drain area of the shelf tapers down slightly from its outer end to where it attaches to the side of the cabinet. This design permits the solvent, draining from the cleaned parts placed there, to return to the main tank for recycling.

The cover itself attaches to the cabinet by hinges, support arm assembly, and a fusible link. This assembly serves two functions. First, the cover acts as a dust protector for the entire tank assembly when it is not in use. Second, if the solvent happens to ignite, the fusible link, attached to the support arm assembly, will melt and cause the cover to close. The fire extinguishes because the closing cover cuts off the oxygen to the flames.

Abrasive Blasters

The abrasive blaster shown in Fig. 1-23 consists mainly of a cabinet-type work chamber with inspection window and glove port openings, blasting gun and hoses, air-control valve, dust collector, and an electric motor and blower assembly. The welded steel cabinet serves as the framework of

Shop Equipment and Tools

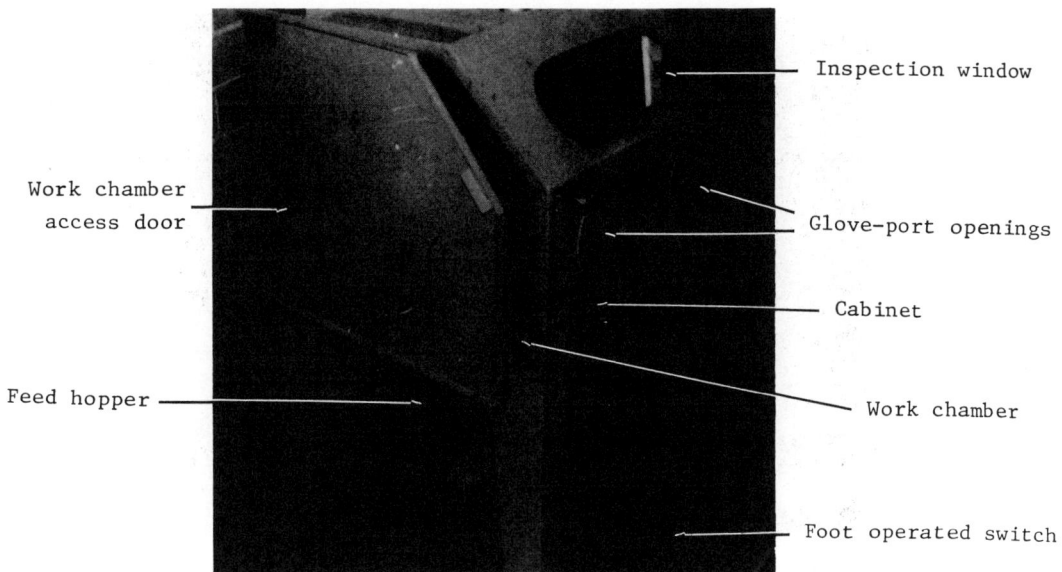

FIG. 1-23 Typical abrasive blaster.

the machine. Inside this cabinet is a well lighted work chamber where the blasting process takes place and a feed hopper that contains the abrasive material.

Situated between the work chamber itself and the feed hopper is a large, steel work stand (Fig. 1-24). This stand has a lattice-style construction heavy enough to support the work. At the same time, however, this design permits the heavier abrasive particles to fall directly back into the hopper after being used to clean the parts.

So that the operator can observe the parts he is blasting, the cabinet has an inspection window. The manufacturer uses safety-type glass for this window, which is 3/16" thick. The window is rubber mounted; therefore, it is replaceable in minutes if the abrasive blast should damage it.

Observe the glove port openings in Fig. 1-23; these openings provide access points through which the operator can work. Each opening is large enough to provide maximum freedom of movement for one of the operator's arms. To protect the operator's hands and arms from abrasive blast, the manufacturer equips both port openings with gauntlet-type sleeves and gloves (Fig. 1-24).

The blasting gun and nozzle, held by the operator through one of these port openings, directs the abrasive blast onto the component being cleaned. This gun has two hoses attached to it. One of these hoses connects the gun to the shop's air supply via the air-control valve (Fig. 1-25). In addition, a foot-operated switch controls the flow of this compressed air from the

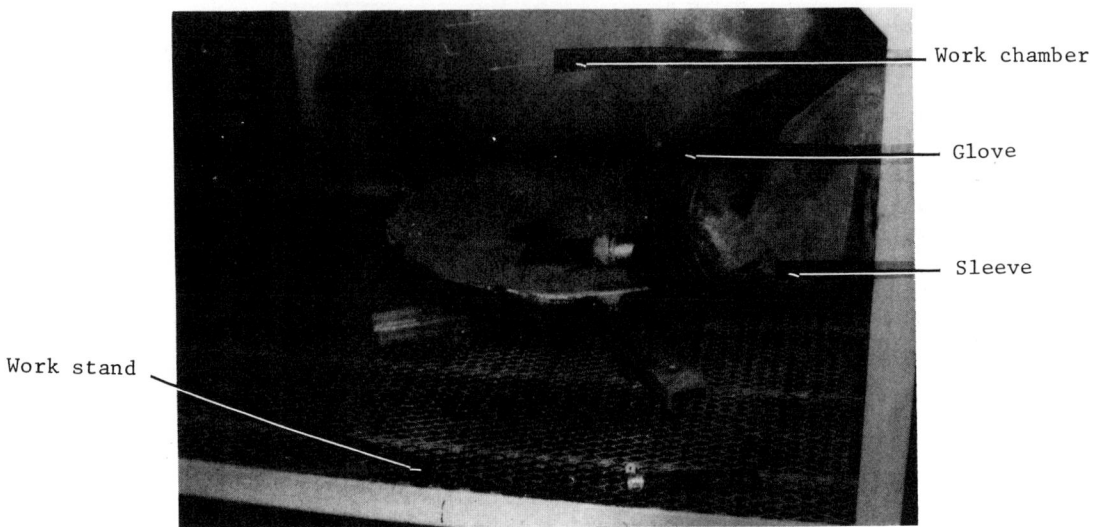

FIG. 1-24 Work chamber and stand area of the abrasive blaster.

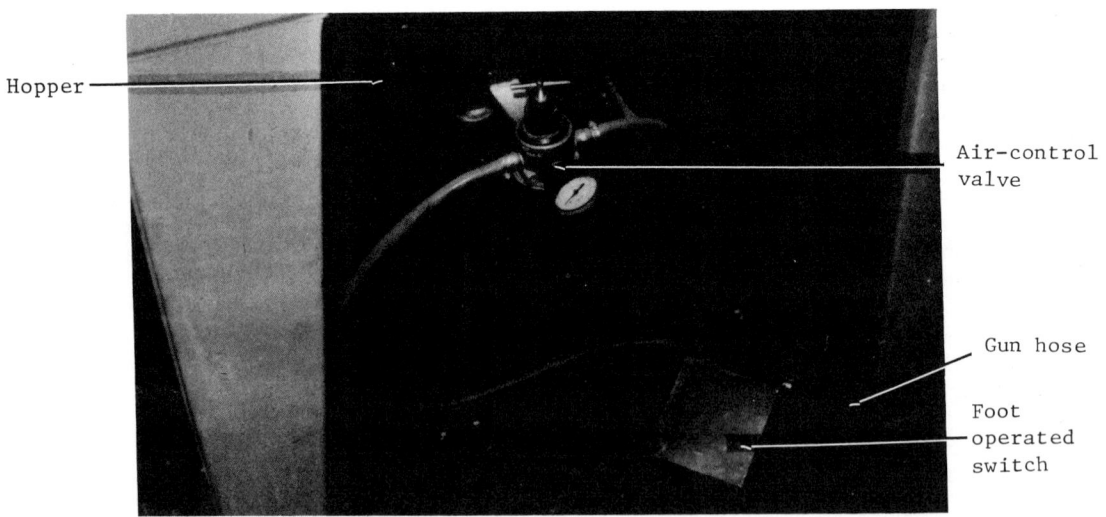

FIG. 1-25 Air control valve and foot switch of the abrasive blaster.

Shop Equipment and Tools

control valve to the blasting gun. Finally, the other hose carries the abrasive material from the hopper to the blasting gun.

The access door (Fig. 1-23) has two functions. First, it provides a large enough opening for the operator to slide components into the work chamber and onto the work stand. Second, through this access door, the operator pours the abrasive material into the feed hopper, which is below the work chamber.

The air control valve (Fig. 1-25) regulates the pressure used by the gun during the blasting process. The valve, in most cases, reduces the amount of air pressure supplied by the shop's air compressor to a value that is suitable for the type of abrasive being used and for the type of work the operator is cleaning. For example, the maximum pressure for glass beads is 70 psi and for silica sand 45 psi, but the operator may have to lower these pressures when cleaning precision parts to avoid damaging them.

The dust collector, motor, and blower assembly (Fig. 1-26) act as a unit to remove and trap dust and dirt from the work chamber. The collector itself attaches to the lower end of the dust bag that fits onto the blower housing. With this construction, all fine particles of dust and dirt pulled from the work chamber by the motor and blower enter the dust bag before reaching their final destination--the collector. This action removes all smaller particles of dust and dirt and permits only heavier abrasive particles to return to the hopper for re-use by the blasting gun.

CLUTCH SPRING COMPRESSORS

Multiple-disc clutch assemblies found in automatic transmissions must have some form of return spring or springs to move the piston back to the base of its bore as the clutch releases. The installation of the springs over the piston is such that they will always exert some force (tension) on the retaining device and the piston itself even when the piston is in the released position. Consequently, with the exception of the diaphragm (bellville) type spring, this initial spring tension must be overcome by some form of compressing equipment in order for the mechanic to remove the spring's retaining device and finally the piston.

Many different types of clutch return spring compressors are now in current use to service open-type clutch assemblies. Figures 1-27 and 1-28 show a few of these units. The two pieces of equipment shown in Fig. 1-27 are universal types. This means that these units are adaptable enough to fit and compress the clutch return springs in most automatic transmission clutch assemblies.

The three devices shown in Fig. 1-28, on the other hand, are factory-type compressors. This means that these units fit and properly compress

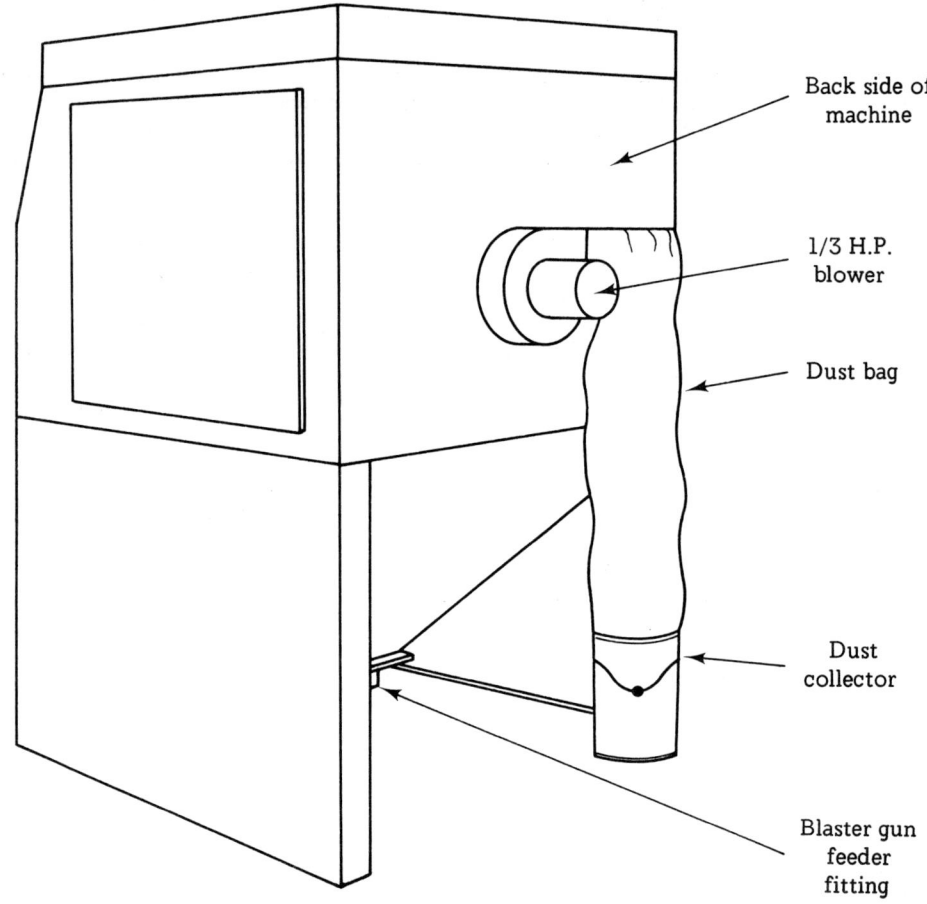

FIG. 1-26 Dust collector, motor, and blower assembly installation on a typical abrasive blaster.

only the springs found in one type of transmission. Therefore, in order to rebuild all the various types of automatic transmissions, the average repair shop must have a rather large quantity of these factory compressors or use the universal type.

Figure 1-29 shows a lever-type universal compressor in position over an open-type clutch drum; this device consists of a base, two vertical posts, and the lever itself. The metal base serves as the framework of the compressor. Each corner of this base has a machined hole that a bolt passes through to hold the base securely to the workbench.

The two vertical posts attach to the base. The left post, Fig. 1-29, attaches to the base via a bolt welded to the end of the post. This post

Shop Equipment and Tools

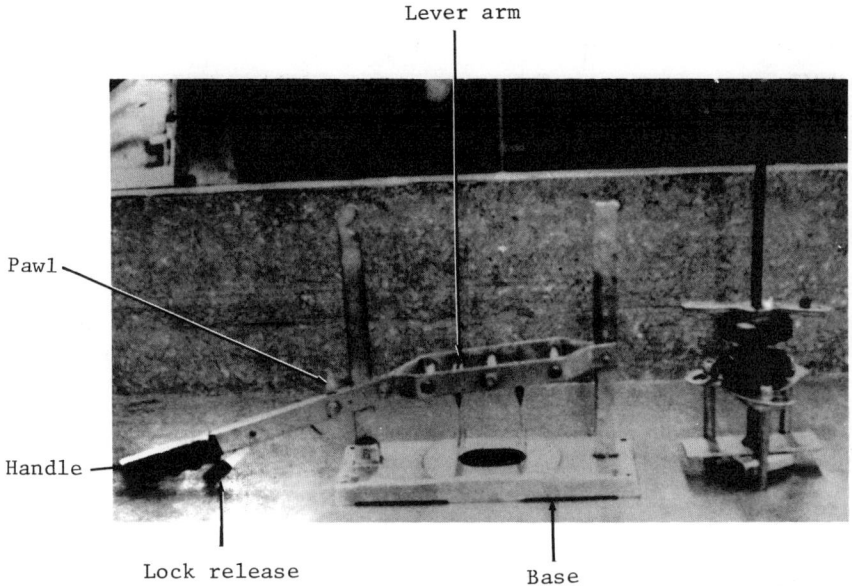

FIG. 1-27 Universal-type clutch return-spring compressors.

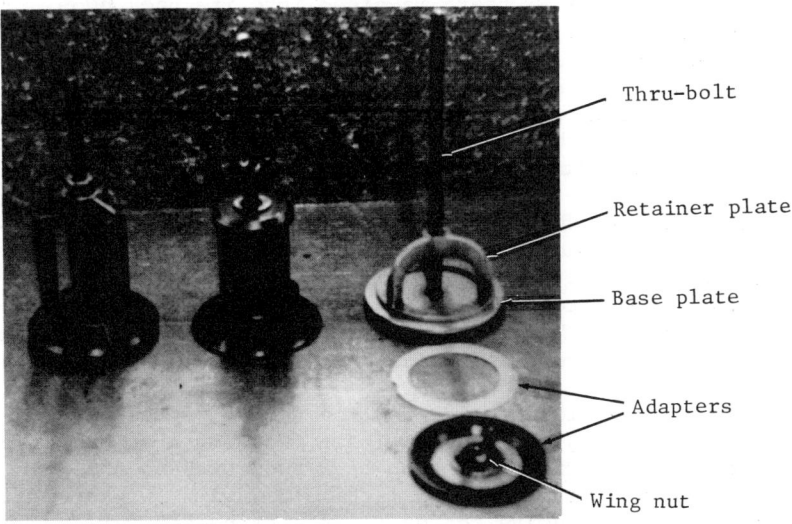

FIG. 1-28 Factory-type clutch return-spring compressors.

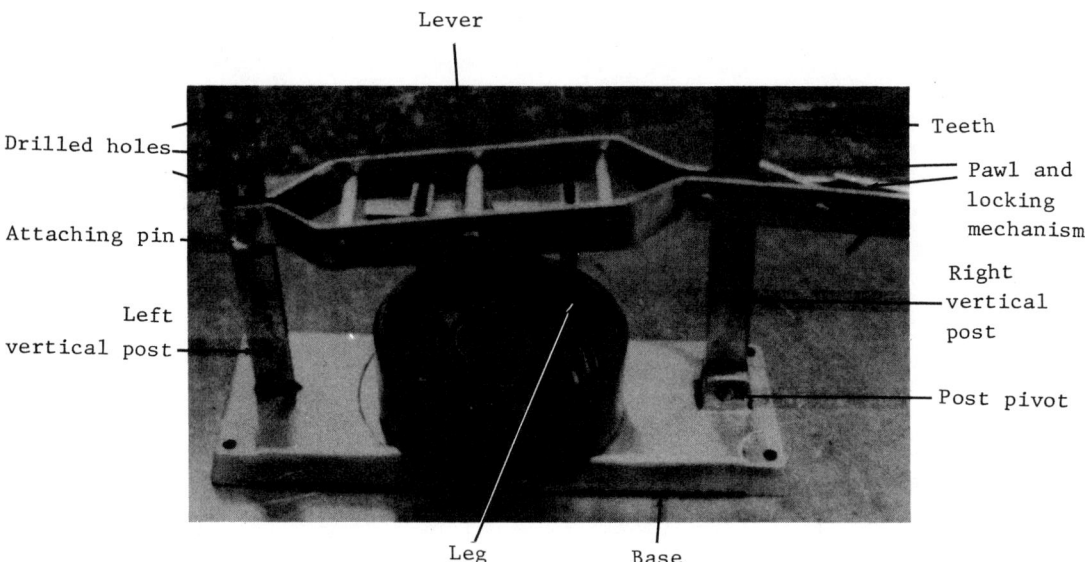

FIG. 1-29 Lever-type universal compressor positioned over an open-type clutch drum.

also has a series of holes drilled into it; these holes accomodate an attaching pin that secures the fulcrum end of the lever to the post. Furthermore, this series of holes permits the mechanic to raise or lower the fulcrum point as necessary to compensate for different drum heights.

The right vertical post pivots back and forth on its mounting bracket, and it has a series of teeth machined into its back edge. The pivot point permits the mechanic to lower this post, making it easier for him to install and remove a clutch drum from the machine. The teeth index with a pawl, which is part of a locking mechanism built into the handle section of the horizontal lever. This mechanism locks the lever to this post which frees the mechanic's hand so that he can remove the spring's retaining device during the compressing procedure.

The design of the horizontal lever is such that it provides the needed mechanical advantage to assist the mechanic in overcoming the tension of the clutch springs. As previously stated, one end of the lever--the fulcrum point--fastens via a pin to the left vertical post. The other end of the lever has a handle assembly where the mechanic applies his downward force which the lever will multiply (Fig. 1-27).

Between the fulcrum point and the handle of the lever are two legs that mount on a pivoting metal bar. These two legs are also adjustable in that the mechanic can slide them both back and forth on the metal bar. This design provides this universal machine now with two adjustments: the legs which move to compensate for different spring and retainer sizes, and a movable fulcrum point to take care of varying drum height differences (Fig. 1-29).

Shop Equipment and Tools

Figure 1-30 shows a factory-type spring compressor installed over a typical open-type clutch drum; this unit uses the principle of the inclined plane to multiply force and overcome spring tension. The machine itself has a base plate, thru-bolt with wing nut, adapters, and a retainer plate. Referring to Fig. 1-28, the metal base plate has a machined hole in its center that acts as the guide for the thru-bolt; and the head of this bolt bears against the bottom of the base plate. When this tool is in service, the base plate rests under the clutch drum; or in the case of a clutch located in the rear of the transmission case, it slips into a recessed area cut in the case.

The thru-bolt and wing nut produce the mechanical advantage, utilizing the principle of the inclined plane, to easily overcome spring tension on the clutch piston. As previously mentioned, the head of this bolt bears against the lower side of the base plate, with its threaded shank passing upward through the clutch drum. This threaded section of the thru-bolt also passes through a hole in a retainer plate that fits over the spring retainer; the wing nut then threads down over the thru-bolt until it contacts the retainer plate (Fig. 1-30). With this arrangement, any additional clockwise rotation of the wing nut forces the retainer plate to move toward the base plate. And since the retainer plate fits over the springs and retainer, this movement compresses them so that the mechanic can remove the snap ring.

The adapters fit over the clutch spring retainer but under the retainer plate of the machine. These adapters are necessary because of the different sizes of spring retainers found in the same model of automatic transmission.

FIG. 1-30 Factory-type compressor installed over an open-type clutch drum.

In other words, with these adapters, this compressing equipment will fit all the different clutch assembly styles found in the same transmission over its years of production.

Some types of clutch drums do not have an open hub like those just discussed. With this design, the transmission's input shaft attaches directly to the drum. As a result, the compressing equipment shown in Figs. 1-27 and 1-28 will not function on these assemblies.

To service the closed-type drums, most shops now use an arbor press and a specially designed retainer press plate (Fig. 1-31). The arbor press supplies the additional force needed to overcome spring force; the hydraulic or rack and gear-type presses are the most common devices used for this purpose.

The retainer press plate can have either the universal or factory design. The universal unit has a slotted press plate with legs that adjust in or out within these slots. Consequently, the mechanic can adjust this assembly to fit several different sizes of clutch spring retainers. The factory plate, however, has fixed legs; therefore, they will conform only to one type of spring retainer. This means that the shop must have a number of these units in order to service the many types of closed-type drum assemblies, or use the universal design.

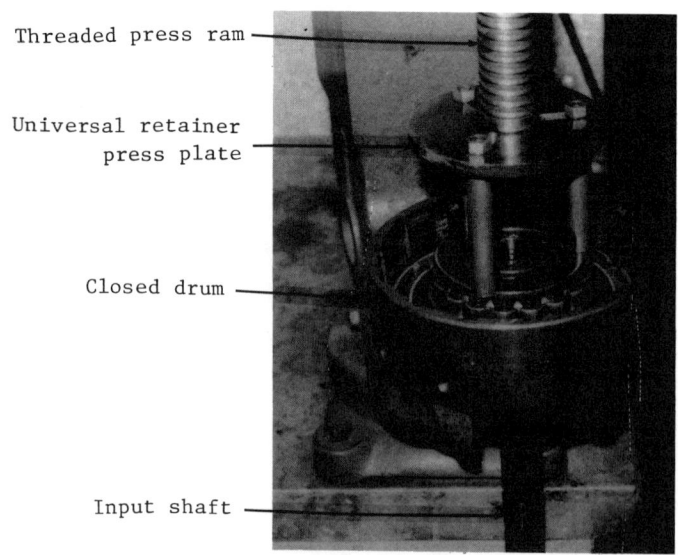

FIG. 1-31 Arbor press and retainer plate in position over a closed-type clutch drum.

Shop Equipment and Tools

BUSHING EQUIPMENT

The average automatic transmission has many sizes and types of bushings installed at various locations within the unit. These bushings are specially designed bearings that support the revolving shafts, planetary gears, and drums of the transmission. The bushing itself has a round, outer steel backing. Cast into the inner circumference of this backing is a softer bearing material such as a babbit metal composed of tin, copper, and antimony or an alloy of copper or aluminum.

To prevent the bushing from shifting its bore once the mechanic installs it, the backing is slightly larger in diameter than the bore it fits into. This provides an interference (press) fit between the backing and the bore. Consequently, in order to remove and replace these interference bushings during a transmission overhaul, special equipment is necessary.

The equipment required to remove an old bushing depends on whether its bore is open or closed. For example, if the bore is open at both ends and has the same diameter all the way through, the bushing will pass out of the unrestricted end during the removal procedure. If, on the other hand, the bushing's bore has a seat or bottoms out (a blind hole), the mechanic must remove the bushing from the open or installation end of the bore.

The pieces of equipment shown in Figs. 1-32 and 1-33 are typical of those used to remove a bushing from a blind bore. The tool illustrated in Fig. 1-32 performs this function by pulling the bushing from its bore. The tool shown in Fig. 1-33, however, splits the backing of the bushing in two pieces so that the mechanic can pry it out easily with a screwdriver or pull it out with needle-nose pliers.

The bushing puller shown in Fig. 1-32 consists of a remover, nut, and cup. On the one end of the hardened steel remover is a tap that cuts a

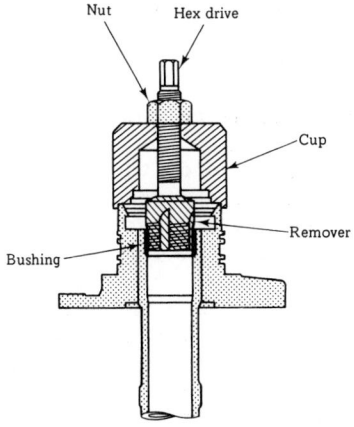

FIG. 1-32 Typical threaded bushing puller assembly.

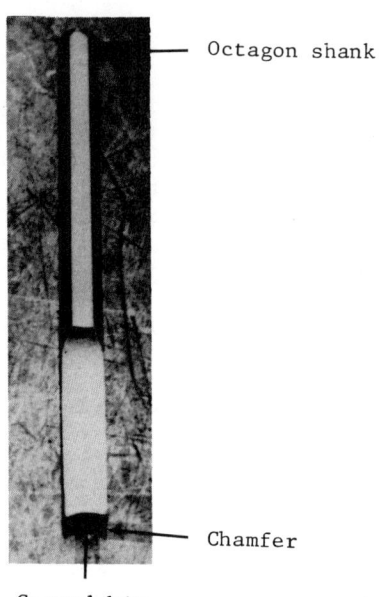

FIG. 1-33 Bushing chisel.

series of threads into the inside circumference of the bushing. Cutting these threads into the bushing is necessary to firmly lock the remover to the bushing itself so that the remover will not pull out of the bushing during the extraction procedure. Finally, flutes or grooves run the length of the tap threads; they permit metal chips to escape, so the chips will not jam and damage the threads of the tap.

On the other end of the remover is a hex drive and a threaded shank. The mechanic uses a wrench on this hex drive to thread the remover into the bushing. The threads of the shank along with the nut provide the pulling force needed to remove (pull) the bushing from its bore.

The cup section of this tool performs two design functions. First, it transmits the pulling force, provided by the nut moving downward on the threaded shank, directly to the surface of the component containing the bushing. Second, the legs of the cup straddle the bushing's bore, thus allowing sufficient clearance for the remover and bushing to clear the cup's inner surface during the pulling process.

The chisel pictured in Fig. 1-33 has a design that permits the mechanic to cut the bushing without damaging the blind bore it fits into. The shank of this cutter has an octagon shape that provides the mechanic with better hand control of the chisel during the cutting operation. Also, the curved bit pushes the bushing material away from the work; therefore, the mechanic can see more easily where to guide the chisel. Lastly, both cutting edges of the tool have a slight chamfer where the bit terminates at the narrow

Shop Equipment and Tools

end of the lower shank. This design tends to reduce bore damage as the chisel cuts through the bushing.

The bushing equipment shown in Fig. 1-34 can remove bushings from open bore installations and can install bushings in both open and closed bores. This universal set consists mainly of various sizes of installing heads, a tool handle, and a set of combination installing head and handle assemblies. The round, multiple-diameter installing head supports a given size bushing as the mechanic installs the bushing in its bore. The smaller diameter of each head fits into and supports the inside surface of the bushing. The outer diameter, its driving flange, bears against the rather thin metal portion of the bushing between its inner and outer diameters. But the outside diameter of the driving flange is slightly smaller than the bushing's outer diameter. This design allows the installing head to pass unrestricted through the housing bore when the mechanic uses it to remove a bushing.

The tool handle of the set is the component that receives the force necessary to remove or install a bushing. This force may be in the form of hammer blows or from an arbor press. The handle itself has a small-diameter machined shank which slides into a hole in each installing head. Furthermore, to protect the head of the handle from the mushrooming effect of hammer blows, the manufacturer grinds a small chamfer around the outer circumference of the head. Finally, to provide a positive gripping point for the mechanic's hand, the manufacturer also knurls the handle section of the tool.

For the installation of small bushings, the combination installing head and handle assembly, as the name implies, combines both the installing head and tool handle into one unit. In other words, on each end of this assembly is an installing head that fits into a given size bushing. With a bushing

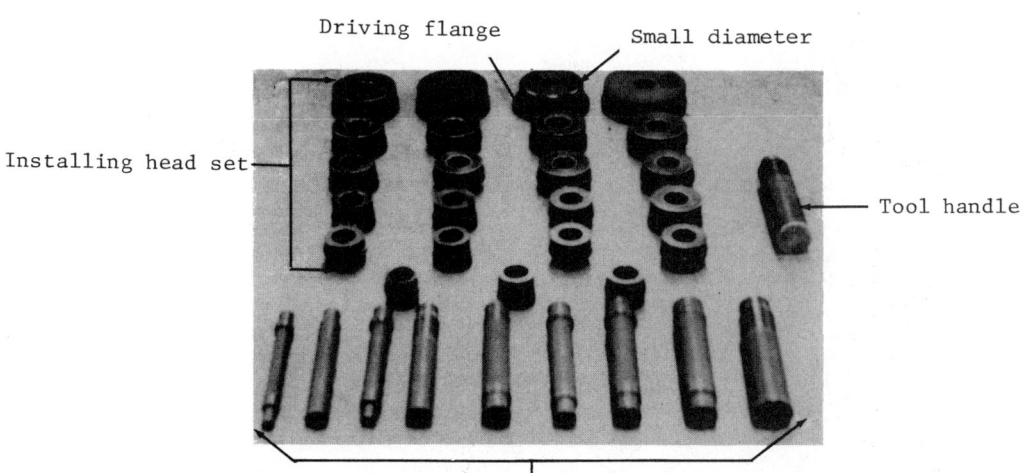

FIG. 1-34 Universal bushing removing and installing set.

installed in one end, the opposite end would then form the driving head where the mechanic applies the pushing force. But in this case, the driving force has to come from a press. <u>A hammer blow would mushroom over the machined surfaces of this assembly; consequently, a bushing would no longer fit over the damaged end.</u>

LIFTING EQUIPMENT

In order to remove, repair, or install an automatic transmission in a motor vehicle, the mechanic must have some method of raising the vehicle before he can work under it. The two most common pieces of equipment used for this purpose are the hydraulic lift or hoist and the floor jack. Of the two, the lift is the most efficient and safest because this unit permits the mechanic to raise the vehicle high enough to work under it standing up. The hydraulic floor jack does not.

Hydraulic Lifts

Two types of hydraulic lifts are in common use today, the single-post and the twin-post lifts both of which usually operate by hydraulic fluid under pressure from a compressed air source. The <u>single post hoist</u> (Fig. 1-35) has a single, hydraulically operated lifting post centered under the vehicle. Attached to the center post are four adjustable support arms and racks which fit underneath the vehicle's frame. With this arrange-

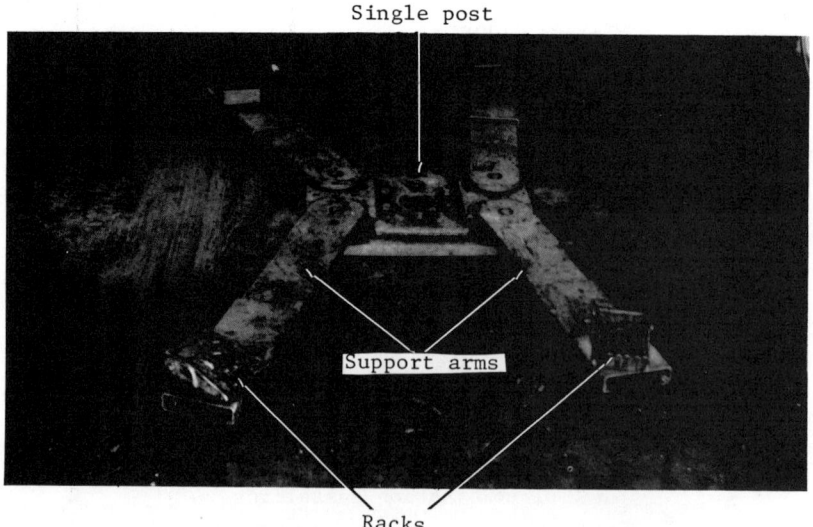

FIG. 1-35 Typical single post lift.

ment, the wheels of the vehicle hang free as the hoist raises it off the shop floor.

Because of its design, the single post lift is not very efficient for transmission repair work. The center post and attaching arms make it difficult for the mechanic to reach and remove certain transmission components. Furthermore, with some transmission and vehicle designs, the single post hoist may block or prevent the removal of the transmission itself.

The twin post lift, on the other hand, has a design that is especially useful in transmission repair work. The unit shown in Fig. 1-36 has two lifting posts which are spaced away from the centerline of the vehicle a given distance. This design provides the mechanic with sufficient working space to repair, remove, or install a transmission. Also, the manufacturer locates the lifting arms and racks to each post in such a manner that this hoist will raise the vehicle by the frame, thus permitting the wheels to hang free.

Another common type of hydraulically-operated twin post lift has a design that raises the vehicle by its front and rear suspension systems. With this design, two individually-operated lifting posts operate near the center of both the front and rear suspension systems. Attached to each of these posts is a lifting arm and pad that is adjustable to fit the various types and sizes of suspension systems.

Manufacturers equip most lifts with some type of safety device that prevents the vehicle from lowering unexpectedly. Figure 1-37 shows one of these devices; it consists mainly of a rack and gear, a locking dog and release handle, and a trip pin. A rack with teeth attaches to the lower side of each booster pad which in turn is fastened to each of the lift posts. The teeth of a locking gear mesh with mating teeth located on one of the racks.

The locking dog, when activated by the trip pin, ratchets on the teeth of the rack as it moves upward; but it wedges between the gear teeth if the

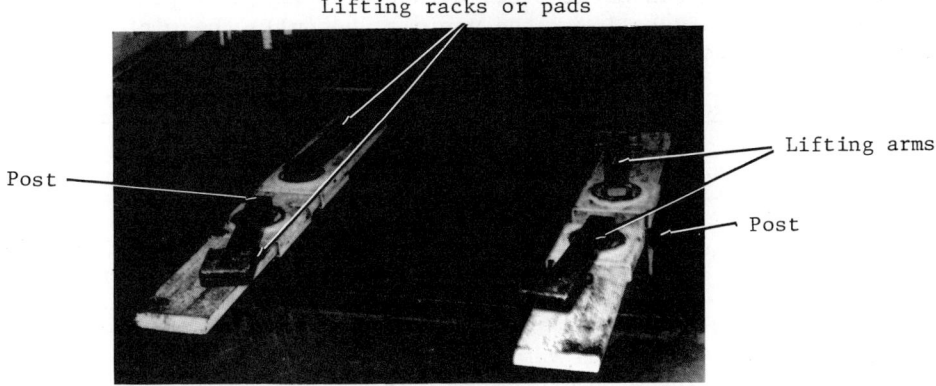

FIG. 1-36 Typical twin post lift.

FIG. 1-37 Typical safety mechanism located on a twin post lift.

rack begins to lower. This action prevents the gear from turning on the rack which, in turn, stops the lift from coming down.

The function of the release handle is to pull the dog away from the gear teeth. With the dog lifted, the mechanic can then lower the hoist in the normal manner.

The trip pin, as previously mentioned, activates the locking dog whenever the lift is in the down position. The pin itself connects to a plate bolted to the bolster. As the lift and bolster come down, the trip pin contacts the release handle and moves it forward. This action sets up the dog mechanism to engage in the gear teeth if the hoist should unexpectedly begin to lower.

Another type of safety device stops the lift from lowering only if the hoist has been raised to its maximum height. This unit uses a weighted lock pin that falls into position in a slot or hole machined into the lower section of a reinforced metal tube, attached to the lift post itself. When the mechanic is ready to lower the vehicle, he pulls the pin out of its slot and lowers the hoist in the normal manner. Finally, to protect himself from injury, <u>the mechanic should install one or more hoist stands under the lift if the unit does not have any safety-type locking mechanism.</u>

Hydraulic Jacks

As previously stated, a hydraulic (floor) jack can raise a vehicle so that the mechanic can perform transmission repair work, and mechanics still use this method of lifting a vehicle when a hoist is not available. But there is a great disadvantage in using a floor jack for this purpose. The jack itself cannot raise the vehicle very high; therefore, the mechanic has little working space under the vehicle and must work under the vehicle on his back. This working position slows down any type of repair work, especially the removal and replacement of the automatic transmission.

Shop Equipment and Tools

The lifting capacity of the average shop floor jack (Fig. 1-38) ranges from 2 to 4 tons. This jack design consists of a jack base, lifting arm with saddle pad, hydraulic ram and pump, and a control handle. The jack base is the framework of the jack and comes equipped with steel wheels in the front and casters in the back. The heavy duty wheels and casters support not only the weight of the entire jack but the vehicle as well. They also provide the means by which the mechanic can position the heavy jack under the load--the vehicle.

The lifting arm with its saddle fastens to the base by means of a bolt that acts as a pivot point for these parts. With this arrangement, as the mechanic pumps the jack handle, the end of the arm moves upward. This causes the saddle pad, mounted on the arm, to contact the vehicle's frame.

The base of the jack also houses the cylinder for the hydraulic ram and pump itself. The hydraulic ram supplies the output force needed to move the lifting arm and pad against the vehicle's weight. The hydraulic pressure required to activate the ram comes from the pump that the handle operates.

The mechanic supplies the input force to the hydraulic pump, housed in the base, via the jack handle; the length of this handle will multiply his efforts. Also, mounted on the handle is the release knob, which is the operating control for the jack. This knob is at the end of the handle, and

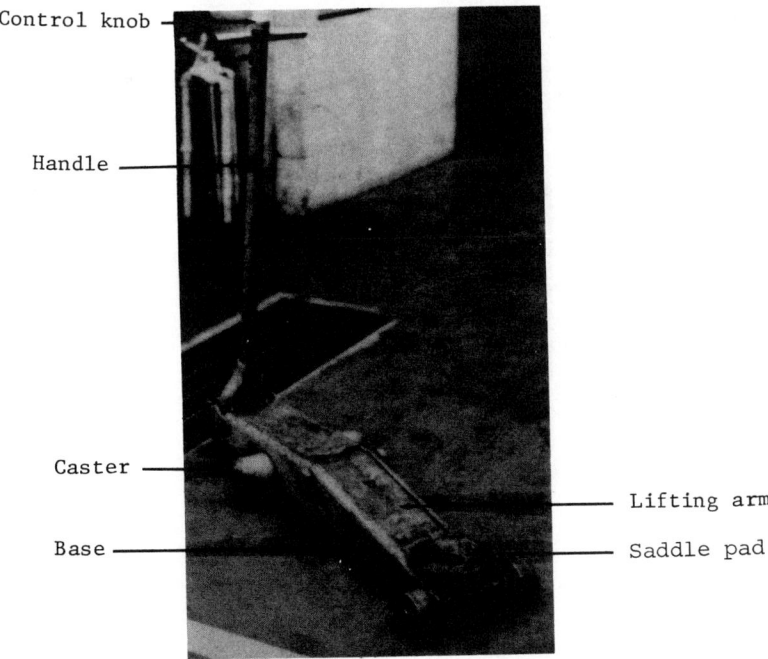

FIG. 1-38 Common hydraulic floor jack.

it controls the release valve located within the jack's hydraulic system. When the operator turns the valve fully clockwise, the valve closes; and his pumping of the handle will raise the lifting arm and saddle pad.

Safety Stands

Whenever a mechanic raises a vehicle with any style of floor jack, <u>he should install safety (jack) stands under the vehicle's frame or suspension system before doing any type of repair work</u>. In service, these stands serve two functions: (1) they support the weight of the vehicle so that the mechanic can remove the jack from under the vehicle (this action also provides him with additional working area); (2) the stands secure the vehicle so that it will not accidentally fall on the repairman while he is working under it.

Most shops now use either the telescoping or ratchet-type stands. The telescoping stand (Fig. 1-39) consists of two strong, steel tubes with different diameters; these tubes slide (telescope) over one another. The stand manufacturer welds one end of the larger tube directly to a round steel base or directly to three steel legs, shaped and reinforced to form a triangle.

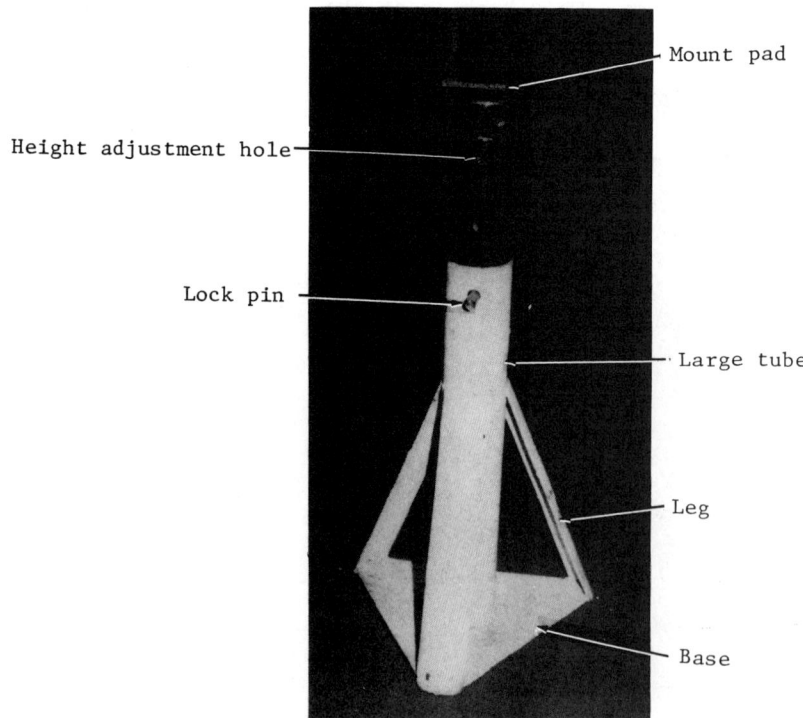

FIG. 1-39 Design of a common telescoping stand.

Shop Equipment and Tools

Onto one end of the smaller tube, the manufacturer welds a vee-shaped mount pad. This pad will fit under the vehicle's frame or suspension system when the stand is in service. The small tube also has a series of evenly spaced holes machined into it that run almost the full length of the tube. These holes index with a hole cut into the base tube.

These holes provide the stand with height adjustments. The holes accomodate a strong steel lock pin, and the mechanic inserts this pin through the holes in both tubes after he adjusts the mount pad to the necessary height.

The ratchet-type stand (Fig. 1-40) consists mainly of a base, a locking dog mechanism, and a rack. This stand manufacturer constructs the square base out of heavy-gauge pressed steel. To increase its strength, the base is rather large in relation to its overall height and has sure-grip type corners.

Built into the upper section of the base is the locking-dog mechanism; this mechanism includes a dog and release handle. The lever portion of the dog engages into the teeth of the rack to lock the rack to the base at various heights for secure load support. The functions of the handle are to pull the dog away from the rack or release it and to act as a carrying handle for the stand.

The heavy steel rack has a series of teeth cut into it on one side. These teeth, as previously mentioned, index with the level section of the dog. This design allows the rack to ratchet freely toward its uppermost position, but the dog locks the rack to the base any time the rack attempts to lower by itself. Finally, a sturdy support bar fastens to the top

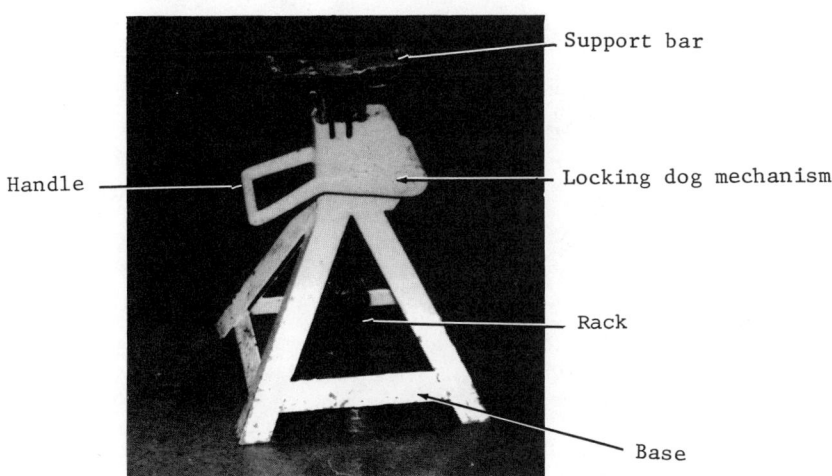

FIG. 1-40 Design of a typical ratchet-type stand.

section of the rack. This bar fits under the vehicle's frame to give it positive support.

Transmission Jacks

Automatic transmissions are very heavy and bulky units and the mechanic should not try to raise, lower, or move them without the aid of a transmission jack. These jacks are available from various manufacturers in different sizes and designs. The overall size and design of the jack will depend on two factors: first, whether the mechanic is going to work under an overhead hoist or beneath a vehicle placed on safety stands; second, the weight of the unit the transmission jack has to support. For example, the heavier units, such as the C-6 and T-400 models, require the use of a hydraulic jack. This jack utilizes hydraulic force to safely raise or lower these large and heavy units, and manufacturers produce hydraulic transmission jacks in both low profile and hoist models.

Shops that must use the low profile jack because of the unavailability of a hoist often have a jack that operates by a worm-screw drive instead of hydraulics. This device safely raises or lowers the smaller transmissions like the C-4 and Powerglide. But they are not very popular because of the time and effort required to raise and lower the transmission with the worm drive.

A typical hydraulic, hoist-type jack with a 1500-pound lifting capacity is shown in Fig. 1-41. This jack assembly includes such components as a base, hydraulic pump, hydraulic ram, and a universal platform or cradle.

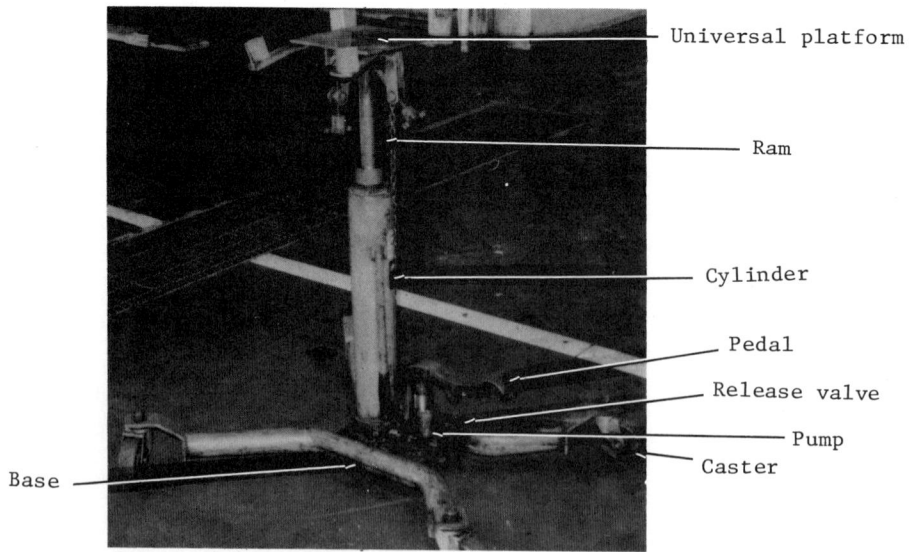

FIG. 1-41 Design of a hydraulic hoist-type transmission jack.

Shop Equipment and Tools

The base itself has four legs made of heavy gauge steel. Attached to the end of each leg is a caster; these casters not only support the total weight of the jack and transmission but also provide the means by which the mechanic can position the jack under the vehicle.

The hydraulic pump and cylinder for the ram both attach to the base. The foot-operated hydraulic pump with its reservoir supplies the fluid pressure necessary to activate the ram. The ram that operates inside a cylinder attached to the base produces the output force actually used to raise or support the transmission. Finally, a foot-operated release valve, located in the hydraulic circuit between the ram and the reservoir, controls the up or down position of the ram. For instance, when the release valve is down, the ram and load lowers because pressurized hydraulic fluid under the ram vents back to the sump.

The universal platform fits on top of the ram and supports the transmission on the jack; the design of a worm-gear arrangement built into this platform is such that it permits the cradle to tilt not only front to rear but also side to side. This tilting of the platform assists the mechanic in aligning the transmission to the engine during the installation process. The average tilt of a typical platform assembly is 50 degrees forward, 15 degrees backward, and 12 degrees sideways.

TRANSMISSION WORK BENCHES

Work benches are very necessary pieces of equipment in any repair shop, but wet benches, as used in automatic transmission shops, are special because they prevent dirty fluid from leaking onto the floor from a transmission the mechanic is tearing down. Therefore, the wet bench assembly reduces the amount of messy clean-up work by collecting dirty fluid and temporarily storing it.

A standard wet-type work bench (Fig. 1-42) consists of the bench, a series of gutters, and a drain-off coupling with an attached sump or container. The manufacturer of these units usually forms this type of bench of heavy-gauge steel with the top (work bench) portion thick enough to withstand the heavy weight of the transmission.

Attached to or made as part of the top is a series of drain gutters, which form a fluid trough around the entire outer edge of the top. Any dirty fluid attempting to leak-off the bench top and onto the floor collects in these gutters. The gutters along the full length of the bench taper down slightly toward one end to allow the dirty fluid to flow by gravity into the sump. Manufacturers accomplish this by installing the gutters slightly lower from the top of the bench on one end than the other or by making the bench legs slightly longer on one end. This latter action, of course, raises one end of the bench a little higher than the other.

The drain-off coupling, an open tube, fastens into the gutter at its lowest point. This design permits the fluid from all the gutters to flow

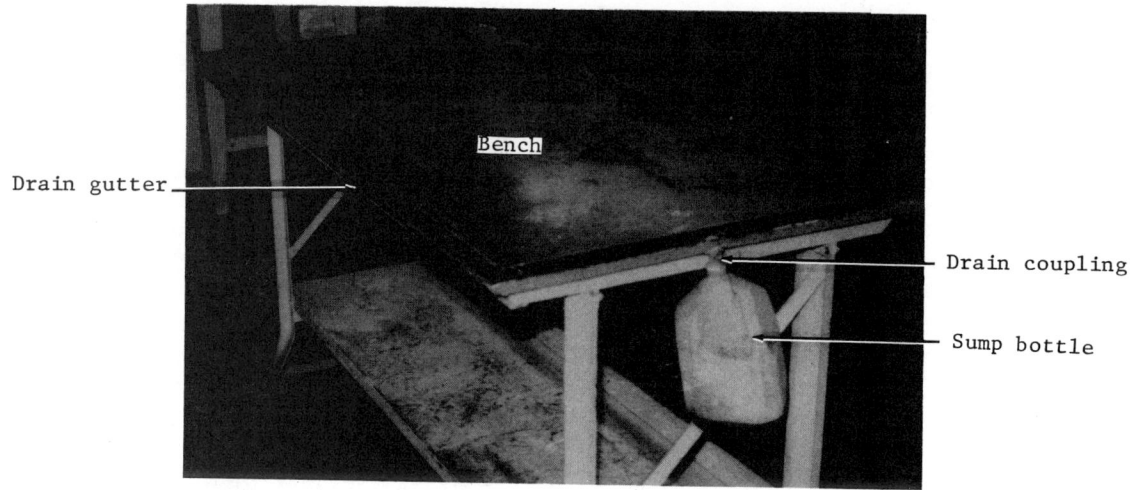

FIG. 1-42 Typical wet-type transmission work bench.

by gravity toward this lowest point and then drain through the open spout and into the sump bottle which stores it. The bottle itself clamps to the under side of the drain coupling, allowing the repairman to easily remove the container for draining as it becomes full.

TRANSMISSION FIXTURES OR STANDS

A transmission holding fixture, like the one shown in Fig. 1-43, is a very useful piece of equipment in any automatic transmission shop. This device supports the heavy and bulky transmission on top of or to one side of the workbench, and the fixture itself is usually adjustable so that the mechanic can rotate the transmission and fixture to a horizontal or vertical position. This feature makes it easier and faster for the repairman to disassemble and reassemble the unit. Although it is not necessary to install all automatic transmissions in a fixture to rebuild them, the fixture is especially helpful when reinstalling heavy or awkward components in a front-loading transmission, like the T-400.

Several manufacturers produce these fixtures in both factory and universal designs. A factory fixture usually fits only one type of transmission, but some units support several transmissions made by the same manufacturer. The design of the universal stand is adaptable enough to accomodate different types of transmissions produced by different companies.

The factory-style fixture assembly, shown in Fig. 1-43, is supporting a T-400 transmission; Fig. 1-44 now shows the assembly by itself. This fixture assembly includes a base and the holding fixture itself. The lower section of the one-piece cast-iron base has a square shape with four holes

Shop Equipment and Tools

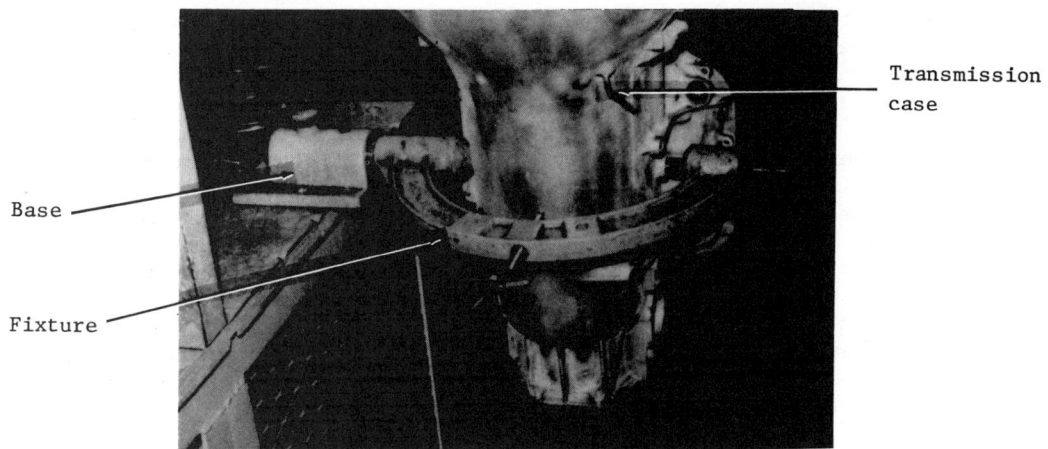

FIG. 1-43 Automatic transmission installed in a holding fixture.

machined into it. These holes accomodate bolts that secure the base to the end of the workbench.

The upper part of the base also has a semicircular shape, and the manufacturer machines a large bore and two smaller holes into it. The large bore acts as a guide and support for the shank section of the holding fixture. The threads, cut into the smaller holes, index with the threaded portion of a set screw and thumbscrew.

The set screw fits into the threaded hole nearest the fixture. The purpose of this screw is to prevent the shank and fixture from accidentally sliding out of its bore in the base.

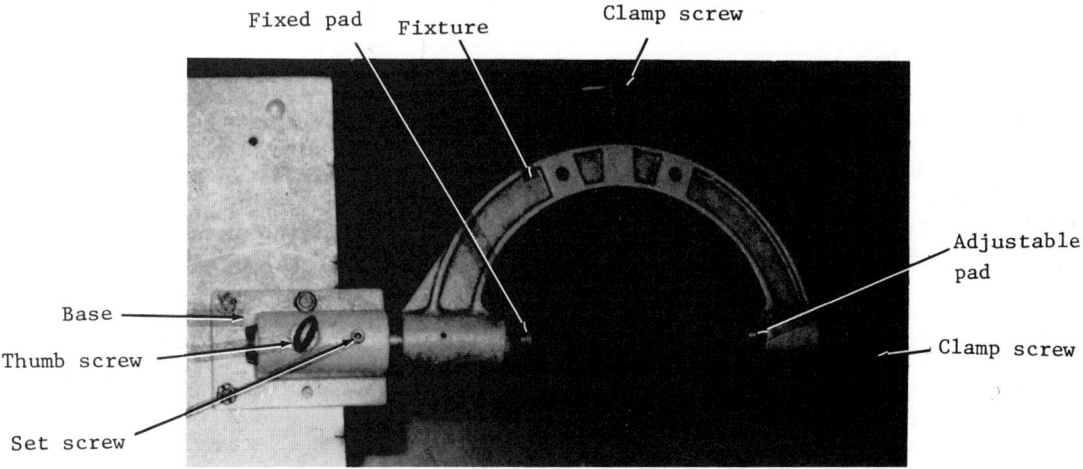

FIG. 1-44 Design of a G.M. transmission holding fixture.

The thumbscrew, on the other hand, threads into the hole farthest away from the fixture. This thumbscrew locks the fixture into either a vertical or horizontal position.

The holding fixture itself attaches to and supports the transmission case in the base assembly; this heavy duty, cast iron fixture has a U-shape along with a solid steel shank and two threaded clamp screws. The solid steel shank presses into a bore machined into one end of the fixture. This shank also has a shallow groove cut into it that aligns with the set screw located in the upper section of the base. As previously mentioned, this set screw, when threaded into the base as far as it will go, prevents the fixture from sliding out of the base unexpectedly. Finally, the shank also has a series of placement holes cut into an area near the end of the shank. These holes index with the end of the thumbscrew to secure the fixture in either the horizontal or vertical position.

The two threaded clamp screws are necessary to hold the transmission case securely into position within the fixture. To understand their function, let's examine how the transmission case attaches to the fixture itself. First of all, the case has two blind mount holes cut into the outside of the case. One of these holes fits over the stationary mount pad on the shank end of the fixture. The other hole accomodates the mount pad fastened to the clamp screw threaded into the opposite side of the fixture.

Now, with the case positioned into the fixture, between the two mount pads, the second (center) screw clamp threads out and bears against the center of the transmission case. This clamp serves as a centering device that maintains the case's centerline almost perpendicular to the centerline of the holding fixture.

THREAD REPAIR TOOL SETS

The mechanic must inspect all threaded holes in the transmission case, front pump, valve body, and extension housing during the rebuilding process for damage, and repair them as necessary, If this is not done, the damaged or stripped threads will not be able to withstand the torque applied to the mating threaded fastener, a bolt or screw. As a result, the fastener will not be tight enough, and the components it holds together can work loose, causing a possible separation of components or an internal or external hydraulic leak.

One method to repair a series of damaged threads is to use the heli-coil thread repair kit (Fig. 1-45). This tool kit includes the heli-coils themselves, a twist drill, heli-coil tap, and installer tool. The heli-coil is a device that the mechanic installs into a specially tapped hole to replace a series of damaged threads. A heli-coil insert is nothing more than a precision formed coil of stainless spring-steel wire, formed into a coil; the wire itself forms the external and internal threads of the insert. The coil manufacturer bends the end of the wire across the coil at one end to

Shop Equipment and Tools

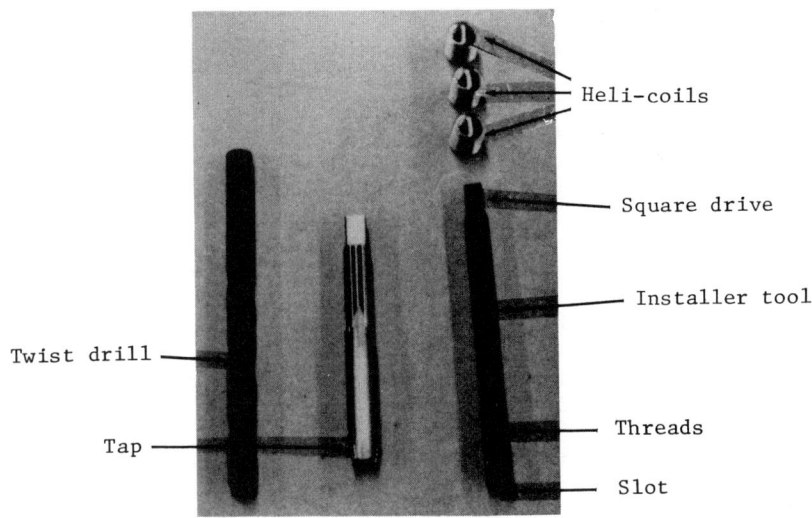

FIG. 1-45 Typical Heli-coil thread repair kit.

form a tang, which is necessary to turn the insert during the installation process. The outer diameter of the heli-coil is also slightly larger than the tapped hole it fits into; consequently, the coil has to compress inward somewhat in order to thread into the tapped hole. Once the mechanic installs the insert into its tapped hole, the coil expands because of its elasticity. This elasticity then helps to lock the heli-coil against the tapped bore.

The twist drill of this kit serves two related functions. First, the drill cuts out all the damaged threads from the bore. Second, the bit resizes the hole in order to prepare it for the heli-coil tap. The function of the tap is to machine a special thread into this oversize hole. The newly cut internal thread has the same pitch as the external threads on the heli-coil insert. But, as previously mentioned, the diameter of the heli-coil's thread is slightly larger than the diameter of the new thread within the tapped hole.

The design of the installing tool is such that it supports and rotates the heli-coil during installation. This tool has external threads that mate with the internal threads of the heli-coil. Furthermore, on the one end of the tool is a slot that engages with the tang located on the end of the heli-coil; this locks the coil to the tool. Finally, on the other end of the tool, the manufacturer machines a square drive. This drive accomodates a tap wrench, used by the mechanic to thread the installer tool and its attached heli-coil into the retapped hole.

POWER TOOLS

Automatic transmission specialists use several types of power tools to service and repair transmissions both in or out of the vehicle. These tools save the mechanic time and energy. Power tools--drills and impact wrenches--increase his productivity during a normal work day and, at the same time, reduce his fatigue.

Normally, the classification of these tools is by the type of power used to make them function: electrically-powered tools and air-powered (pneumatic) tools. Most popular tool producers now offer power tools that do the same job with either electricity or air as their power source.

The drill motors, shown in Fig. 1-46, are samples of one type of power tool. The transmission technician uses this tool to drive a twist drill which, in turn, machines a hole in a given component. For instance, in transmission repair and service work, the drill motor and twist drill together cut the holes necessary to insert heli-coils and alter valve bodies for the installation of shift kits.

Figure 1-46 pictures both the electric and air-type motors. The electric drill motor is not as popular today as the air drill for transmission and general automotive-type repair for several good reasons. For instance, although electricity is probably the most widely used power source today, it can be very dangerous under shop conditions. With the drill motor, normal brush arc can ignite any nearby volatile fumes. Also, if a grounded condition happens to occur within the motor itself, and the cord attached to the drill is not the three-wire type with a three-blade plug, the mechanic can receive a severe electrical shock.

Air drills, on the other hand, do not have these inherent, negative characteristics and are easier to use than the electric type because they are smaller and lighter and run cooler. The air drill, with the same overall

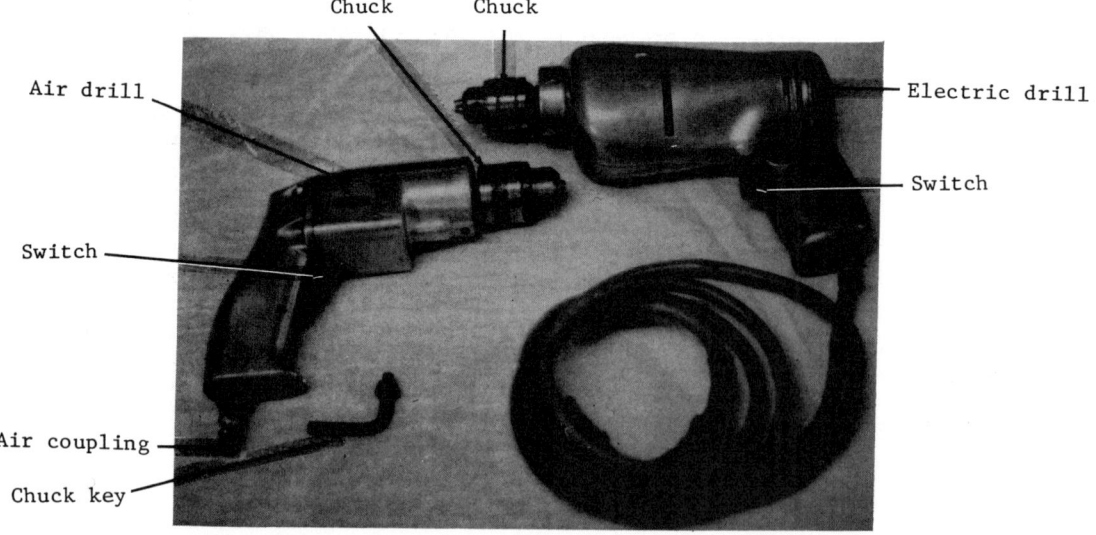

FIG. 1-46 Typical electric and air drills.

Shop Equipment and Tools

chuck size and power rating as the electric type, is smaller and quite a bit lighter because of the difference in design between the drive motor found in each unit. Furthermore, since electricity is not the power source to rotate the motor itself, the operating temperature of the air drill, even under loaded conditions, remains relatively low.

Both drill motor designs do have some of the same features. For example, both units have a three-jaw chuck that tightly holds the twist drill in place and a chuck key to tighten and loosen the chuck. Furthermore, manufacturers of both drill designs designate the size of their units by the maximum capacity of the chuck, and the chuck sizes that most automatic transmission shops use have a capacity of 1/4 inch or 3/8 inch. Finally, both motor types are available with switches that allow them to operate with varying speeds, which is very helpful when cutting holes in different types of materials.

Figure 1-47 illustrates other types of air-powered tools--the impact wrench and the air ratchet. The impact wrench, as its name implies, loosens or tightens most bolts or nuts that hold the transmission itself together. With this device, the mechanic squeezes a trigger to activate the working mechanism of the wrench, which applies partial to full impact force on the fastener to loosen or tighten it. Manufacturers do produce electric impact wrenches, but they are not very popular today for the same reasons that inhibit the use of the electric drill.

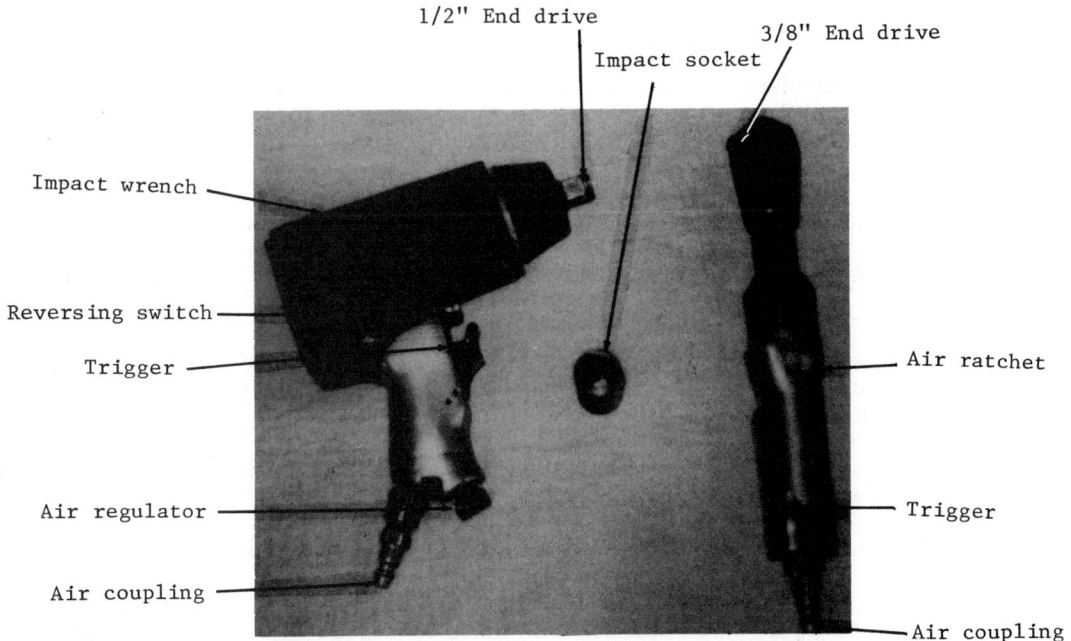

FIG. 1-47 Air impact wrench and air ratchet.

A common method of rating air-impact wrenches is by the size of its end drive. The most common wrenches used by transmission technicians have either a 3/8- or 1/2-inch end drive. These end drives, in turn, mate with 3/8- or 1/2-inch drive impact sockets. These sockets are special in that the manufacturer reinforces them so that the socket can withstand the impact (hammering type) loads placed on them by the air wrench.

Most air impact wrenches also have some method of regulating input (shop) air pressure to the driving impact mechanism inside the wrench. The position of this regulator then determines not only the speed of the end drive but the torque applied to the impact socket. A typical 3/8-inch air wrench, for example, has a regulator that controls torque output from zero to a maximum of 75 pounds-foot (10.35 Kg-m) while the working torque range of a regulated 1/2-inch impact will be between zero and about 200 pounds-foot (27.6 Kg-m).

When using any size air impact wrench, follow these recommendations, and the tool will provide long, maintenance-free service along with a reduction in fastener damage:

1. Always use the special impact sockets with an impact wrench.

2. Always utilize the proper size socket for the nut or bolt.

3. Always put to use with the wrench the simplest assembly of a socket, extension, and universal joint.

4. Where possible, always use a deep socket in place of a standard-length socket and extension.

5. Always hold the wrench in a position so the socket fits squarely on the nut or bolt. As necessary, apply a slight forward pressure on the wrench to hold the socket in place.

6. Once a bolt or nut is tight, never impact it beyond an additional one-half turn on the socket. Continued tightening beyond this point will strip the threads or break the bolt, and <u>never use the impact wrench to tighten down fasteners which call for a given torque specification.</u>

7. Never utilize an impact wrench beyond its rated capacity. If the fastener stays tight after impacting with the tool for 5 seconds, use a larger wrench.

8. Always presoak large, rusty bolts and nuts with penetrating oil before attempting to remove them.

The air ratchet, shown in Fig. 1-47, is another type of device that saves the repairman time and energy. This reversible 3/8-inch tool provides a small amount of output torque, about 45 pounds-foot, (6.21 Kg-m) for speeding up the removal and replacement of nuts and bolts. When in service, the mechanic uses the tool as a common ratchet first to break the fastener loose; then he depresses the switch to spin the nut or bolt off. Of course, during the installation of hardware, he reverses this process.

This wrench, due to its unique design, is very versatile. It fits easily into limited work areas because of its smaller size. This increases the mechanic's productivity in removing and disassembling the transmission. Also, because of its low torque output, the serviceman can use the regulated air ratchet to run in smaller fasteners when assembling the transmission. Finally, due to its design and output, this wrench does not require the use of the special impact sockets; therefore, the mechanic can use all the sockets and attachments of a standard 3/8-inch set with this tool.

For proper performance, any type of air powered tool requires a regulated supply of clean, dry compressed air. The recommended air pressure necessary to operate most common air tools is about 100 psi (7.03 Kg/cm^2). This rated air pressure is the pressure at the tool while the tool is running under a no-load condition, and it is measurable by attaching a pressure gauge as close to the tool as possible. The pressure that registers on this pressure gauge is the pressure <u>at the tool</u> and not the output pressure of the shop's air compressor.

Several problems will occur if the serviceman uses an air tool operated by too low or too high air pressures. If the air pressure is less than recommended, the overall efficiency of the tool is reduced. On the other hand, excessively high air pressure causes the tool to operate above its rated capacity. This action shortens the life of the tool and can damage or break fasteners.

The mechanic should lubricate the air tool each day before using it, by applying three or four squirts of recommended oil into the tool's air inlet, connecting it to a source of air pressure, and operating the tool. This action flushes any moisture, dirt, and gum out of the air motor and lubricates its moving parts. The injected oil removes deposits from the tool by carrying them out the air exhaust.

When performing this lubrication procedure, observe these safety precautions: (1) do not flush out the tool around an open flame; (2) always point the tool's air exhaust port away from your skin or clothing.

PULLERS

Gear Type Pullers

The mechanic uses several types of pullers when disassembling an average automatic transmission. These tools, as their names imply, remove gears, seals, or pumps using either a pulling-type or impact-type force. For example, Fig. 1-48 pictures two pullers used to remove the pressed-on speedometer gear from the output shaft of two different transmission types.

54 SECTION 1

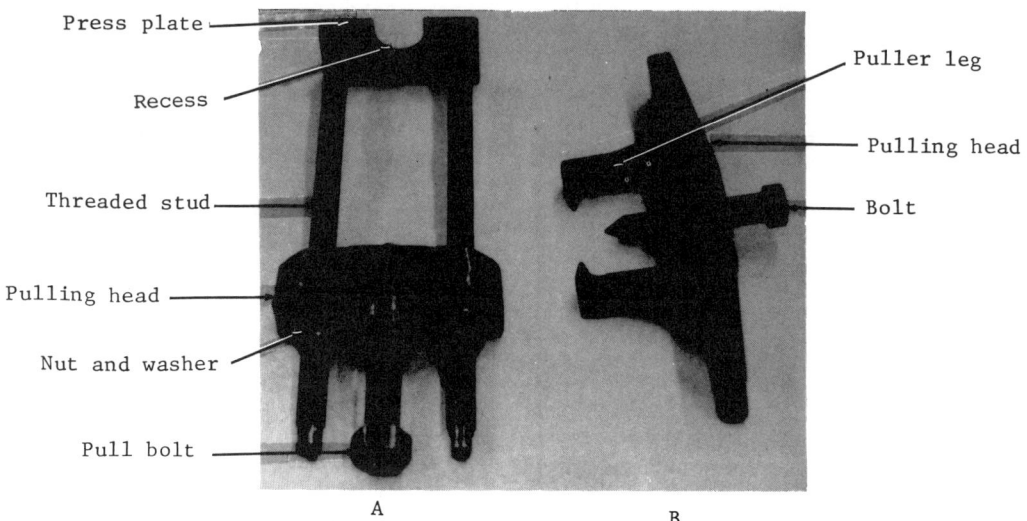

FIG. 1-48 Two typical speedometer gear pullers.

Both of these devices use the incline-plane principle, via the threads on the large bolt, to supply the force necessary to remove the gear.

The Powerglide transmission puller shown at A is a unique device in that the tool not only pulls the gear off, but it also presses the gear back on the output shaft as well. This tool consists of a remover plate, two long threaded studs with washers and nuts, and a puller head and bolt. The remover plate is nothing more than a piece of steel plate with a recessed opening and two tapped holes. The opening is large enough so that the remover will fit around the output shaft and under the gear. The recessed area that extends from the opening conforms to the gear's shape and provides adequate support to this area during the removal of installation procedures.

The ends of the two studs thread into the tapped holes in the remover plate; these studs along with their washers and nuts serve several functions. First, they act as a rigid but adjustable connection between the remover and head. By turning the two nuts one way or the other, the mechanic can shorten or lengthen the distance between the remover and head to match the length of the output shaft (Fig. 1-48).

Second, the technician can use these two studs and their hardware to push or press the gear back onto the output shaft. For a pressing operation, the two nuts fit under the puller head with the washers against the head itself. Now with the puller head in position over the output shaft, the repairman can easily press the gear on by turning each nut out a little at a time until the gear moves fully into position on the shaft.

The pulling head and bolt also serves two functions. First, the two components together provide the force necessary to remove the gear from the

Shop Equipment and Tools

shaft. In this situation, the bolt threads through the head until it contacts the end of the output shaft. Any further rotation of this bolt causes sufficient force on the head, studs, and remover to pull the gear off the shaft.

Second, the head along with the aid of the bolt acts as a reaction area during the gear pressing operation. To accomplish this function, the head must lock to the output shaft. The tool manufacturer accomplishes this by machining a retaining collar into the lower side of the head; this collar, in turn, fits into a groove in the output shaft. Now, with the head attached to the shaft and its bolt threaded in against the end of the shaft, the head can act as the reaction area (stop) upon which the threaded nut and washers react while pushing the gear in place over the output shaft.

The design of one type of T-400 transmission puller, shown at B of Fig. 1-48, is such that it removes a gear only from its shaft. This tool consists basically of a pair of removers, bolts of varying lengths, and the puller head and bolt. The pointed end of each remover fits under the gear, and they are adjustable to fit any gear size via slots machined into the puller head.

Two bolts attach the removers to the puller head. The actual length of the bolts the mechanic will use depends on the length of the shaft the gear is on. In Fig. 1-49, the bolts are short, which brings the two removers in direct contact with the head.

The puller head and bolt supply the force necessary to remove the gear. The bolt itself threads through the center of the head and bears against the shaft. Now, if the removers are in place around a gear, and the serviceman continues to turn the bolt clockwise, the tool will pull the gear off the shaft.

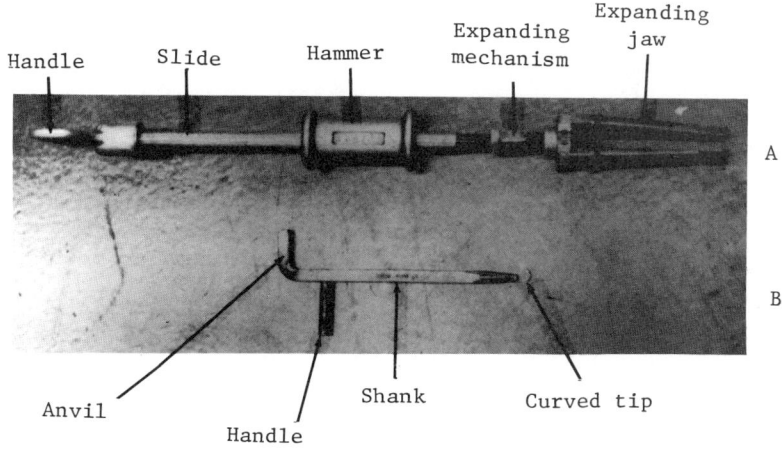

FIG. 1-49 Two common types of universal seal pullers.

Seal Pullers

It is also necessary, in most cases, for a mechanic to use another type of puller to remove metal-clad seals from the front-pump or extension housing. The puller normally used for this purpose employs impact force to loosen and remove the seal from its housing bore.

Figure 1-49 shows two kinds of universal--impact seal pullers. The tool shown in A is a slide-hammer type that includes a set of expanding jaws, a slide and handle, and the slide hammer itself. The set of jaws fits inside the seal with each jaw bearing against a steel lip formed into the backing. Furthermore, an expanding mechanism threaded onto the end of the slide not only adjusts the set of jaws to fit various seal sizes, but also keeps the individual jaws in firm contact with the steel backing.

The slide with its attached handle acts as a guide for the movable slide hammer. The hammer fits over the slide and moves back and forth on the unthreaded portion of the slide between the expanding mechanism and the stop built into the handle. The handle therefore acts as a stop for the slide hammer and as a place where the mechanic can support the end of the puller.

The heavy slide hammer supplies the tool with the impact force necessary to loosen and remove the seal. The hammer itself, when moved from the jaw end of the slide toward the handle, imposes an impact blow on the handle stop. This blow attempts to force the stop, slide, and jaw set in the direction of the hammer blow. The end result of one or more hammer blows is the removal of the seal from its housing.

The design of the universal tool shown at B, Fig. 1-49, is such that it also uses the impact energy of hammer blows to remove a seal, but in this case, the blows come from a hand-held, ball-peen hammer. The tool itself is made out of square-steel stock with one end of the tool bent at a 90-degree angle to the shank. This end now forms an anvil that receives the impact force from the hammer blows.

On the opposite end of the shank from the anvil is a curved tip made of hardened steel. This tip fits inside the seal and against the lip formed into the steel backing. With this design, any impact force from a hammer blow on the anvil passes from the anvil to the shank and finally to the tip and seal.

The knurled handle threads into the shank. The technician uses this handle to support and guide the tool during a seal pulling operation.

Pump Pullers

Certain transmission models also require the use of a puller to remove the pump from the transmission case; this puller is necessary for several reasons. First, the internal component design of the transmission is such that it is impossible to get behind the pump to push it out. Second, the

Shop Equipment and Tools

pump's outer seal, which exerts some tension on its bore inside the case along with a stuck pump gasket, usually makes it impossible for the mechanic to remove the pump by hand.

Figure 1-50 shows the most common type of pump puller. This puller consists of two hand-held slide hammers and two long bolts. Each weighted slide hammer fits over one of the bolts, and it slides back and forth on the unthreaded section of the bolt. If the hammer strikes the head of the bolt, its impact energy transmits to the head. This action results in a pulling-type force on the bolt itself in the direction the hammer was moving.

The two bolts which act as the guides for the hammers each thread into a tapped hole within the pump housing. Each bolt also has a nut to lock the threaded end of the bolt in place, so it cannot come loose during the pulling operation and damage the threads. Now, with these bolts in position, the mechanic can easily remove the pump by firmly pulling both hammers back until they contact the bolt head; the resulting impact force will pull the pump from the case.

SNAP-RING PLIERS

The factory installs several types of snap rings when assembling an automatic transmission to lock or hold a component in a given position. Some snap rings, like those used in a clutch drum, the mechanic can remove easily with a screwdriver. Others require the use of special snap-ring pliers for both their removal and installation.

Snap-ring pliers are of two types, inside and outside. The jaws of the inside snap-ring pliers close and grip an internal snap-ring when the mechanic closes the handles. The jaws on the outside snap-ring pliers open to expand an external snap ring as the handles come together.

Figure 1-51 pictures several sizes and types of snap-ring pliers.

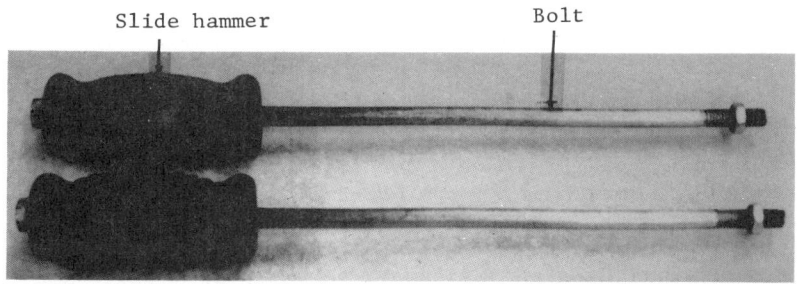

FIG. 1-50 Set of slide hammers, used to remove the front pump from a G.M. transmission.

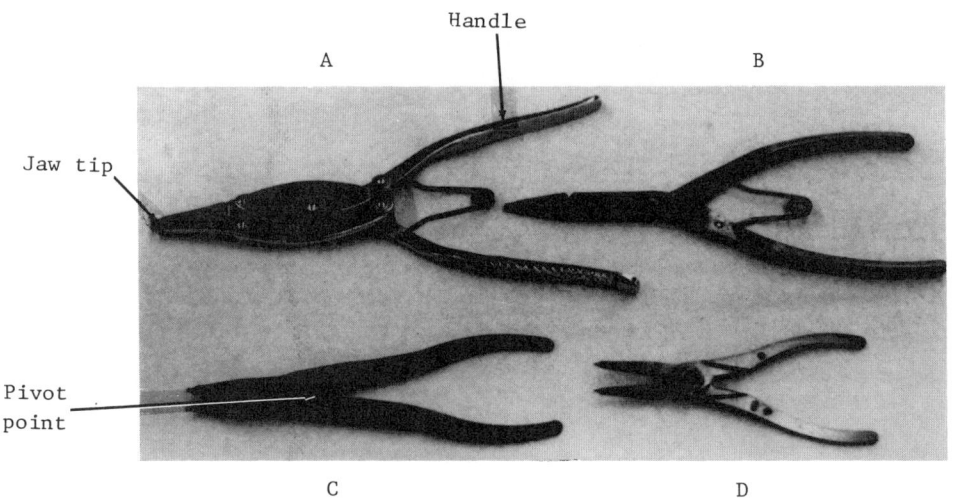

FIG. 1-51 Several sizes and designs of snap-ring pliers.

Pliers A and B are both outside-type tools which fit plain snap rings. The main difference between these pliers is in the design, shape, and free opening of the jaw tips.

Pliers C and D are outside Truarc pliers which have a specially designed jaw tip. The manufacturer in this case machines a round pin on the tip of each jaw. These pins fit into the holes formed into the lip ends of a Truarc snap ring. With this design, there is less possibility of the pliers slipping off the ring during removal or installation.

Some Truarc pliers also have detachable tips, while others have several pivot points. The detachable tip design permits the mechanic with the help of one pair of pliers and various sizes of interchangeable tips to service a wide variety of different-sized snap rings. If the tool has two interchangeable pivot points, the mechanic can convert an inside pair of snap-ring pliers to an outside pair.

OTHER HAND TOOLS AND SPECIAL EQUIPMENT

On the market today are other multipurpose hand tools that make the automatic transmission mechanic's job a lot easier and faster. Figure 1-52 shows three of these devices, the scribe, the pencil magnet, and the modulator wrench. A mechanic uses the pointed scribe for removing small snap rings or seals in places where a small screwdriver, for example, will not fit. The pencil magnet is very useful in removing check balls from fluid passages in the transmission case or valve body, and it is also quite helpful for removing valves and springs when overhauling a valve body.

Shop Equipment and Tools

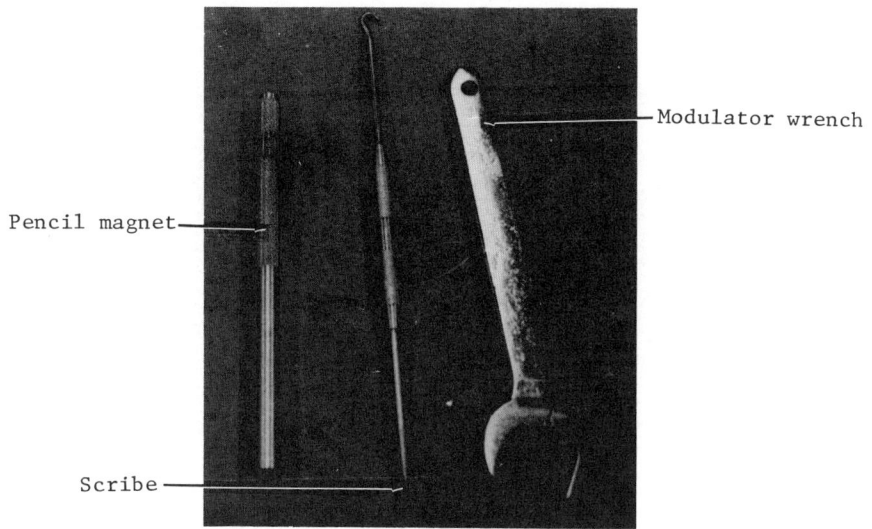

FIG. 1-52 Typical scribe, pencil magnet, and modulator wrench.

These two tools have, of course, other service applications, but the important thing to remember about them is that the technician can use them on all transmission models.

The modulator wrench is necessary, in most cases, to remove and install a vacuum modulator without damaging the can, but this special tool is not as universal as the scribe and magnet. Most modulator wrenches are made for a specific transmission, others for a family of transmissions, and only a few are universal.

All automatic transmission manufacturers recommend the use of certain other special tools when rebuilding their particular units. All of these tools are nice to own because they make the rebuilding job easier and faster. But they are very expensive, sometimes hard to get, and not always necessary.

Figure 1-53 shows some examples of the special tools utilized to rebuild the T-400 transmissions. The two devices shown at A are seal protectors, used when overhauling the forward and direct clutches. The device pictured at B is a center support alignment tool. Item C is an accumulator piston installer, used to remove and install the accumulator piston in the valve body. The three items at D are the rear band apply fixture with its two apply pin gauges; these tools are necessary to determine the correct selective apply pin to install with the rear servo piston. And finally, tool E is a speedometer gear installer, used to drive this gear onto the output shaft.

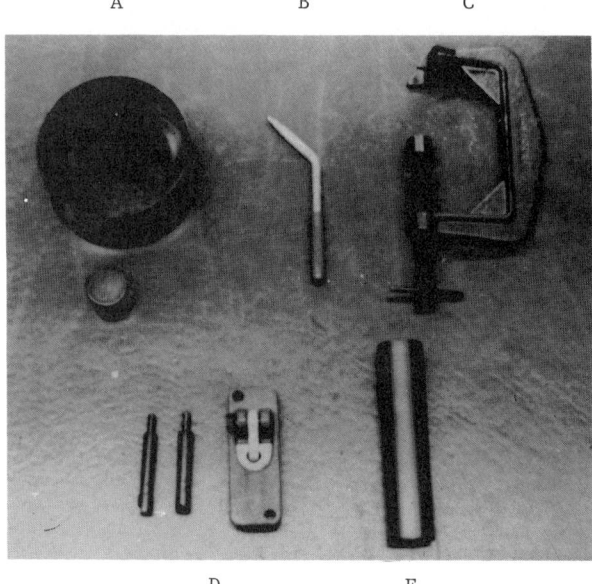

FIG. 1-53 Some of the special tools needed to rebuild a T-400 transmission.

CHECK-UP QUESTIONS

The questions listed below will assist you in determining how well you remember the material contained in this section. Read each question carefully before adding the word or words necessary to complete the sentence. If you can't complete the sentence, review that portion of the section which covers the question.

1. The function of a torque converter flusher is to _____ used converters and eliminate costly comebacks.

2. The flushing machine, described in this section, utilizes the _____ method to circulate the cleaning agent inside the converter.

3. Many sealed converters come from the factory without a _____ _____; this makes servicing the converter much more difficult.

4. The dynamometer _____ _____ the rebuilt or malfunctioning transmission without the need of installing the transmission into the vehicle.

5. The speed-control cylinder of the dynamometer has the responsibility of actually changing the width of the _____.

6. The dynamometer load-application gauge shows the amount of _____ applied to the machine's load section.

Shop Equipment and Tools   61

7. The dynamometer's 12-volt system tests the _____ _____ circuit on the T-300 and 400 transmissions.

8. The hydraulic gauge set tests the _____ _____ of an automatic transmission that is still in the vehicle.

9. Transmission repair shops can use _____ types of cleaning equipment.

10. The heating coil of a steam cleaner changes tap water into _____.

11. The oscillating jet assemblies of a jet cleaner are _____ the cleaning chamber.

12. The safety-type parts washer, mentioned in this section has an _____ agitation system.

13. A clutch spring compressor is not usually necessary to compress a _____ type return spring.

14. The factory-type, clutch-spring compressor mentioned in this section uses the principle of the _____ _____ to multiply force.

15. Bushings are a special type of _____ that supports revolving transmission components.

16. A mechanic can use a bushing chisel to remove a bushing for a _____ bore.

17. Because of its design, the _____ lift is not very efficient for transmission repair work.

18. The lifting capacity of the average shop floor jack ranges between _____ and _____ tons.

19. Most shops now use either the _____ or _____ safety stands.

20. The _____ type work bench is a necessary piece of equipment in a transmission shop.

21. A factory-type transmission holding fixture will usually fit only _____ transmission type.

22. One method the mechanic can use to repair damaged threads is through the use of a _____ _____ _____.

23. Mechanics use the air-type drill because it is _____, _____, and runs cooler.

24. Along with the impact wrench, the _____ _____ also saves the repairman both time and energy.

25. Pullers, used on automatic transmission, primarily utilize _____ or _____ force to remove a gear.

26. The factory installs several types of _____ _____ to lock or hold a component in a given location.

27. The special tool that is very useful in removing valves from the valve body is the _____ _____.

For the answers to these check-up questions, turn to the Appendix in the back of the book.

SECTION

2

Measuring Devices and Fasteners

REFERENCES: Manufacturers' instructions on the operation of each type of measuring device.
Automatic Transmission Fundamentals, Chapters 5, 6, and 8.

One of the more important jobs of the technician performing service or repair work on the automatic transmission is the taking of accurate measurements. The purpose behind this particular task is to make sure that the transmission itself, along with all of its components, is correctly put together and will operate with given tolerances, called factory specifications. Sometimes this means measuring the torque on bolts or nuts. In other cases, it means checking the clearance in clutch assemblies or the end play of rotating-type shafts. But in order to check items against the specifications, the repairman must be familiar with the function of various kinds of measuring tools like the torque wrench, micrometer, feeler gauge, and dial indicator.

MEASURING DEVICES

Torque Wrenches

The torque wrench is a measuring tool that is essential to properly overhaul the automatic transmission. This wrench is basically a measuring device that indicates the amount of torque or twist applied to the fasteners, bolts, or nuts of the transmission during reassembly. <u>Remember that the automatic transmission manufacturer establishes a given torque value for all threaded fasteners, and these specifications require exact measurements. Therefore, do not depend on experienced feel or a regulated impact wrench to torque these fasteners</u>.

   Failure to use a torque wrench or failure to follow factory specifications can cause damage to transmission components or cause the unit to malfunction. For example, over-torquing a valve body cap screw can result in a broken housing, pulled threads, or a warped valve body. Under-torquing of the same hardware can cause a hydraulic leak, resulting in a transmission malfunction or total failure.

   Manufacturers produce torque wrenches not only in several shapes and sizes, but also with different methods of reading or calibrating the torque value applied by the wrench. Figure 2-1 shows several examples of different types and sizes of wrenches. Wrenches <u>A</u> and <u>B</u> are micrometer- or clicker-type wrenches. Wrenches <u>C</u> and <u>D</u> are both beam-type torque wrenches.

   The clicker-type torque wrench has a particular torque value calibrated into it by the mechanic. When the mechanic tightens a fastener to this value with this wrench, it makes a load click, and the wrench handle feels as though the tool has broken loose from the fastener itself.

   In order to calibrate the torque limit of this wrench, the mechanic loosens the lock nut and rotates the hand grip one direction or the other. The hand grip itself is part of a micrometer mechanism built into the

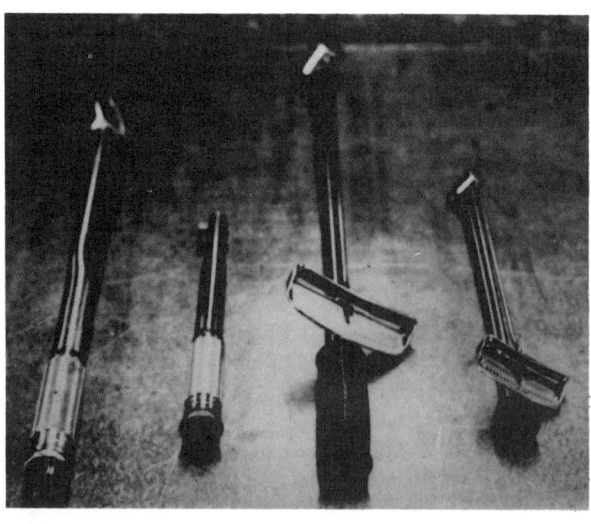

  A   B   C   D

   FIG. 2-1   Examples of different types and sizes of torque
              wrenches.

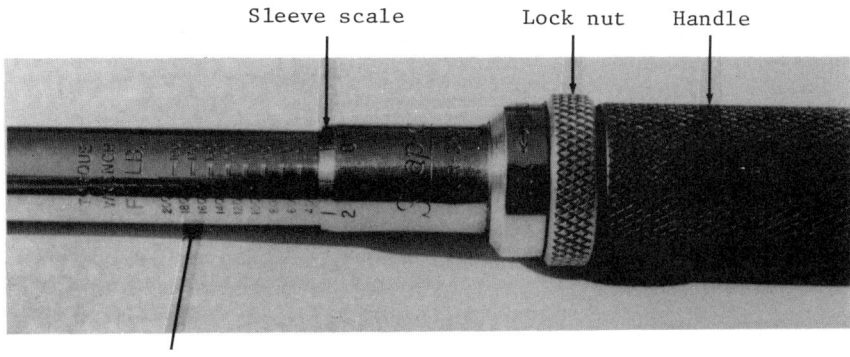

FIG. 2-2 Micrometer mechanism built into the handle section of the clicker-type torque wrench.

handle end of the wrench (Fig. 2-2). By twisting the hand grip and observing the micrometer scale, the mechanic can set whatever torque value he desires into the wrench; and the wrench will automatically click and break free at this torque setting. This provides a very accurate method of torquing nuts or bolts.

The design of the deflecting-beam torque wrench (Fig. 2-3) is such that the amount of torque applied to the fastener registers on a scale as the mechanic pulls on the handle. With this wrench design, the beam itself bends (deflects) in proportion to the amount of torque applied to the fastener. On the free end of the beam is a pointer that moves across the indicator scale mounted on the arm just above the handle. Since the arm does not bend with the beam, the position of the pointer on the scale indicates the amount of applied torque by measuring the amount of bend. Thus, the mechanic can determine the actual amount of torque applied at any given time by simply observing the pointer's position on the scale.

Torque wrench manufacturers produce the clicker and beam wrenches with various sizes of square end drives and overall torque ranges. For example, a common 1/4-inch-drive micrometer torque wrench measures applied torque from 30 to 200 pounds-inch (34.5 t0 230 Kg-cm). A 3/8-inch drive wrench of the same design has settings from 5 to 75 pounds-foot (.69 to 10.35 Kg-m).

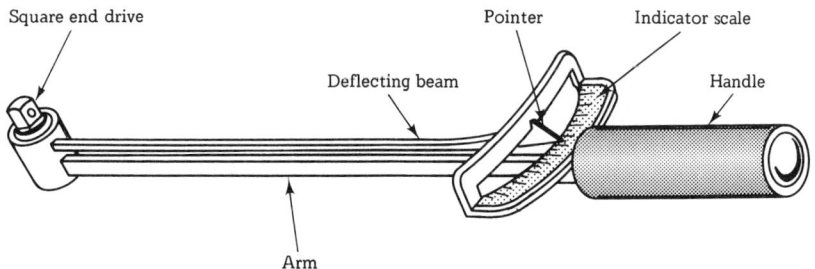

FIG. 2-3 Deflecting beam torque wrench.

A typical 1/2-inch-drive beam-type torque wrench has a torque range from 0 to 200 pounds-foot (0 to 27.6 Kg-m).

Three factors determine the size torque wrench the mechanic should use on a given job. The first factor is the diameter of the bolt or screw; the second is the torque specification for the fastener; the third is the torque range of the wrench itself. In most cases, the torque on the bolt or screw does relate to its diameter because fasteners of different diameters have, of course, varying shear points under torque load. However, many nuts, bolts, or screws receive a torque value way below their shear point.

This reduced specification protects the components attached together by the fasteners from distortion or breakage. Thus, if a torque specification for a given bolt is 30 pounds-inch, a mechanic must use a small torque wrench similar to the 1/4-inch drive, mentioned earlier, to tighten the bolt. If he happens to use a larger wrench that cannot accurately measure this amount, the hardware may break or the components they fasten together may distort. To prevent damage to the torque wrench itself, the mechanic should select a wrench with a torquing capacity high enough so that the specified torque value will fall near the midpoint on the wrench's scale.

When torquing bolts or nuts in obstructed areas, it is sometimes necessary to add certain extensions and or adapters to the torque wrench in order to provide it with operating clearance. As long as the extension or adapter fits onto the end drive at a 90-degree angle to the wrench handle (Fig. 2-4), the amount of torque, indicated on the scale or set into the clicker-type wrench, will be accurate. Whatever torque reading the mechanic observes on the scale or sets into the wrench will be what the wrench applies to the fastener.

If the extension or adapter adds length that is parallel to the handle of the wrench (Fig. 2-5), the torque applied to the fastener will be increased. The amount of torque showing on the scale or set into the wrench <u>will no longer be accurate</u>. Consequently, whenever adding length to the torque wrench, be prepared to calculate a new specification before torquing the hardware.

FIG. 2-4 Extension installed on the end drive of a torque wrench at a 90 degree angle to the wrench handle.

Measuring Devices and Fasteners 67

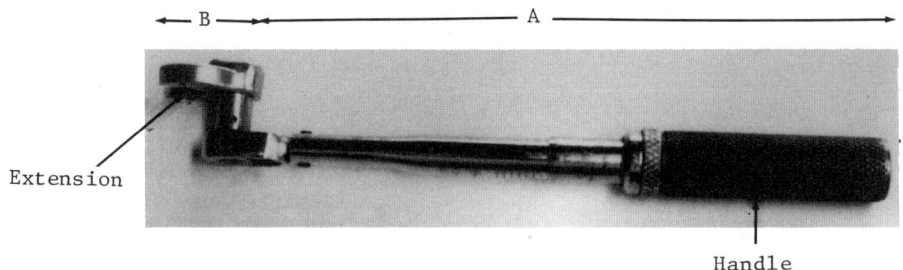

FIG. 2-5 Adapter unit installed on the end drive of a torque wrench, parallel to the wrench handle.

To determine the proper torque when length extensions are necessary, use the following simple formula, $D = \frac{A \times T}{A+B}$: D is the new dial reading or amount of torque to be preset into the wrench; A is the original length of the torque wrench from the center of the square end drive to the center of the hand grip; B is the length of the extension from the center of the end drive to the center of the adapter, parallel with the wrench handle; and T is the factory torque specification for the fastener.

Now let us put this information to work on a practical torquing application. Suppose that, for example, a given inch-pound torque wrench has a length (A) of 8 inches. The adapter needed to clear the obstruction (B) has a length of 1 inch; and the specifications call for an applied torque (T) of 45 pounds-inch. Referring to the formula above:

$$D = \frac{8 \text{ inches} \times 45 \text{ pounds-inch}}{8 \text{ inches} + 1 \text{ inch}}, \text{ or } \frac{360 \text{ pounds-inch}}{9 \text{ inches}}, \text{ or } 40 \text{ pounds-inch}.$$

The correct torque reading on the scale or set into the wrench must now be 40 pounds-inch instead of 45 pounds-inch--the amount it would be without the added adapter or extension.

In conclusion, when using the torque wrench observe the following general rules:

1. A torque wrench is a precision measuring device; do not use it as a general purpose turning tool.

2. Never use the torque wrench to tighten a bolt or nut to a higher value than the maximum capacity of the wrench.

3. Never use a torque wrench on a bolt or nut tightened with a wrench, socket set, or impact tool. For an accurate torque reading, the mechanic must use a torque wrench for the final tightening sequence.

4. On a clicker-type torque wrench, never position the handle of the micrometer below the minimum torque setting on the scale. However,

when <u>storing</u> this tool, <u>always adjust the handle to its lowest setting</u>.

5. Always have the clicker-type wrench calibrated at the interval periods, specified by the manufacturer.

6. Always clean and lubricate the threads of the fasteners before torquing; this action reduces thread friction that can cause an inaccurate torque reading.

7. Always pull on the torque wrench with an even and steady movement; a fast or jerky motion will cause an incorrect application of torque on the nut or bolt.

8. Always follow the manufacturer's torque specifications for each fastener used on the transmission.

Micrometers

Another kind of measuring device frequently used by transmission mechanics is the micrometer (mike), a precision instrument that the technician uses to determine the thickness of such items as selective thrust washers and snap rings. The micrometer itself measures these parts much more accurately than a ruler because the micrometer reads in tenths, hundredths, thousandths, and sometimes ten-thousandths of an inch instead of fractions of an inch like most rulers.

Figure 2-6 shows a 0 to 1-inch micrometer used in automatic transmission service work. This standard mike consists of a frame, anvil, spindle, hub, thimble, and ratchet stop. The frame, as the name implies, is the U-shaped housing or framework of the micrometer.

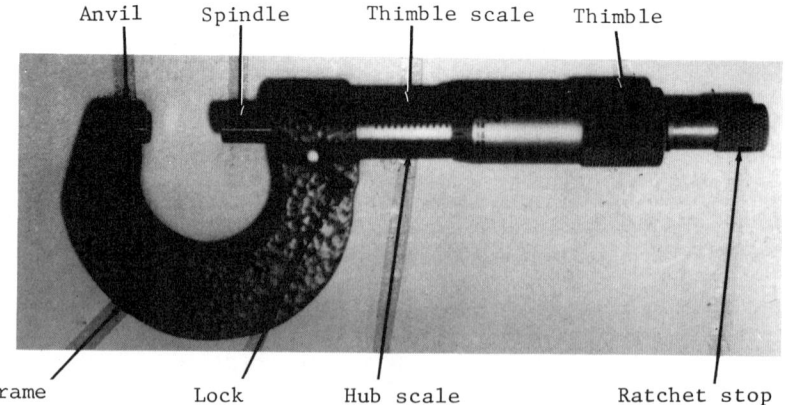

FIG. 2-6   Typical 0-1 inch outside micrometer.

Measuring Devices and Fasteners   69

The anvil fits into the frame on one end. The highly polished surface of this anvil forms one of the two measuring faces of the tool. The precision measurements taken by the mike are from the face of the anvil to the other measuring face located on the spindle.

The spindle, with its polished measuring face, attaches on the inside to the thimble. The part of this spindle that the thimble and hub conceal has threads which fit into a nut in the opposite end of the frame from the anvil. Since the frame is stationary, any rotation of the thimble causes the attached spindle to revolve with it and move through the nut in the frame. This action causes the spindle's measuring face to approach or recede from the face of the anvil.

To measure a part, place the part first against the anvil. Then by slowly turning the thimble clockwise, the thimble and the spindle move toward the anvil, wedging the part between the two measuring faces. The technician determines the actual measurement of the opening between the anvil and spindle, the thickness of the part, by observing the lines and figures located on the hub and thimble.

The manufacturer marks off the hub with 40 lines to the inch (Fig. 2-7). These markings correspond to the number of threads on the spindle, 40 per inch. Accordingly, one complete revolution of the spindle moves it longitudinally 1/40 of an inch (.025 inch). To indicate this movement on the mike, each small vertical line on the hub indicates a distance of .025 inch; and every fourth line that is longer than the others has number designations from 0 to 9. Each of these long numbered lines indicates a distance of .100 inch, or one-tenth of an inch.

The inside end of the rotating thimble also has a series of markings. Each marked line represents .001 inch, and there are 25 of these markings. Consequently, every time the mechanic turns the thimble one complete revolution, it moves exactly .025 inch. If he moves it four complete turns, the thimble travels .100 inch.

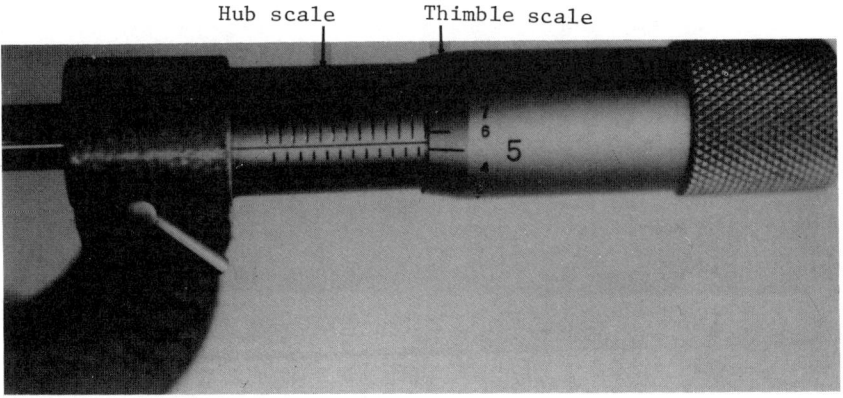

FIG. 2-7   Hub and thimble markings of a typical micrometer.

The ratchet stop (Fig. 2-6) is a friction clutch build into the micrometer that allows faster and more accurate mike readings. By rotating the ratchet stop instead of the thimble, the mechanic applies the correct pressure to the work, which is wedged between the spindle and anvil. In other words, when the force of this wedging action reaches a given point, the ratchet stop will slip. This slippage prevents the spindle from turning any further and, at the same time, indicates to the mechanic the work is held with enough tension to take an accurate reading.

To properly read the micrometer itself, follow this simple procedure. First, multiply the number of vertical divisions (lines) visible on the hub by 25. In Fig. 2-8, 16 vertical lines are visible; therefore, the reading on the hub is equal to 16 x 25 = .400 inch. Second, add to this the number of divisions located on the **bevel** end of the thimble, from 0 to the line that coincides with the horizontal line of the hub. In Fig. 2-8, this is .002 inch. The complete micrometer reading now is .400 inch plus .002 inch or .402 inch.

The micrometer is a precision tool and costly to replace if damaged. To avoid damage to this device, the mechanic should follow these general rules covering its use and care:

1. Never clamp the micrometer down hard on the part being measured. Tighten the thimble only enough to cause the micrometer's measuring faces to drag slightly as they slide over the part or use the ratchet stop, if so equipped. Excessive clamping distorts the spindle threads and frame and could ruin not only the part being measured but also the micrometer.

2. Never open or close the micrometer by holding the thimble and spinning the frame. This action can throw the micrometer out of adjustment.

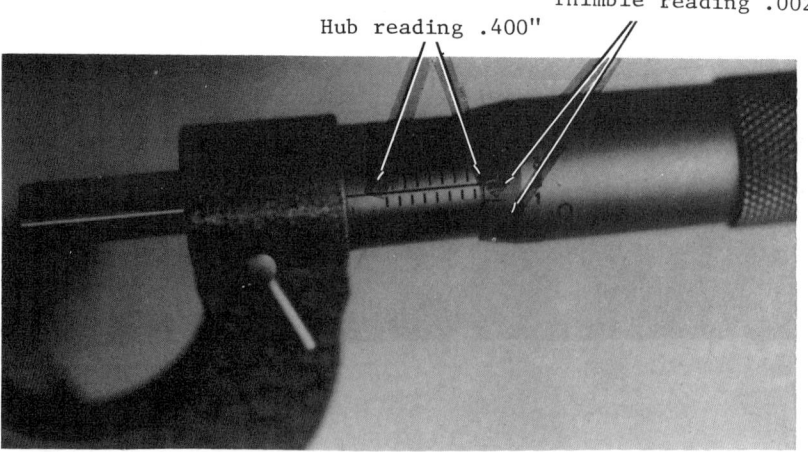

FIG. 2-8   0-1 inch micrometer that reads .402 inch.

Measuring Devices and Fasteners   71

3. Never throw the micrometer down or leave it lying unprotected on the workbench. Wipe it clean after every use, and place it in a special drawer or container that protects it from dirt and impact damage from other tools.

Feeler Gauges

Automatic transmission mechanics use the feeler gauge (Fig. 2-9) for such measuring jobs as checking the operating clearance within clutch assemblies and for the amount of wear in hydraulic pumps. A feeler gauge is a strip or blade of hardened steel or other metal which is ground or rolled to the proper thickness with extreme accuracy. The manufacturer marks on the blade its thickness in thousandths of an inch, thousandths of a millimeter, or both. The blade shown in Fig. 2-9 is a .010 gauge. Note also the markings for its metric equivalent of .25 mm.

For the sake of convenience and compactness, the manufacturer assembles a series of individual gauge strips into a set. The measuring range of a typical set is from about .0015 to .040 inch (.037 to 1.02 mm). With this arrangement, the mechanic can easily locate and pull out the thickness blade he desires. Furthermore, he can combine two or more blades together to measure a space for which the set does not contain a gauge of the correct size. In this latter situation, the total thickness of all the blades used to fill this space is the actual measurement between the surfaces.

Manufacturers also make feeler gauges in several shapes. The three shapes most commonly employed by transmission technicians are the tapered, bent, and round. The tapered gauge (Fig. 2-10) as its name implies has a blade with one end that tapers down nearly to a point. This particular gauge design is very useful for checking clearances in gears and pumps, especially, when these components operate in restricted areas.

The bent feeler gauge (Fig. 2-11) has a rounded tip, but the manufacturer bends the blade at a given angle near the measuring tip. With this design, the mechanic can use the gauge to measure clearances between parts located in recessed locations. For instance, to measure clutch plate operating clearance in many multidisc assemblies, the technician must use a bent-type gauge because the plates and their retaining snap ring fit below the level of the opening in the drum, which makes it impossible to insert a flat strip between the plates.

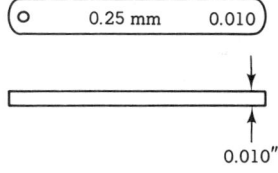

FIG. 2-9   .010 inch (.25mm) feeler gauge strip.

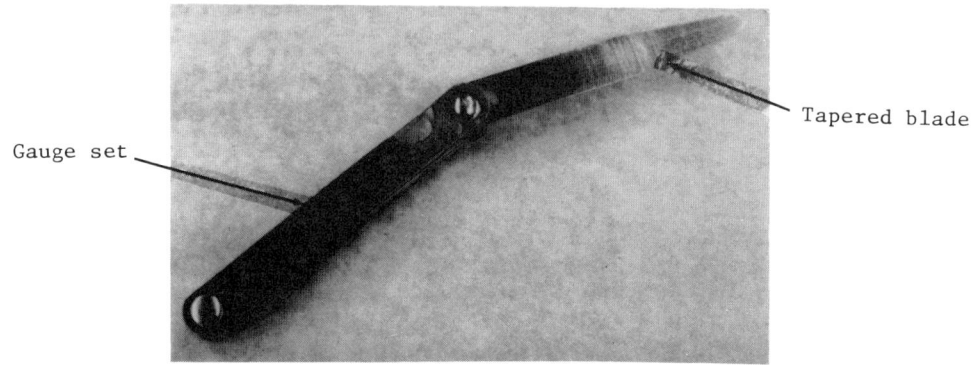

FIG. 2-10 Typical tapered feeler gauge set.

The round feeler gauge, as its name implies, has a perfectly round cross section. It is made of carefully calibrated steel wire of the proper thickness. This gauge design is especially useful in checking the clearance between a bushing and the shaft it supports.

Feeler gauges are precision measuring devices. In order to preserve their accuracy, the mechanic should follow these general rules for their use and care:

1. Never force a feeler gauge into an opening that is too small for the size of the strip. Some blades are very thin; this action can easily bend, dent, or tear the gauge strip.

2. Before using a feeler-gauge strip, wipe the blade with a clean, oiled cloth. This action removes any dirt from the blade and prevents an inaccurate reading.

3. After using a feeler gauge and before returning it to its storage area, wipe the blades and the holder with a clean, oiled cloth. The

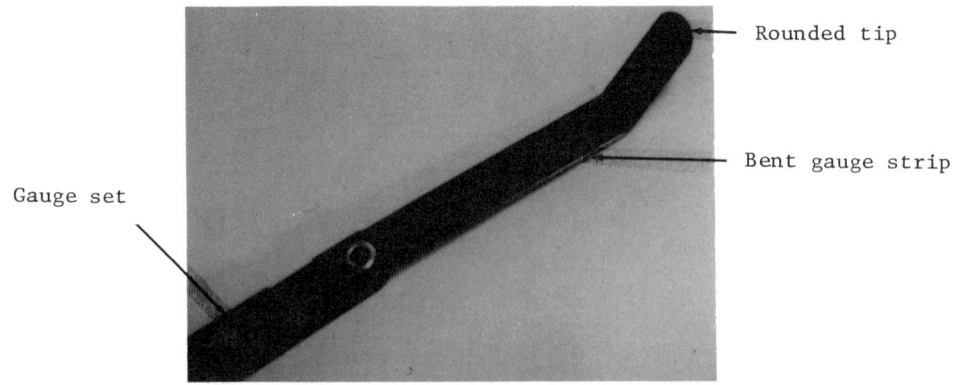

FIG. 2-11 Typical bent-type feeler gauge set.

Measuring Devices and Fasteners

oil from the rag prevents the acids and moisture on your hands from rusting the blades.

Dial Indicators

The last measuring device that this section will consider is the dial indicator (Fig. 2-12). The transmission mechanic utilizes this precision instrument to determine if the end play of the input shaft, output shaft, or converter turbine is within factory specifications. The indicator measures the amount of end play (back and forth movement) of these components, and it does so with an accuracy of one thousandth of an inch.

The dial indicator set shown in Fig. 2-12 includes the indicator itself, C-clamp with attaching rod, and a hole-lever attachment. The indicator itself mounts in a housing and has a dial face and needle to register the end-play measurements. The capacity of this device is from .0 to .050 to .0 inch, in .001 increments; therefore, one full revolution of the needle will measure a total plunger travel of .100 inch (Fig. 2-13).

A movable, spring-loaded plunger activates the needle. When a shaft or other object contacts this plunger and moves it inward, the needle rotates on the dial face to indicate the plunger's travel in thousandths of an inch. If the shaft then moves away from the plunger, the spring pushes the plunger outward, and the needle turns in the opposite direction.

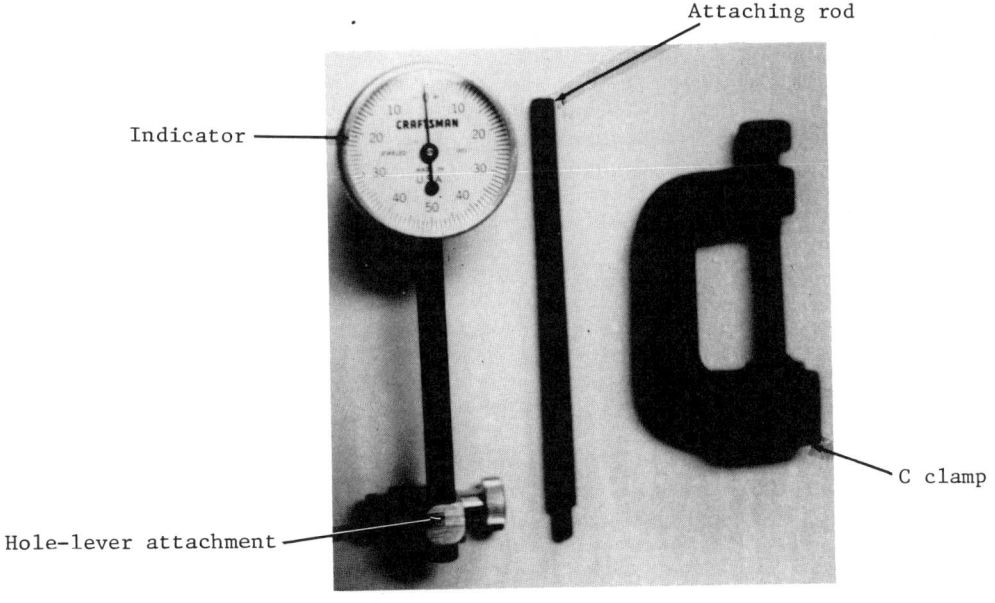

FIG. 2-12 Universal dial indicator set.

FIG. 2-13 Face of a typical dial indicator.

Finally, the dial face of the indicator is adjustable, a design permitting the mechanic to zero the indicator dial after he positions the plunger against the component he wants to check. In addition, a clamp screw located on the indicator housing locks the dial face in this set position while the mechanic performs the necessary measurements with the indicator.

The C-clamp and attaching rod, via the hole-lever attachment (Fig. 2-14), are necessary to fasten the indicator housing to the transmission or turbine end-play attachment. The C-clamp itself can fasten directly to a lip or flange on the transmission itself if its design permits; but in most cases, the mechanic will tighten down the clamp over a threaded stud installed in a tapped hole. For instance, when checking the end play of most transmission input shafts, the stud threads into one of the front-pump attaching bolt holes. In the case of the turbine end-play attachment, the rod or stud is a permanent part of the fixture itself.

The hole-lever attachment provides a great deal of flexibility necessary to mount the indicator in different locations. This attachment, when used in conjunction with the C-clamp and rod, allows the mechanic to turn or twist the indicator to various positions in order to align the indicator plunger with the work. Then by tightening the thumb screw on the hole-lever attachment, the indicator is solid enough in its mounting to take an accurate measurement.

In conclusion, the dial indicator is a delicate measuring instrument, and the mechanic must treat it with care. The technician should be careful not to drop or repeatedly push hard on its activating plunger because the internal mechanism is subject to damage or wear if treated roughly. When finished with the indicator, the mechanic should clean and return it and its attachments to their storage case.

Measuring Devices and Fasteners

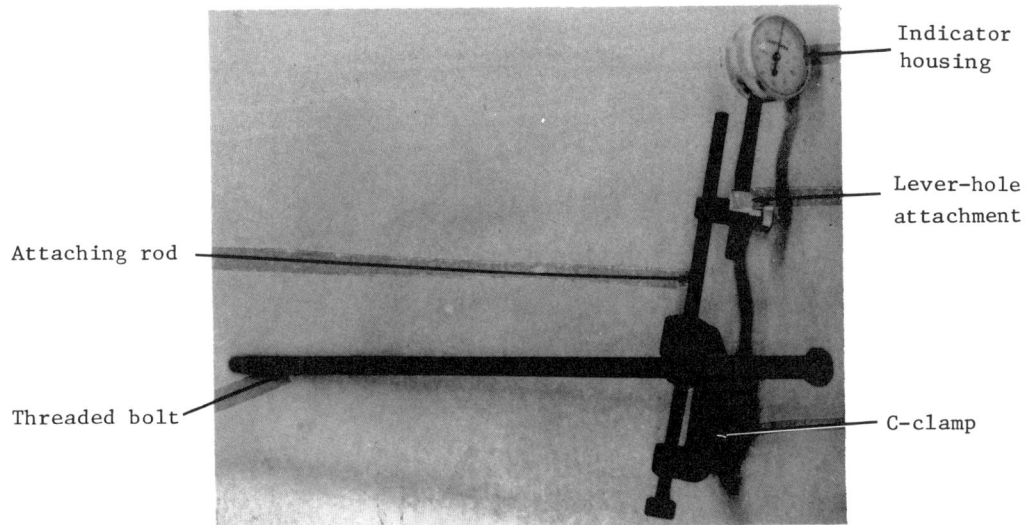

FIG. 2-14 C-clamp and attaching rod connected by the hole-lever attachment to the indicator housing.

FASTENING DEVICES

Many types of fasteners are necessary to hold the large number of transmission components together. It is extremely important, therefore, that the mechanic be familiar with the various types, designs, and functions of these devices before actually working on a transmission. The term "fastener" is the name given to any device that secures several parts together. The most common fasteners used in the automatic transmission are the machine screw, bolt and nut, washers and cotterpins, keys, splines, and snap-rings.

Machine Screws

The machine screw (Fig. 2-15) is a device that threads into a drilled and tapped hole in another metal component, like the transmission case or valve body. Manufacturers use the machine screw more extensively in automatic transmission build-up than any other single type of fastener. The actual function of the machine screw is to hold another part in place. The part fits between the base of the screw's head and the tapped hole.

Since the manufacturer does utilize the machine screw so extensively in transmission assembly, they use screws with different head designs, sizes, and tensile strengths, in addition to different thread pitches and series, to meet certain design and load requirements. The most common machine screw head designs, for example, are the Phillips, slotted, Allen, and hexagonal (Fig. 2-16). Manufacturers commonly install the Phillips and

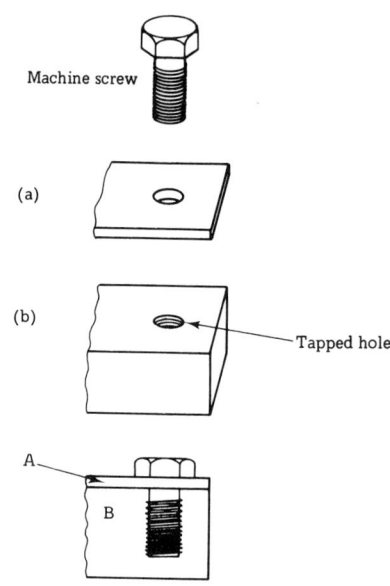

FIG. 2-15 Machine screw attaching part A to part B.

slotted as structural components of such items as the valve body and governor assembly. The Allen screw, on the other hand, performs well in holding some types of internal linkages together; but in this type of installation, the screw requires a nut to hold the parts together. Finally, the machine screw that secures the larger transmission components together has a hexagonal (six-sided) head. This head design can withstand larger torque loads on it with less tendency of ruining the area where the driver, socket, or wrench normally fits.

Manufacturers designate the size of a screw by a number or by the outside diameter of its threads. For instance, a certain valve body will have twelve number 10 Phillips machine screws holding it together. A

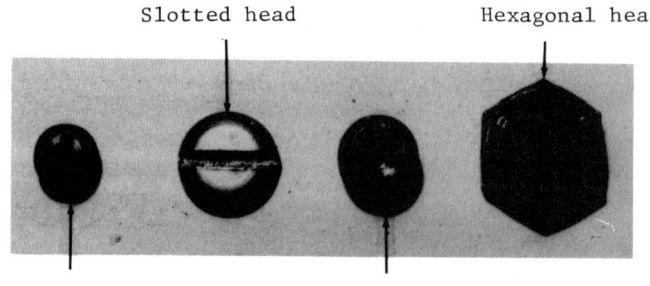

FIG. 2-16 Various head designs for machine screws.

Measuring Devices and Fasteners

Minimum tensile strength P.S.I.	64,000	105,000	133,000	150,000
Material	Low-carbon steel	Medium-carbon steel	Medium-carbon steel	Alloy steel
Quality	Indeterminate	Minimum commercial	Medium commercial	Best commercial
Head markings	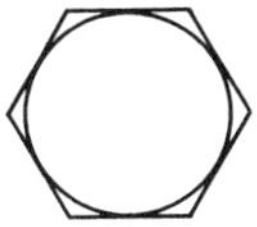 ⬡	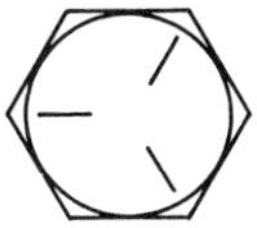 ⬡	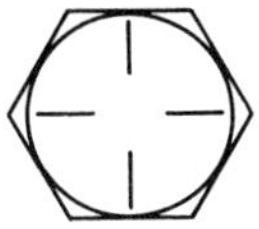 ⬡	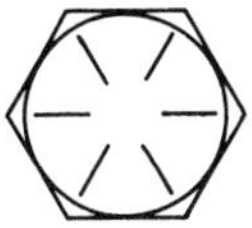 ⬡

FIG. 2-17 Common tensile strength markings on machine screw or bolt heads.

typical servo cover attaches to the transmission case by four 5/16-inch hexagonal machine screws.

Manufacturers also produce hexagonal machine screws of different materials, having different strengths. Figure 2-17 illustrates the head marking of several hexagonal screws; these markings indicate the quality, or tensile strength or the screw. Tensile strength is the amount of pull in pounds a screw can withstand before it tears apart or breaks. The stronger the screw, the more expensive it is to produce; and the transmission manufacturer installs them only where additional strength is necessary. Therefore, if a hexagonal machine screw is defective, <u>always examine its head markings, and replace it with one of equal or superior strength</u>.

Designers use two factors in determining the length requirements of a given machine screw (Fig. 2-18). The first factor is the thickness of the part the machine screw holds in place. This thickness determines length <u>B</u> of the screw. The second factor is the type and the amount of load placed on the thread. If torquing loads are high, the screw should have more pitch (threads per inch) and the threaded section <u>C</u> must be longer as well as the depth of the tapped hole. In other words, the overall length <u>A</u> is a combination of length <u>B</u> and <u>C</u>. Finally, the mechanic must consider these

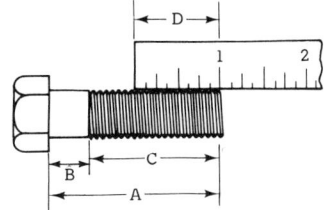

FIG. 2-18 Measurements for length and pitch on common machine screws or bolts.

facts also before choosing a replacement machine screw, or it may fail, resulting in damage to or malfunction within the transmission.

The most common designation for threaded fasteners is by thread pitch or series (Figs. 2-18 and 2-19). Pitch, as previously noted, is the actual number of threads per given inch of fastener (D of Fig. 2-18). Thread series, on the other hand, is a designation by the coarseness or fineness of the thread itself. This information is important in that thread damage will occur if the mechanic attempts to install a replacement machine screw not mated by thread pitch or series with the original into a tapped hole.

Bolts

A bolt requires a nut to attach several parts together (Fig. 2-20). The bolt, in this situation, passes through the hole in both parts. A nut threads over the end of bolt to secure the components together.

Bolts and machine screws are really the same devices; therefore, both have the same classifications for head designs, sizes, tensile strengths,

| | | Threads per inch ||
| | Diameter | Coarse | Fine |
Size	(decimal)	NC	NF
0	0.0600	...	80
1	0.0730	64	72
2	0.0860	56	64
3	0.0990	48	56
4	0.1120	40	48
5	0.1250	40	44
6	0.1380	32	40
8	0.1640	32	36
10	0.1900	24	32
12	0.2160	24	28
1/4	0.2500	20	28
5/16	0.3125	18	24
3/8	0.3750	16	24
7/16	0.4375	14	20
1/2	0.5000	13	20

FIG. 2-19 Chart showing size, diameter, pitch, and series of commonly used machine screws.

Measuring Devices and Fasteners   79

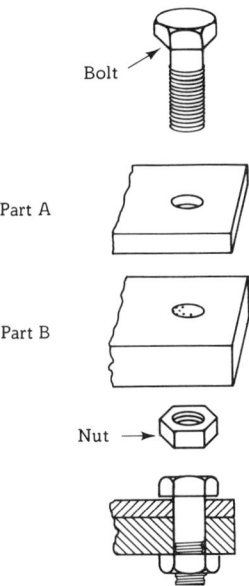

FIG. 2-20 Several parts held together by a bolt and nut.

and thread pitch and series. The main difference between the two is that
the bolt requires a nut; the screw threads into a tapped hole. Finally,
it is a general shop practice to refer to any hexagonal machine screw as
a bolt even if it does, in fact, thread into a tapped hole.

Nuts

Manufacturers produce nuts in many sizes and designs, but the two types
of common nuts used in automatic transmission installation or assembly are
plain hex and slotted hex (Fig. 2-21). The factory uses the plain hex nut,
usually along with a locking type washer, more extensively than the slotted
hex to tightly hold several parts together. The slotted hex nut, cotter
pin, and bolt, on the other hand, connect together linkages or similar
components where the components may require some rotational movement in
order to function.

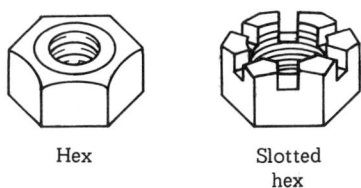

FIG. 2-21 Plain-hex and slotted-hex nuts.

Washers and Cotter Pins

During assembly of the transmission, the factory can use several types of washers. The two main washers used are the flat and the lock (Fig. 2-22). The flat washer fits under the head of the machine screw or bolt and serves two functions. First, it protects the head-surface contact area from being scratched or marred as the mechanic torques the fastener. Second, the washer extends the head-surface contact area which aids in preventing the head from distorting or tearing the metal surface as the mechanic tightens the fastener.

Some manufacturers form the flat washer as part of the screw head itself. This design accelerates the assembly process and prevents the loss of the washer when a mechanic services or rebuilds the transmission. A fluid pan attaching screw is a good example of a fastener where the factory usually combines the head and washer together into a single unit.

Finally, some flat washers used on the transmission have a gasket or sealing material bonded to both of its mating faces. The factory installs this type of washer on a screw or bolt to prevent hydraulic leakage. A front pump-to-transmission attaching screw will usually come equipped with this type of washer to prevent leakage of fluid from the area where the head seats against the pump housing. When the technician removes and replaces a machine screw or bolt equipped with a sealing washer, he must replace the washer or a leak may develop.

The factory installs lock washers (B and C of Fig. 2-22) under nuts or the heads of bolts or machine screws to prevent them from working loose from vibration or road shock. This spring-steel washer has a split with edges that cut into the nut or head as the mechanic torques the fasteners. This action keeps the fastener from turning or loosening, and the mechanic should never reuse a lockwasher once the split edges have flattened out because the washer will have lost its locking characteristic.

The design and installation of the cotter pin (Fig. 2-23) is such that it forms a locking device for a bolt and slotted nut. The bolt, in this

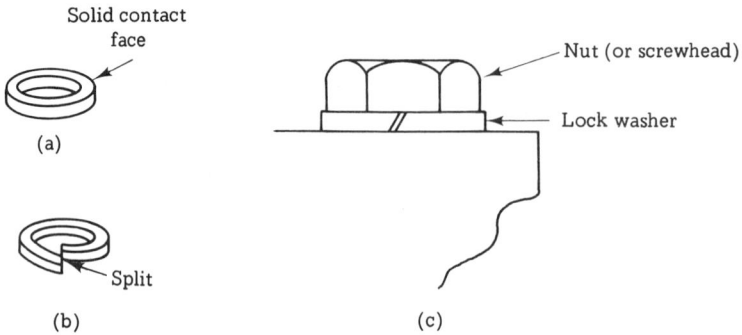

FIG. 2-22 (a) Typical flat washer; (b) plain lock washer; (c) plain lock washer installed under a nut, screw, or bolt head.

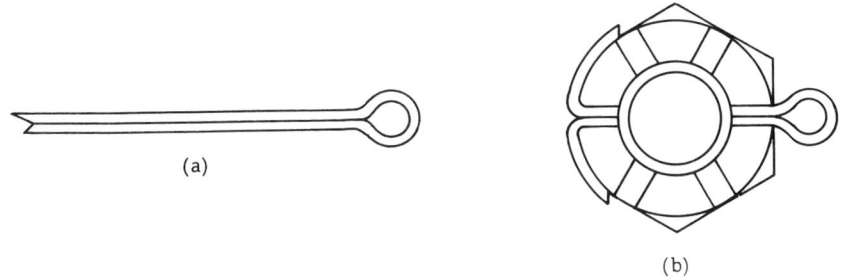

FIG. 2-23 (a) Cotter pin before installation; and (b) its installation through the hole in the bolt and slot in the nut.

situation, has a hole drilled into its threaded end large enough to accommodate the pin. The pin also passes freely through the slots machined into the nut.

To install the cotter pin, first thread the slotted nut snugly down on the bolt. Next, tighten or loosen the nut until one of its slots aligns with the machined hole in the bolt. Finally, insert the cotter pin and bend its two exposed legs (ends) over the sides of the nut so it cannot fall out. Note: <u>Once a cotter pin has been used, never reinstall it because the ends may later break, permitting the pin to fall out</u>.

Drive Keys, Balls, and Pins

Woodruff Keys, steel balls, and pins are also forms of fastening devices used to assemble a typical automatic transmission. These special fasteners lock a gear, like the rear-pump drive and speedometer, to the output shaft of the transmission so they both can rotate together. The Woodruff Key (Fig. 2-24) is nothing more than a half-round piece of hardened steel that fits into a half-round slot (keyway) machined into the output shaft.

The gear also has a slot. But, in this case, the slot has a flat, square design. When the mechanic slides this slotted gear over the Woodruff Key located in the output shaft, the key extends through the gear, thus locking it to the output shaft.

The round, steel ball has the same basic function as the key. The ball, in this situation, also fits between a gear and the output shaft. But instead of fitting into a slot in the output shaft, the ball rests in a seat machined to conform with the ball's curvature. However, the drive slot in the gear does resemble the one used with a Woodruff Key.

Manufacturers have also used a T-shaped pin to lock the rear-pump drive gear to the output shaft. The shank of this pin fits into a hole drilled into the output shaft, leaving the head portion of the pin resting on the surface of the output shaft. The square drive slot of the gear then slides over the head during assembly, locking the gear to the shaft.

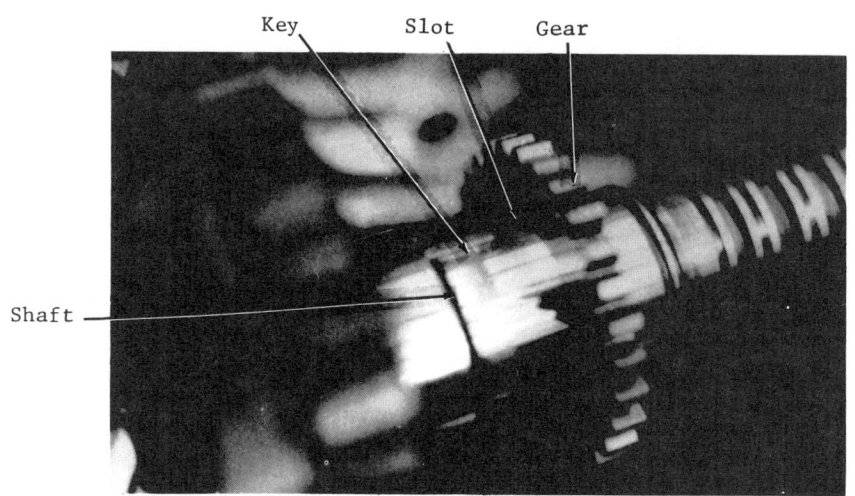

FIG. 2-24   A key locks a gear to a rotating shaft.

Splines

Transmission manufacturers also use splines on several shafts of the transmission for assembly purposes. The splines are external, and internal teeth cut into the shaft and the component the mechanic will install. The splines permit the installed part to move back and forth on the shaft, but they force the installed part to rotate with the shaft when it turns.

Figure 2-25 illustrates a typical splined output shaft and installed yoke. The external output shaft splines mate snugly with the internal splines of the slip yoke. The yoke, with its attached drive shaft, can easily slide over the output shaft, but the splines will force the yoke to turn if the output shaft is rotating.

Snap Rings

The factory installs several types of spring-steel snap or retaining rings when assembling an automatic transmission to lock or hold a component in a

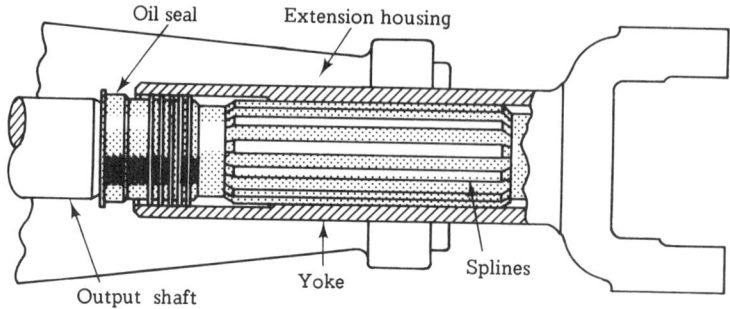

FIG. 2-25   Typical splined output shaft and slip yoke.

Measuring Devices and Fasteners

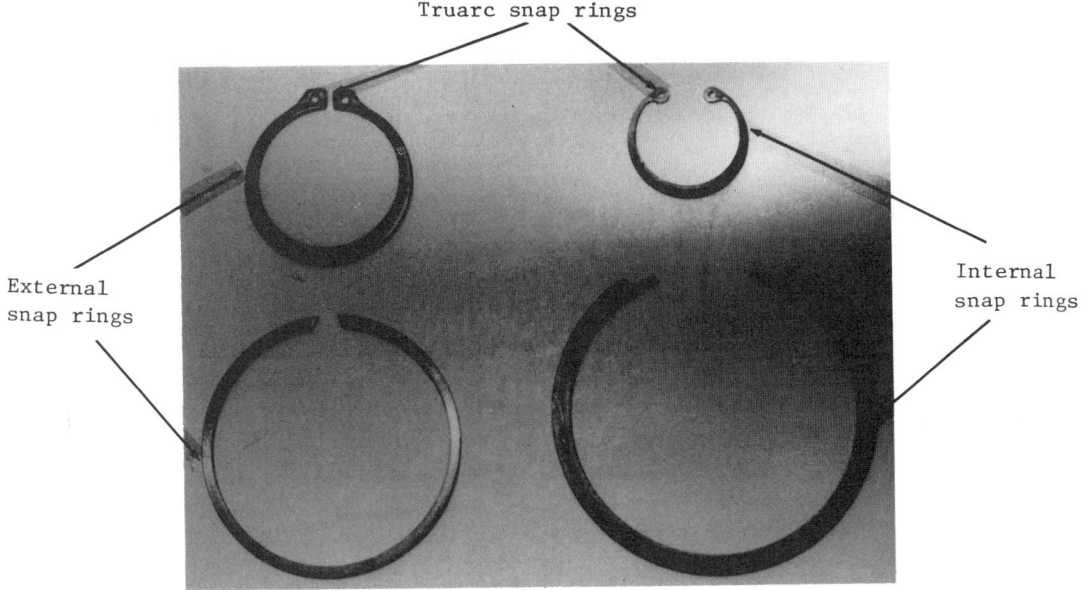

FIG. 2-26 Typical snap rings used to assemble a transmission.

given position. The two general types of rings used for this purpose are the *external* and the *internal* (Fig. 2-26).

The external snap ring in one type of installation prevents the end-to-end movement of a gear on a shaft. Another use for this type of snap ring is to

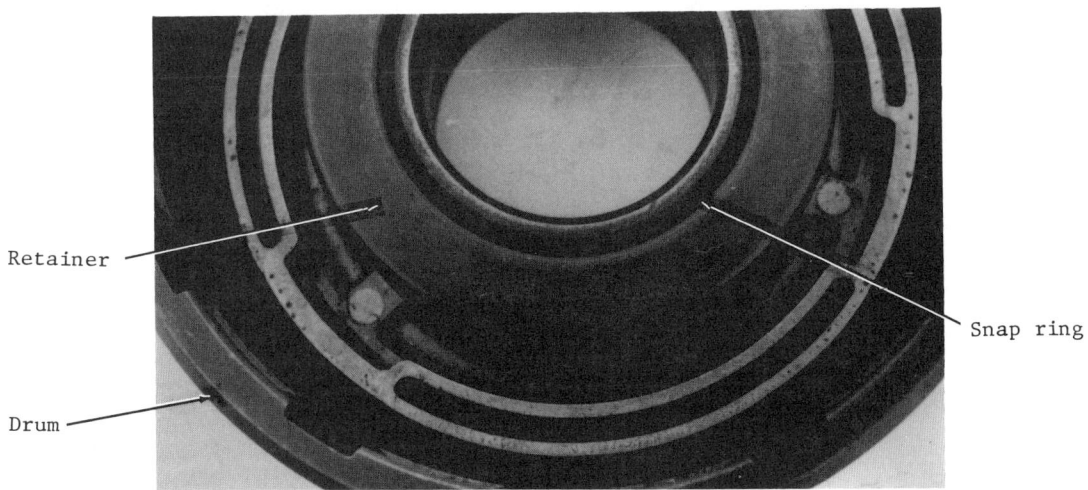

FIG. 2-27 External-type snap ring installed in its groove in a clutch piston guide.

FIG. 2-28 Internal-type snap ring installed in its groove in a governor housing.

secure the return-spring retainer to a clutch drum (Fig. 2-27). In either installation, the mechanic has to stretch or expand this ring with special snap-ring pliers before slipping it over the shaft or clutch piston guide and into its groove.

The internal snap ring, on the other hand, holds a shaft or component in a given position (Fig. 2-28). The mechanic, in this case, has to compress or squeeze the ends of this ring together with special pliers before installing it inside its bore, and finally down into a recessed groove.

The Truarc retaining ring is a special type of snap ring made for both internal and external ring applications (Figs. 2-26 and 2-28). This ring has two lips, each machined with a hole. These holes accomodate the pin ends of a special pair of snap-ring pliers. With this design, there is less possibility of the pliers slipping off the ring during removal or installation.

CHECK-UP QUESTIONS

The questions listed below will assist you in determining how well you remember the material contained in this section. Read each question carefully before adding the word or words necessary to complete the sentence. If you can't complete the sentence, review that portion of the section that covers the question.

1. A _____ _____ measures how tight a bolt or nut is.
2. Pulled threads can result from improper _____.

Measuring Devices and Fasteners

3. The wrench that has a micrometer mechanism built into its handle is a _____ type _____ wrench.
4. When an extension or adapter adds length to a _____ wrench, the applied torque will be _____.
5. To determine the thickness of a thrust washer, the mechanic should use a _____.
6. The hub section of a _____ has _____ lines to the inch.
7. Each marked line on the thimble represents _____ inch.
8. The _____ stop is a friction clutch built into the _____.
9. To check the operating clearance in a clutch assembly, the mechanic should use a _____ _____ _____.
10. The mechanic should use a _____ _____ to check pump clearances.
11. The mechanic checks input shaft end play with an _____ _____.
12. The _____ face of the indicator is adjustable.
13. The term _____ is the name given to any device that screws several parts together.
14. The device that threads into a tapped hole is a _____.
15. _____ is the actual number of threads per inch of fastener.
16. A _____ requires a nut to attach several parts together.
17. The _____ washer protects the contact surface from scratching or marring by the bolt head.
18. A _____ pin is used with a _____ nut.
19. The manufacturer can lock a gear to a shaft through the use of _____, _____, or _____.
20. The name given to the external teeth machined on a shaft is _____.

SECTION

3

Automatic Transmission Problem Diagnosis

REFERENCES: <u>Automatic Transmission Fundamentals</u>, Chapters 3, 4, 5, 6, 8, 9, 10, 11, 12, and 13.
<u>Automatic Transmission Service</u>, Sections 1, 4, 5, 7, and 8.
Vehicle or transmission service manual.

Automatic transmission problem diagnosis (troubleshooting) is probably the greatest challenge to the transmission technician; a challenge even greater, in many instances, than overhauling the unit itself. To understand why this is such a hard task, let us examine for a moment just what the serviceman must know and do to locate a customer complaint. First of all, before a mechanic can figure out what is wrong with the transmission, he must have a working knowledge of how the transmission and its systems operate. He must know, for example, the function of hydraulically operated units, clutches, and bands; basic hydraulics and the function of the many hydraulic valves and components; transmission powerflow; and basic engine operation, engine systems that relate to transmission performance, and basic engine tune-up procedures.

The technician must be familiar with the various tools and charts used in diagnosing transmission malfunctions. The tools include such items as the hydraulic gauge, tachometer, vacuum gauge, and portable vacuum pump. The hydraulic gauge is necessary to measure the amount of transmission internal-hydraulic pressure. The tachometer and vacuum gauge measure engine speed (rpm) and engine vacuum. Finally, the portable vacuum pump is a tool necessary to check a transmission's modulator system and modulator-diaphragm condition.

There are many different types of charts the serviceman may need to refer to when troubleshooting a transmission. These charts provide specifications and procedures for checking stall speed, shift point, control pressure, governor pressure, and throttle pressure. Not all of the charts mentioned above will be available on every transmission model, mainly because the manufacturer either does not desire the technician to do a specific test on the unit or does not provide a test plug or method of actually performing the check. For instance, many transmission manufacturers do not recommend stall testing their units while others do not provide test points for checking governor or throttle valve pressures.

Regardless of what is available, the main thing that a mechanic must always do to be successful in transmission troubleshooting is follow a logical, specific, and thorough test procedure. Hopefully, if he does this, he will quickly locate and correct the cause of the specific transmission problem or malfunction. By following a given procedure (guide), the technician will locate what started the problem (the main cause), and by doing so, he will prevent this particular problem from reoccurring.

Several factors determine the type and scope of any diagnosis procedure. The first is the type of transmission itself, and the actual troubleshooting procedures recommended by its manufacturer. The second factor is the type and severity of the malfunction. Since there is no one simple trouble diagnosis procedure to cover all transmissions and all complaints, this section will attempt to provide only an overview of the check and procedures commonly used by the transmission repair industry with emphasis placed on understanding what the test is for, how to perform it, and what to look for.

PRELIMINARY FLUID LEVEL AND CONDITION CHECKS

The first and easiest preliminary check to perform when there is a complaint of a malfunctioning transmission is the fluid level and condition.

Fluid Level

The correct fluid level is important because low fluid level can cause a number of complaints, such as delayed engagement and slip by starving the transmission's hydraulic system. This condition can also permit air, which is compressible, to enter the system, causing fluid foaming and mushy application of clutches and bands. If the transmission has too high a fluid level (overfilled), the gear train can churn up the fluid, again causing foaming and mushy operation. Therefore, the transmission must have proper fluid level before the technician can perform other diagnostic tests.

Before examining the actual fluid level checking procedure, it is a good time to mention one particularly misleading fluid level problem caused by converter drainback. This condition occurs when fluid drains back from the converter to the transmission case via a faulty check valve or worn bushing in the front pump (Fig. 3-1). The problem itself becomes apparent only after the vehicle sits idle for several hours. Then, when the driver starts the engine and shifts the transmission into gear, the vehicle will not move until the hydraulic pump refills the converter. Finally, after the engine has operated for a period of time, the converter and transmission fluid levels stablize and the vehicle will move normally in either direction.

The easiest and fastest way to diagnose this malfunction is to check the transmission fluid level after the vehicle has sat for a few hours. If the fluid level on the dipstick is quite high with the transmission cold, an excessive amount of fluid has leaked back from the converter to the case. This converter drainback generally does not harm either the converter or transmission. However, it is a definite nuisance and appears to most drivers to be a major transmission problem. This condition is correctable only by removing the transmission, repairing or replacing the check valve, installing a new bushing into the pump, and reinstalling the transmission.

<u>normal checking procedure</u>

To perform a fluid level check, follow this procedure:

1. Position the vehicle on a level floor, and operate the engine at fast idle until the fluid reaches its normal operating temperature.

2. With the engine idling, move the shift selector lever momentarily into each driving range to fill all the clutches and servos.

3. Leave the selector level in either Neutral or Park position, as specified by the manufacturer, and with the engine running at idle take a dipstick reading. <u>Note</u>: Before removing the dipstick, wipe the area around the dipstick cap clean to be sure no dirt will get into the transmission fluid. Then pull the dipstick out, wipe it clean, and push it back in until the dipstick seats in the tube.

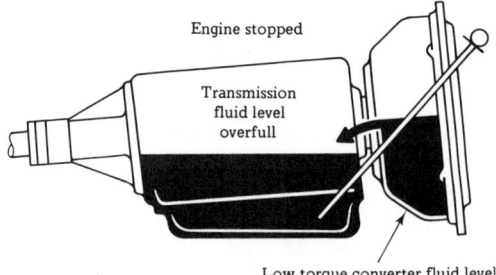

FIG. 3-1 Fluid level in the transmission will be too high if the converter begins to drain back due to a defective bushing or check valve.

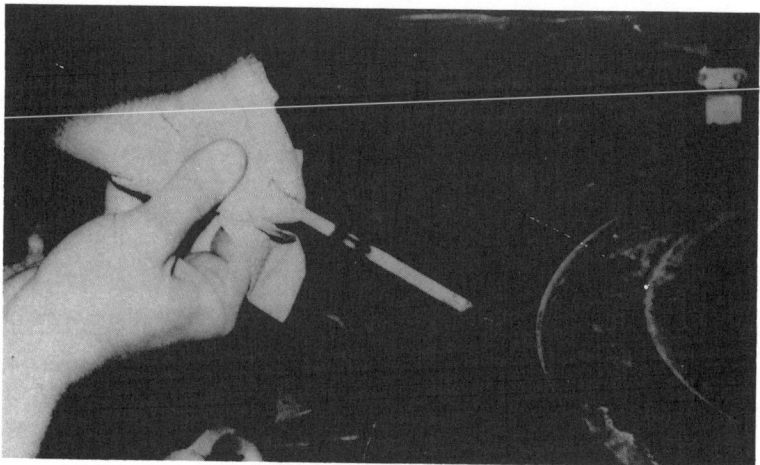

FIG. 3-2 Taking a normal dipstick reading.

4. Pull the dipstick out again and check to see if the fluid is between the add and full marks, never lower than add or above fill (Fig. 3-2). If the fluid is below the add marks, slowly pour in the correct fluid until the level on the stick is between add and full. Never overfill the transmission with fluid. Note: If the fluid level was low, perform a fluid leakage check as outlined later in this section.

5. If the fluid level is too high, insert a length of clean hose into the dipstick, attach the free end of ths hose to a suction gun, and pull the excess fluid from the transmission (Fig. 3-3). Then, with the engine idling in Neutral or Park, recheck the fluid level.

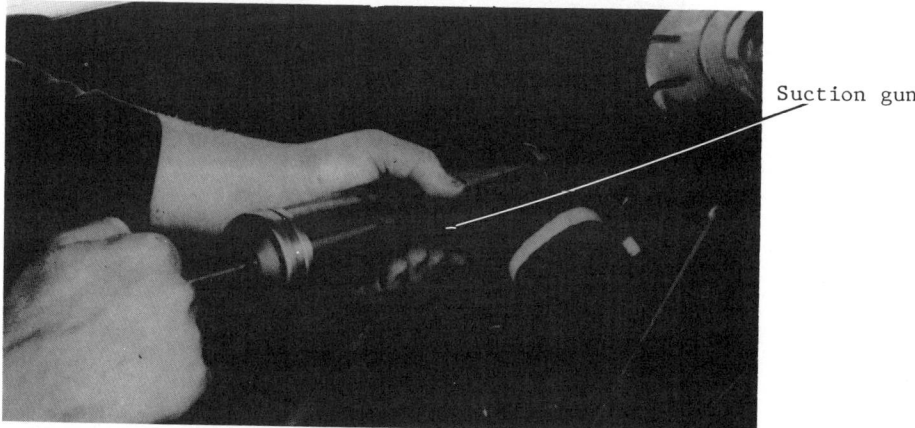

FIG. 3-3 Removing excess fluid with a suction gun.

Fluid Condition

The condition and color of the transmission fluid can also provide the mechanic with clues to specific transmission problems. The fluid should normally always be clean and be the color of transmission fluid--red. However, dirty fluid, fluid containing solid material, or discolored fluid are good indications that the transmission may have overheated or has internal damage such as a clutch or band failure.

To check the fluid's condition and color while the engine is operating, pull the dipstick and wipe it on a clean white paper towel or tissue. Once the fluid liquid soaks into the towel or tissue, any solid particles will be clearly visible on the wiping towel (Fig. 3-4). If the fluid sample appears discolored, smell the fluid sample. Fluid that is burnt has quite a distinguishable odor similar to that of a burned-out electrical motor coil.

<u>results and indications</u>

As previously mentioned, the results of the fluid condition check will provide a good indication of the general condition of the transmission. Therefore, carefully examine the fluid sample and compare the results with the typical ones listed below:

1. The fluid sample appears clean and has a normal color. If the fluid has been in the transmission for some time, this is a good indication that the unit is still in good mechanical condition and the cause of the problem is elsewhere.

2. The fluid sample appears slightly brown, but it has no burnt smell and contains no particles. In this case, the transmission may just

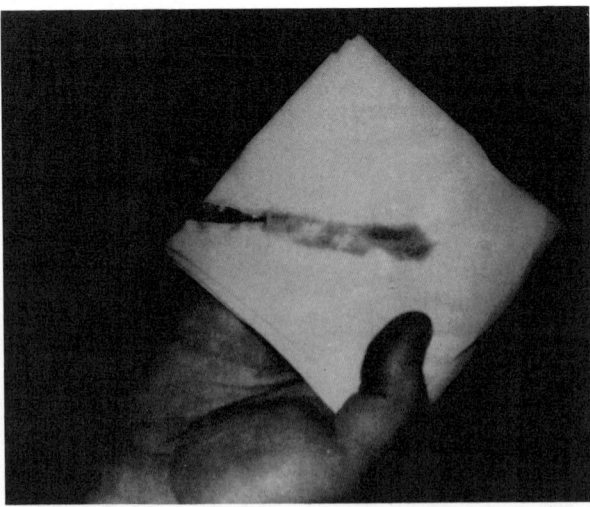

FIG. 3-4 Checking the fluid's condition using a clean paper towel or tissue.

be overdue for a fluid change. Change the filter and transmission fluid and inspect the pan for residue.

3. The fluid sample is dark brown or is dirty (containing solid material). This is a very good indication that the transmission has overheated or has some internal damage. Smell the fluid for the burnt odor. Then, check the sample for particle content. Dark particles may be from dislodged clutch or band material while silvery or shining particles may be from a bushing or thrust washer. In either case, remove the fluid pan for further damage inspection.

4. A dark fluid sample without the burnt odor or presence of solid material, and varnish on the dipstick are good indications that there is antifreeze in the transmission fluid. When this problem is evident, the transmission cooler requires inspection and repair in addition to a complete transmission fluid change.

5. A black fluid sample that has a strong burnt odor and contains particles of solid material, along with a dark varnish stain on the dipstick, usually indicates prolonged transmission overheating or a burned out clutch or band. In almost all cases, any transmission with fluid such as this will require a complete tear down, inspection, and overhaul.

FLUID LEAKAGE CHECKS

If the first preliminary fluid level check indicated a fluid shortage in the transmission and this low-level condition caused the malfunction, the next logical step is to determine where the fluid is going. there are three possibilities: the system was low on fluid to start with, the engine is pulling the fluid through its vacuum system, or the fluid is leaking to the outside of the transmission or into the engine cooling system. A quick question to the vehicle owner will verify if the transmission had a fluid change or service recently. It is not uncommon for some transmissions to start malfunctioning a short time after a fluid change just because it is a quart low on fluid.

Fluid that is being burned inside the engine is another very common cause of fluid loss in transmissions using a vacuum operated modulator. This condition is not very hard to detect because there are usually other symptoms other than a loss of fluid. For example, the engine usually smokes excessively, especially when developing high vacuum. Engine idle will be erratic and higher than normal. Finally, depending on the type of transmission, the defective modulator will cause either harsh up-and-down shifting, delay upshifting, or no upshifting at all.

There are two checks a mechanic can make to verify a defective modulator. The first is to just disconnect the vacuum hose at the modulator itself. If fluid drips out of the modulator or the vacuum hose, the modulator diaphragm

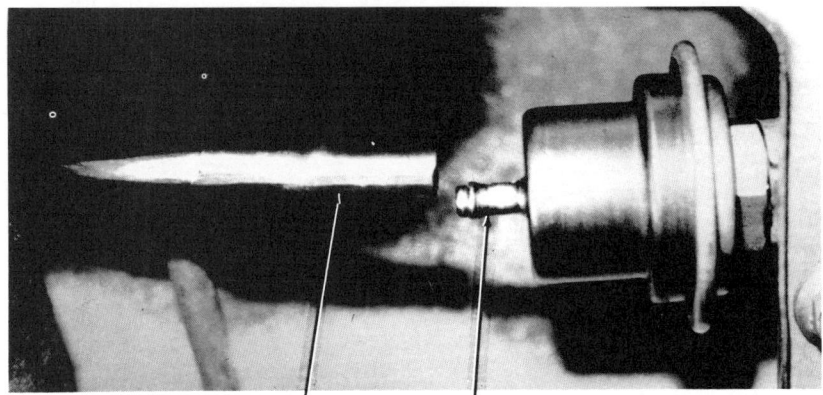

Vacuum hose   Modulator vacuum port

FIG. 3-5 If evidence of fluid is found in the modulator's vacuum line or vacuum port, the diaphragm has a hole in it.

has ruptured, and engine vacuum is sucking the fluid from the transmission (Fig. 3-5). The second check is to connect the hose from a portable vacuum pump (refer to Section 1) to the modulator, and attempt to apply 18 inches of vacuum to the modulator diaphragm (Fig. 3-6). If the vacuum pump cannot build up this amount of vacuum in the modulator or it won't hold that amount, the unit is defective.

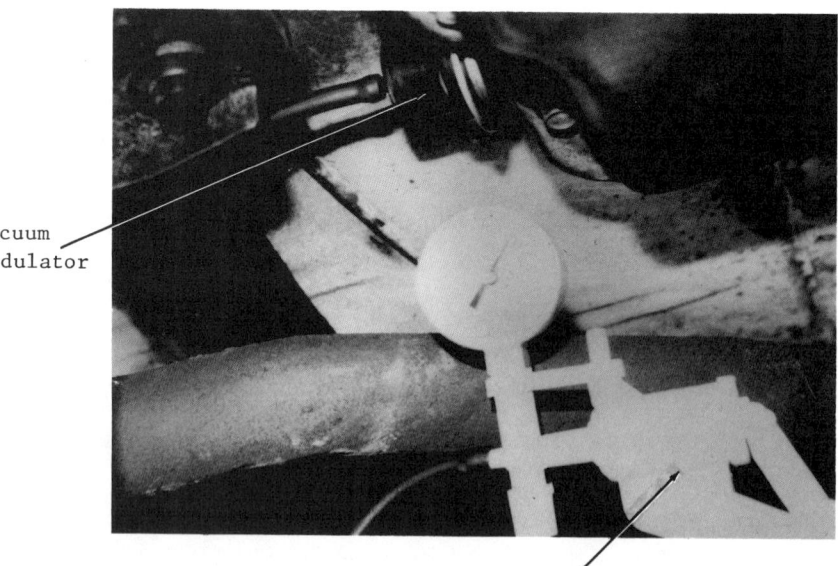

Vacuum modulator

Portable vacuum pump and gauge assembly

FIG. 3-6 Checking a modulator with a hand-operated vacuum pump.

Automatic Transmission Problem Diagnosis

The third possibility for fluid loss and unfortunately the most common is through external fluid leakage. This simply means that the fluid is leaking to the outside of the transmission from one or more points. This situation is unfortunate for both the vehicle owner and the mechanic in that it is costly to the owner to have leaks diagnosed and repaired, and it is sometimes very difficult for a technician to locate and correct many types of leaks.

External diagnosis can be a very trying experience for a mechanic for several reasons. First, the vehicle may have more than one component leaking fluid into the same general area. For example, there may be an engine or power steering leak along with a transmission leak. The airstream, passing under the vehicle, compounds the problem by carrying the leaking engine or power-steering oil back where it deposits onto the transmission case and leaks off along with the transmission fluid at various points. Second, a transmission leak itself may also originate in one place, but the force of gravity along with the air stream will cause the fluid to move to another location before it drips off.

Typical Checking Procedure

With these facts in mind, let's examine the common procedures used to locate the various types of leaks you may have to diagnose. To determine the source of a leak:

1. Position the vehicle on a hoist.
2. Wipe off the underside of the transmission and engine carefully to remove all traces of old fluid leakage. Steam clean these two areas or use a degreaser if necessary and then allow the areas to dry thoroughly.
3. Place a piece of white cardboard or heavy paper on the shop floor under the vehicle (Fig. 3-7). Make sure the cardboard or paper is large enough to encompass the area from the front of the engine to the back of the transmission.
4. Start the engine and allow it to run for about 15 minutes.
5. Shut the engine off and examine the cardboard or paper for traces of fluid. The traces will appear on the cardboard at a point approximately below where the fluid has leaked out of the transmission, engine, power-steering component, or cooling system.
6. If under the transmission, check the general area where the leak appears to originate. <u>Note:</u> The leak may have started at a higher point, followed the curvature of the case, and then dripped off the transmission in another location.
7. Once you have located the general area where the leak originated, it may be necessary to wipe this area off again and observe it carefully with the engine running.
8. If fluid traces appear on the paper in front of the converter and

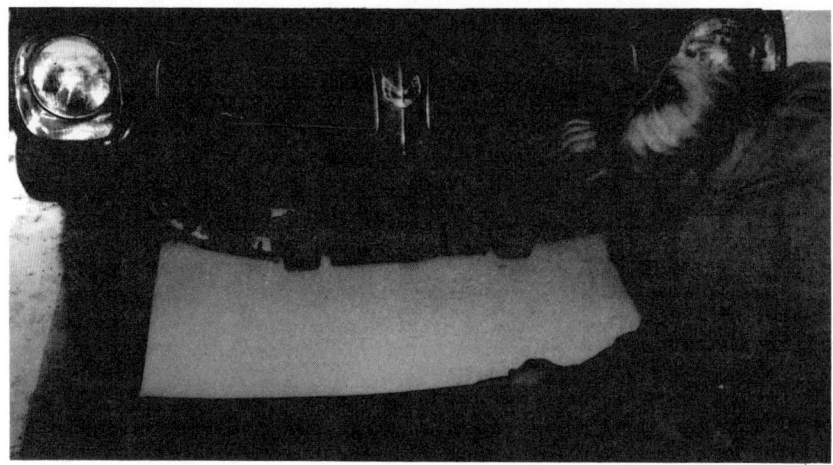

FIG. 3-7 Placing a piece of clean, white cardboard or paper under a vehicle to locate the source of a leak.

transmission, examine its color. Engine oil is usually brown unless really dirty; engine coolant is green unless contaminated; and power steering fluid is usually yellow-green or blue. If any of these traces do appear, the leakage is coming from the component indicated by the fluid color and not the transmission.

<u>results and corrective actions</u>

If the leak originated at any of the following areas of the transmission, take note of the usual cause and perform the corrective action as specified.

1. Leakage at the governor housing or cover is usually due to a defective gasket, warped cover, or loose attaching bolts. Replace the defective parts or retorque the bolts as necessary to stop the leak.

2. Leakage at the fluid pan usually results from loose attaching bolts, defective gasket, or warped or damaged pan. Retorque the bolts or replace the defective parts in order to prevent further leakage.

3. Leakage coming from the fluid filler tube is usually caused by a loose fitting, defective seal, or loose attaching bolts. Retorque the fitting or attaching bolts or replace the seal as necessary to stop the leak.

4. Leakage at the cooler lines and fittings usually results from loose or damaged fittings in addition to worn, damaged, or leaking lines. Retorque any loose fitting or replace any damaged or worn fittings and cooler lines. <u>Note:</u> Refer to Section 8 for repairing a leaking cooler line.

5. Leakage at either the throttle-lever shaft or manual-lever shaft is usually due to defective seals. Replace the seals as necessary to

stop the leakage. <u>Note</u>: Refer to the shop manual for this procedure because it may be necessary to remove and disassemble the transmission in order to replace the leaking seals.

6. Leakage at any of the test plugs usually results from a loose or worn plug. Retorque the plugs to specifications; if tightening the plugs does not stop the leak, replace the plugs as necessary.

7. Other components than the transmission itself can cause a fluid leak from the converter housing. If the color of the fluid does not indicate its source, there are two other methods commonly used by technicians to locate the leak, a visual inspection and the use of black light.

Converter-Housing Visual-Leak Inspections

To locate the source of a leak in the converter housing area by visual inspection, do the following things:

1. Remove the lower converter housing from the transmission.

2. Thoroughly clean off any fluid residue from the top and bottom of the converter housing, front of the transmission case, and rear face of the engine and its oil pan. Clean these areas by washing them with a suitable nonflammable solvent, and blow the areas dry with compressed air.

3. Wash out the inside of the converter housing, the front of the flywheel or flexplate, and the converter drain plug. Clean these areas by washing them out using a suitable nonflammable solvent and a squirt-type oil can. Next, blow all of these washed areas dry with compressed air.

4. Start and run the engine until the transmission reaches its normal operating temperature. Shut the engine off, and observe the back of the engine block and top of the converter housing for evidence of engine oil leakage.

5. Raise the vehicle on a hoist and then have an assistant run the engine at a fast idle, then at idle, while occasionally shifting to both the drive and reverse ranges to increase the pressure within the transmission.

6. Shut the engine off and observe the front of the flywheel or flexplate, back of the block (in as far as possible), and inside the converter housing and front of the transmission case. <u>Note</u>: Run the engine as long as necessary until oil leakage is evident and the probable source of leakage is apparent.

<u>results of the converter housing visual inspection</u>

The actual source of the leak is determined by observing the paths the fluid takes to reach the bottom of the converter housing (as shown in Fig. 3-8):

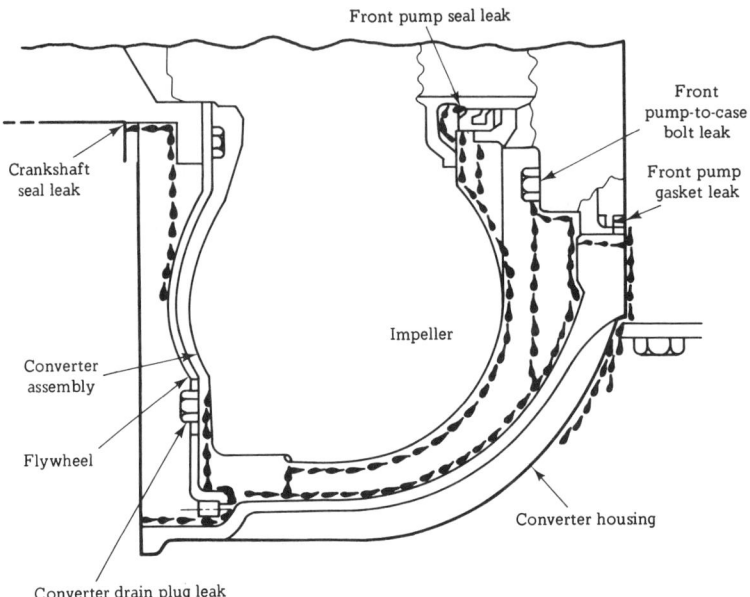

FIG. 3-8 Sources of fluid leakage from the torque converter housing.

1. Fluid that leaks by the lip of the front-pump seal tends to move along the pump-drive hub and onto the back of the impeller housing. In the case of total seal failure, fluid leakage by the seal's lip also deposits on the inside of the converter housing itself, near the outside of the housing.

2. Fluid leakage past the outside diameter of the front-pump seal will follow the same path as leakage by the seal's lip.

3. Fluid which leaks by one or more front pump-to-case attaching bolts deposits onto the inside of the converter housing only; fluid will usually not be found on the back of the converter.

4. Leakage past the front pump-to-case gasket usually causes fluid deposits on the inside of the converter housing, or it may seep down between the front of the transmission case and the back of the converter housing. Fluid on the front of the transmission case above the level of the pan gasket is evidence that the front pump-to-case gasket is leaking.

5. Fluid leakage from the converter drain plugs appears at the outside diameter of the converter, on the rear face of the flywheel or flexplate, and inside the converter housing only, near the flywheel or flexplate.

6. Engine oil-gallery plug leaks will also allow oil to flow down the rear face of the engine block and to the bottom of the converter housing.

7. Leakage by the engine-crankshaft seal will work its way back to the front of the flywheel or flexplate, and from there into the converter housing.

8. Leakage at the rocker arm covers may allow oil to flow over the converter housing or to seep down between the block and converter housing, causing oil to be present in or at the bottom of the converter housing.

Black-Light Tests

With some transmission designs, it is impossible to perform a good visual inspection of the converter housing area. To determine if the leak coming out of the housing is transmission fluid or not, perform the following procedure before starting any repair work.

1. Mix a solution of an oil soluble fluorescent dye, at the rate of 1/2 teaspoon of dry powder to 1/2 pint of transmission fluid, and pour the solution into the transmission filler tube. Note: This dye is necessary to determine, in some cases, if the leaking fluid is from the engine or transmission.

2. Run the engine for about 10 minutes to mix this solution into the transmission fluid and circulate it throughout the transmission.

3. View the converter housing area with a black light (Fig. 3-9). The dye in the transmission fluid, when under a black light, is clearly visible and indicates whether or not the leak from the housing is actually transmission fluid.

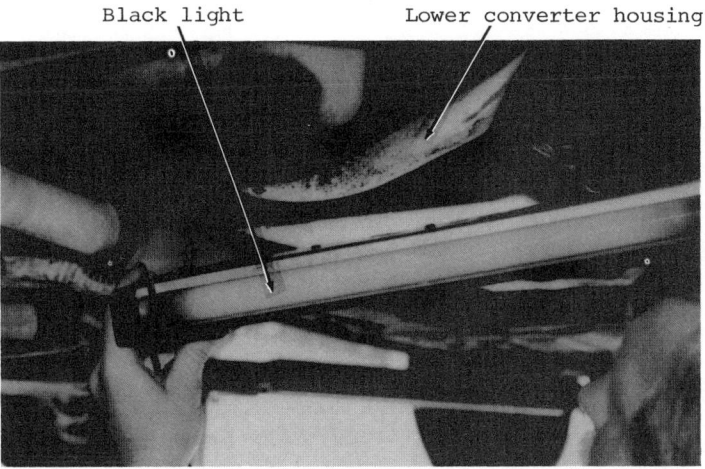

FIG. 3-9 Checking the converter housing area with a black light to determine if the fluid leaking from the housing is from the transmission itself.

Cooler Leakage Tests

Probably the least likely cause of low-fluid level in the transmission is a fluid leak into the engine cooling system. In this situation, the fluid leak is from a crack or hole in the fluid cooler located inside the radiator. As the fluid circulates through this cooler under pressure, small amounts of fluid pass into the coolant.

This type of leak causes two problems. First, of course, is the loss of fluid from the transmission itself; which in time causes the unit to malfunction because of low-fluid level. Second, the transmission fluid entering the radiator overfills and contaminates the coolant. This multiple condition can cause cooling system deterioration, engine overheating, coolant venting out the radiator overflow tube, and coolant to be forced into the transmission.

To check for this problem, shut the engine off and allow it to cool sufficiently before removing the radiator cap. Then check the engine coolant in the radiator. If transmission fluid is present in the coolant, the cooler in the radiator is most likely leaking.

In order to check the cooler itself, do the following things:

1. Disconnect both cooler lines at the radiator and plug one of the cooler fittings.

2. After attaching an air fitting to the other cooler fitting, apply a pressure of 25 to 50 psi into the cooler. <u>Do not exceed a pressure of 50 psi; higher pressure may damage the cooler</u>.

3. Check the coolant in the radiator for the presence of bubbles. If the cooler is leaking, remove the radiator for reconditioning or replacement.

OTHER PRELIMINARY CHECKS

If the fluid level and condition are satisfactory and the transmission does not leak, the next items the technician should check are engine idle speed, linkage adjustments, and the vacuum or electrical circuits to the transmission. All of these items influence the manner in which the transmission operates. Consequently, whenever the owner complains of a transmission malfunction, the mechanic should check them <u>all</u> before performing a stall and road test.

Idle-Speed Checks

The idle speed of the engine if not in proper adjustment will cause several problems. If the idle is set too low, the engine will stall or run

Automatic Transmission Problem Diagnosis

roughly whenever the transmission is in gear. This condition results from the load placed on the engine by the converter. If the engine idle is too high, the vehicle will creep excessively when placed in gear or the initial engagement will be very harsh.

There is no one simple method of adjusting engine idle speed. The actual adjustment does vary from one vehicle manufacturer to another because of the differences in such things as linkages, carburetors, and smog controls. Therefore, when setting engine rpm at idle, use an accurate tachometer and follow the exact procedure set forth in the appropriate service manual.

Linkage Checks

All automatic transmissions have at least one cable- or linkage-type connection to the transmission, while others will have two. This linkage provides control input into the transmission from one or more sources. For instance, there is always a mechanical link between the gearshift-selector lever and the manual valve in the valve body. When the driver moves the selector lever into a given position, the linkage or cable moves the manual valve which, in turn, sets up the actual operating conditions within the transmission.

Because of its very important function, proper manual-valve linkage adjustment is important. If this linkage is out of adjustment, the transmission may creep in Neutral or Park, not engage immediately into any gear, prevent the engine from starting in Neutral or Park, not lock in Park, or have low hydraulic pressure and premature clutch or band wear.

In order for some automatic transmissions to respond to the varying loads placed on the engine, there is also some linkage installed from the carburetor down to the transmission. This linkage or cable operates both the throttle and kickdown valves but sometimes only the kickdown valve. The Powerglide, T-200, and the Torqueflite are examples where the linkage operates both the throttle and kickdown valves. The C-4 is an example of a transmission in which the linkage activates the kickdown valve only.

Several transmission malfunctions can occur if the combination throttle- and kickdown-valve linkage or cable is out of adjustment. For example, the transmission may upshift too soon or too late in relation to engine load or vehicle speed. Also, a forced downshift or kickdown may not occur when the driver depresses the accelerator pedal enough to move the linkage through the detent position, or the downshift may even occur before reaching the detent position. Finally, this premature or no kickdown condition can also occur in a transmission that has only a kickdown linkage if this linkage is out of adjustment.

The actual procedures for checking and adjusting the manual valve, throttle valve, and kickdown valve linkages sometimes vary greatly between one vehicle manufacturer and another. Always refer, therefore, to the linkage adjustment section of the service manual for the make and model of vehicle you are repairing before checking or altering any adjustments. In addition, Section 5 of this manual provides samples of typical linkage checks and adjustments.

Vacuum System Checks

Some transmissions have a vacuum-operated modulator unit in place of or in addition to a linkage-operated throttle-valve system. If the vacuum modulator system replaces the mechanically operated system, the modulator circuit will then vary shift points and control pressure to match engine load. On the other hand, if the transmission has both a modulator and mechanically operated circuit, the modulator system will vary only control pressure to match engine load.

If the modulator system itself does not function properly or does not receive the proper amount of engine vacuum, the following malfunctions can occur on transmissions without a mechanically operated throttle valve:

1. No upshift.
2. Early or late upshifts.
3. Slipping.
4. Harsh upshifts and downshifts.
5. Low or high control pressure.
6. Losses of transmission fluid.
7. Excessive engine smoking.

If the vacuum modulator system fails or does not receive sufficient engine vacuum on a transmission with a mechanically operated throttle valve, the following malfunctions can occur:

1. Harsh upshifts and downshifts.
2. Slipping.
3. Low or high control pressure.
4. Losses of transmission fluid.
5. Excessive engine smoking.

If a transmission does have a vacuum operated diaphragm or modulator unit, it must, as previously mentioned, receive sufficient engine vacuum to function properly. To check the amount of engine vacuum at the modulator unit, perform the following procedure:

1. Disconnect the vacuum hose at the diaphragm. Note: If transmission fluid drips out of the line or modulator, the diaphragm inside the modulator is probably ruptured, and the modulator unit will require replacement.

Automatic Transmission Problem Diagnosis

2. Using a tee, connect a vacuum gauge, via a hose, between the disconnected vacuum hose and the modulator's vacuum port (Fig. 3-10).

3. Start the engine and check the vacuum reading on the gauge. The reading should be relatively steady and be within the limits specified by the manufacturer.

4. Accelerate the engine briefly and then let it return to idle. The vacuum gauge reading should drop rapidly, then return to a steady idle vacuum reading.

5. If no vacuum is available at the gauge, inspect the hose and line leading to the engine for kinks, restrictions, or breaks. Repair or replace the line or hose as necessary.

6. If the vacuum gauge reading is below specifications, check the engine's condition. <u>A badly worn engine produces low vacuum, and as previously mentioned low vacuum can cause abnormal transmission operation.</u>

Some Ford automatic transmissions come equipped with a dual-port vacuum modulator. The front hose on most of these models connects to the exhaust gas recirculation (EGR) system. The second (rear) port connects directly to engine intake manifold vacuum and receives normal engine vacuum.

The EGR tube, on the other hand, receives ported vacuum. Ported vacuum in this case means simply that at engine idle there should be no vacuum at this hose connection. As the driver opens the throttle to about the 1/4 to 1/2 open position, there should be a specified amount of vacuum routed to the EGR connection on the modulator. To check for EGR ported vacuum, perform the following steps:

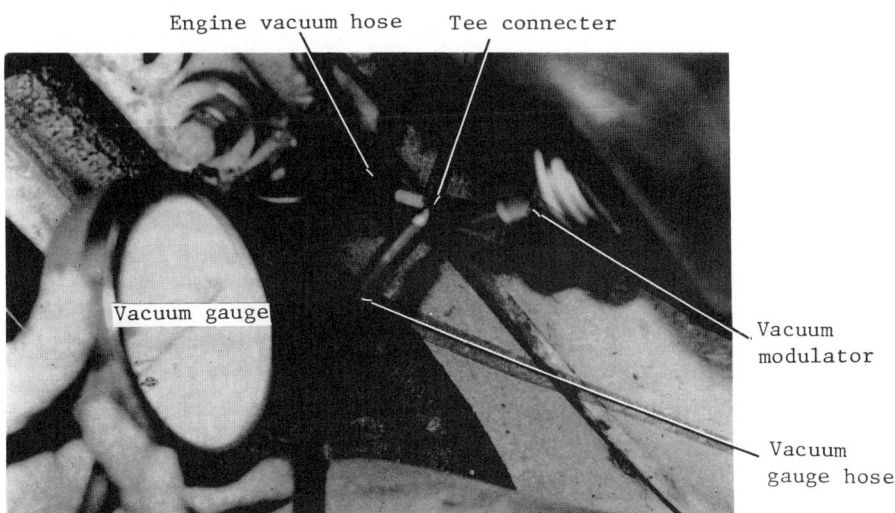

FIG. 3-10 Checking engine vacuum at the modulator with a vacuum gauge.

1. Remove the EGR vacuum hose at the modulator unit.
2. Connect an accurate vacuum gauge to the hose.
3. Start the engine and check the vacuum gauge for a reading. There should not be any vacuum gauge reading at idle.
4. Open the throttle until it reaches the 1/4 to 1/2 position. The vacuum gauge should now show a reading. If there is no gauge reading, the EGR valve may be defective or the hose or line may have kinks, restrictions, or breaks. Note: When checking any dual-port modulator system, always check the hose routing to the modulator itself against the manufacturer's specifications. If the hose connections are incorrect, the transmission will receive the wrong vacuum signal which will cause the transmission to malfunction and eventually fail.

Along with checking the condition of the engine, vacuum lines, and hoses, check the modulator itself for leaks. To test the modulator, connect a hand operated vacuum pump and gauge assembly (refer to Section 1) to the modulator inlet ports. Apply 18 inches of vacuum to the modulator diaphragm (Fig. 3-11). The gauge reading should hold steady for at least 30 to 60 seconds. If the vacuum bleeds down, the diaphragm inside the modulator can is leaking, and the entire unit requires replacement. Later on, this section will also provide additional information on modulator testing when the section describes the checking of transmission control pressure.

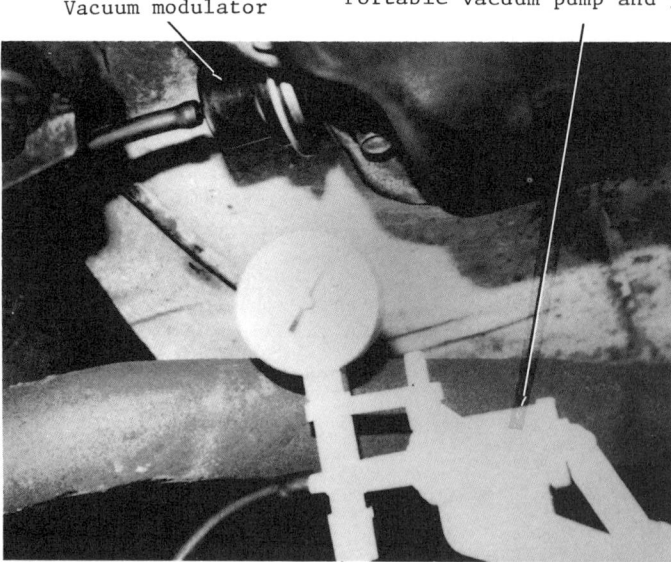

FIG. 3-11 Checking the vacuum modulator for leakage in the vehicle using a hand-operated vacuum pump and gauge assembly.

Automatic Transmission Problem Diagnosis

Electrical Transmission Circuits

Some automatic transmissions have one or more electrical leads connected to switches or terminals on the transmission case. These wires, switches, and terminals are parts of electrical circuits that function during various phases of transmission operation. As the transmission operates at different times, these units will complete the electrical circuits, through the assistance of the transmission, to such items as the back-up lights, starter, kickdown system, and transmission controlled spark system (TCS).

Because of the differences in the way manufacturers construct their vehicles, these circuits and their components are all slightly different and so are their servicing and testing procedures. If one of these circuits malfunctions, therefore, always refer to the appropriate service manual for the make and model of vehicle you are repairing for the proper troubleshooting procedure.

OPERATIONAL TESTS

Now that the preliminary inspections are out of the way, the next usual step in diagnosing transmission malfunctions is a series of operational tests--the stall, the road, and the hydraulic pressure.

Stall Tests

A technician uses the stall test to find the highest engine speed (rpm) at wide open throttle with the wheels locked and the transmission in gear. This test quickly checks engine performance, stator one-way clutch operation, and the holding ability of bands and clutches.

Some manufacturers do not specify the stall test as part of their recommended diagnostic procedures. The reason for this is two-fold. First, if a clutch or band is marginal before a stall test, or if the design of the transmission is such that controlled slippage of a clutch or band is normal, the strain of the test can ruin them. Second, if the technician does not follow the factory procedures, the test can damage the transmission, engine, and brakes. Consequently, before performing a stall test on a given transmission, always refer to the vehicle or transmission service manual to see if stall test is permissible and, at the same time, review the recommended procedure.

To conduct a typical stall test, follow these steps:

1. Check the engine coolant level, transmission fluid, and band adjustment.
2. Connect an accurate tachometer to the engine. Position the gauge so that you can easily view it from the driver's seat (Fig. 3-12).

FIG. 3-12 Installing a tachometer inside a vehicle in preparation for a stall test.

Mark on the gauge with a grease pencil the maximum stall speed (rpm) for each driving range listed in the service manual for the particular engine and transmission configuration you are testing (Fig. 3-13). This will provide you with a quick reference during the test.

3. Start the engine and operate it at fast idle until the engine reaches its normal operating temperature.

4. Apply fully the service and parking brakes. Note: <u>Never allow anyone to stand in front of the vehicle during the test. If the vehicle should break free during the stall test, the impact could injure him severely.</u>

5. Shift the gearshift selector lever into the range being tested. Push the accelerator pedal rapidly to the floor, and hold it just

Selector lever position	Clutch applied	Band applied	Stall speed (rpm)
D2	Front	Front	1350-1700
D1	Front	One-way clutch	1350-1700
L	Front	Rear	1350-1700
R	Rear	Rear	1350-1700

FIG. 3-13 A typical stall test chart.

Automatic Transmission Problem Diagnosis

long enough until a stable rpm reading registers on the tachometer. <u>Do not keep the accelerator pedal depressed longer than 5 seconds because this will severely overheat the transmission</u>.

6. Release the accelerator pedal immediately if engine rpm goes over the specified amount previously marked on the tachometer; this is a good indication a clutch or band is slipping.

7. Record the maximum rpm achieved during the test.

8. Run the engine at fast idle (about 1000 rpm) for 60 seconds with the transmission in Neutral to cool the transmission fluid.

9. Repeat step 5 for each driving range and record the results. Cool the fluid after each check as outlined in step 8.

10. Compare all the maximum stall speed readings taken during the test with the manufacturer's specifications.

<u>high stall speed--results and indications</u>

If the stall speed is more than 200 rpm above the manufacturer's specifications, this indicates a possible clutch or band slippage. For instance, if this situation occurs in a C-4 transmission in <u>all</u> forward driving ranges, the forward clutch is probably slipping because this is the only friction-type clutch applied in these ranges. Consequently, a high stall reading in <u>all</u> forward drive ranges is a good indication of forward clutch slippage.

If on the other hand, the high stall speed only occurs in second gear of a C-4 transmission, the clutch is probably not at fault. In this range, the intermediate band and forward clutch are both on. Therefore, if the excessive stall speed is obvious only in Drive Two (second gear), the intermediate band is slipping.

<u>low stall speed--results and indications</u>

If the stall speed is about 250 to 400 rpm below manufacturer's specifications and the engine is in tune, the torque converter is usually at fault. The stator's one-way clutch has failed and is no longer holding the stator against counterclockwise rotation. This causes the returning turbine fluid to strike the forward faces of the impeller blades, which slows the engine down. The end result of this failure is a low rpm reading, indicating an early stall condition.

You can also confirm this condition during the road test. The vehicle should operate satisfactorily at cruising speeds, but it will have poor low-speed acceleration. If the road test confirms a failure of the one-way clutch, replace the converter.

<u>normal stall speed--results and indications</u>

A normal stall speed does provide a positive indication of the serviceability of clutches and bands, but the converter can still be malfunctioning. Therefore, a road test should follow the stall test to check stator operation during cruise conditions. If low-speed acceleration is good but the vehicle drags or requires a high throttle opening at highway speeds, the stator has seized or locked up. In this situation, the transmission will usually run hotter than normal.

Before replacing the converter to correct this problem, check the vehicle's exhaust system. A partially blocked exhaust system produces the same symptoms as a seized stator clutch. An almost completely restricted exhaust system will affect both low-speed and high-speed vehicle operation.

Road Tests

A properly conducted road test, a test in which the technician checks the operation of the transmission under various vehicle driving conditions, is a valuable diagnostic tool for several good reasons. The road test verifies the owner's complaint, and this verification is very important in locating and repairing the <u>cause</u> of the problem. There will be times when the owner's explanation of what the transmission is doing or not doing will be unclear just because he cannot explain in technical terms just what the problem really is. In some situations, there will be no real transmission defect at all, just a case of improper driving techniques or poor engine performance.

The road test verifies the results of a stall test, and at the same time, it checks items that the stall test could not. For instance, suppose that while stall testing a Ford FMX transmission, the stall speed was too high in Reverse <u>only</u>, indicating a possible slipping low-reverse band or rear clutch. The stall test, in this case, cannot really tell you which of the two units is slipping. The reason for this is that while the band holds the carrier by itself in Reverse, the band along with a one-way clutch will hold the carrier in Manual Low. Consequently, if the band slips in low, the one-way clutch automatically stops and holds the carrier.

The rear clutch, on the other hand, also engages in third, direct drive. Therefore, a road test should certainly verify a slipping clutch condition when the transmission upshifts to direct. A well-executed road test also provides a good indication of how the throttle valve, governor, valve, and shift valves are functioning.

<u>preliminary steps</u>

Before road testing any vehicle for a transmission malfunction, do the following:

1. Check the engine's coolant and oil levels along with the transmission fluid level.
2. Check engine performance. The engine has to idle properly and otherwise perform relatively well, or the transmission can operate erratically.
3. Look up the manufacturer's shift-speed specifications in the service manual (Fig. 3-14). The shift points will differ according to the

Engine cu. in.	225	318	360-4	360-2	400-2
Axle ratio	2.76	2.45	3.21	2.45	2.71
Tire size	6.95 × 14	E78 × 14	H78 × 14	GR78 × 15	HR78 × 15
Throttle minimum					
1-2 Upshift	9-16	8-16	8-15	9-16	9-16
2-3 Upshift	15-25	15-25	15-23	17-25	15-25
3-1 Downshift	8-13	9-14	8-13	9-14	8-13
Throttle wide open					
1-2 Upshift	31-43	39-54	43-56	41-57	37-52
2-3 Upshift	63-76	79-95	78-93	83-100	77-92
Kickdown limit					
3-2 WOT downshift	60-73	76-92	75-90	79-96	73-89
3-2 Part throttle downshift	46-61	30-56	34-57	31-58	30-56
3-1 WOT downshift	28-35	30-44	34-47	31-46	29-43

FIG. 3-14 A typical chart showing shift-speed specifications.

type of engine and transmission, axle and tire sizes, and model-year combinations. Write these specifications down on a work sheet for easy reference during the test.

4. As needed, connect a tachometer and vacuum gauges to the engine in addition to a fluid-pressure gauge to the transmission. Position the gauge set inside the vehicle at a convenient place so you can view them when necessary during the test (Fig. 3-15).

conducting the road test

When performing the road test, check the transmission for the following:

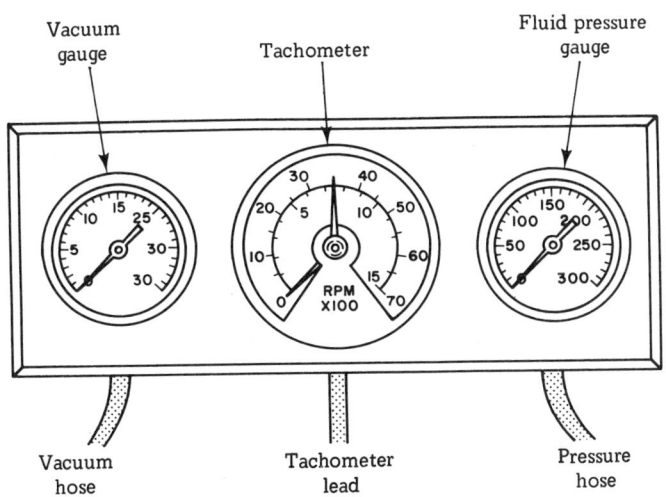

FIG. 3-15 A tachometer, vacuum, and fluid-pressure gauge set positioned inside a vehicle in preparation for a road test.

1. Slippage -- operate the transmission in all ranges to test for variation or signs of slippage during all the shifts. Note: A slipping condition is verifiable by observing the tachometer. If the transmission is slipping, the gauge will register an engine overspeed resulting from a slipping clutch or band.
2. Shift quality -- note the quality of each upshift or downshift. Are they harsh, normal, or mushy?
3. Premature, late, or erratic shift timing -- check and note the actual speeds at which all automatic upshifts and downshifts take place. Check light-throttle upshifts, full-throttle upshifts, closed-throttle downshifts, and engine braking in Manual Low and Intermediate ranges.

typical test procedure -- initial engagement checks

Perform the initial engagement checks as follows to determine if initial band and clutch engagements are smooth:

1. Run the engine until it reaches normal operating temperature.
2. With the engine at the correct idle speed, shift the selector lever from Neutral to Drive, Neutral to Intermediate, Neutral to Low, and Neutral to Reverse. Band and clutch engagements should be smooth in all positions.

typical drive-range tests

After the initial engagement check is complete, begin the road test from a stationary position with the gearshift selector in Drive. Check the light-throttle upshifts as the transmission starts the vehicle off in first, then upshifts to second, and finally to third. Note the results of this check.

Again start the vehicle off in Drive, allowing the transmission to normally upshift to third. While the transmission is in third gear, depress the accelerator pedal to the floor (through the detent). The transmission should shift from third to second or third to first, depending on vehicle speed. Note the results of this check.

Accelerate the engine once more until the transmission upshifts to third gear. Then, check the closed throttle downshift from third to first by allowing the vehicle to slowly coast down from about 30 mph in third gear. The downshift should occur as specified in the service manual. Note: You may experience a 3-2-1 downshift under these conditions; this is not abnormal. Note the results of this test.

Check the partial-throttle downshifts in Drive using the service brakes as a load. To do this, accelerate the engine until the transmission upshifts into third gear and vehicle speed reaches 30 mph. Then depress the accelerator pedal to the half-throttle position while applying the service brakes to the point where road speed slowly reduces. The third to second and then second

Automatic Transmission Problem Diagnosis

to first shifts should occur as road speed decreases. Note the results of this check.

typical drive two (intermediate) tests

Accelerate the vehicle from a stationary position with the selector lever in Drive Two, 2, or Intermediate range. Note: General Motors and Chrysler transmissions will start the vehicle off in first ratio as they would normally do in Drive, and then upshift to second. Check and note the vehicle speed as the 1-2 upshift occurs. There will be no 2-3 upshift in this driving range. In other words, these transmissions should remain in second ratio regardless of engine or vehicle speed and provide engine braking on deceleration.

A Ford transmission, on the other hand, will start the vehicle from a stationary position in second gear with the gearshift-selector lever in Drive Two, 2, or Intermediate. On some models there will be a 2-3 upshift at a given road speed and throttle setting while on still others the transmission remains in second gear. But in either case, the transmission provides engine braking on deceleration. Note these results.

manual low checks

Accelerate the vehicle from the stationary position with the gearshift selector lever in Manual Low. The transmission must start the vehicle off in first ratio and remain in this gear range. Note: Some transmissions will upshift to second at a given road speed; this protects the engine from operating in excessively high rpm.

After accelerating the vehicle for a few moments, release the accelerator pedal. The transmission should provide engine braking on deceleration. Note the results of this test.

With the transmission in Drive and upshifted into third gear, move the selector lever back into Manual Low at about 25 mph. The transmission should downshift from third to first and provide engine braking in Manual Low. Note: At speeds above 25 mph, most transmissions when manually downshifted from Drive to Manual Low will downshift into second and provide engine braking before finally downshifting to Low. Note the results of this test.

reverse tests

With the vehicle stationary, place the gearshift-selector lever into Reverse. Release the brakes, and check transmission operation in Reverse under part- and full-throttle settings. Note the results of this test.

road test results and indications

After completing the above tests and noting the results, it is time to analyze the results and what they indicate. A handy reference for analyzing the results of the road test is a clutch-and-band application chart, found in the appropriate vehicle or transmission service for the unit being tested

(Fig. 3-16). This chart tells which application unit, clutch or band, is functioning in each ratio, and this information is necessary to analyze what is the probable cause of a malfunction.

Let's compare the results of a typical road test with a clutch-and-band application chart to analyze a transmission malfunction. Suppose you have road tested a vehicle with a Torqueflite transmission and found the unit slipped in Second gear only. By observing the chart (Fig. 3-16) note that there are two friction devices applied in both Drive, second ratio, and 2, second ratio--the rear clutch, and the kickdown band. Note that the transmission utilizes the rear clutch in all forward ratios, but the kickdown band only applies when the transmission shifts into second. Since the transmission slips only in second gear, in Drive and 2, and in no other forward ratio, the rear clutch is functioning satisfactorily, so the kickdown band is most likely the cause of the slipping condition.

Pressure Tests

The stall and road test can assist you in determining which clutch or band is not functioning properly. These tests, however, cannot identify the actual <u>cause</u> of the problem. The cause may be mechanical, a burned out

Selector-lever position Drive ratio	Front clutch	Rear clutch	Front (kickdown) band	Rear (low-rev) band	Overrunning clutch
N-Neutral	Disengaged	Disengaged	Released	Released	No movement
D-Drive (first) 2.45 to 1	Disengaged	Engaged	Released	Released	Holds
(second) 1.45 to 1	Disengaged	Engaged	Applied	Released	Over runs
(direct) 1.00 to 1	Engaged	Engaged	Released	Released	Over runs
Kickdown (to second) 1.45 to 1	Disengaged	Engaged	Applied	Released	Over runs
(to low) 2.45 to 1	Disengaged	Engaged	Released	Released	Holds
2-Second 1.45 to 1	Disengaged	Engaged	Applied	Released	Over runs
1-1 Low 2.45 to 1	Disengaged	Engaged	Released	Applied	Partial Hold
R-Reverse 2.20 to 1	Engaged	Disengaged	Released	Applied	No movement

FIG. 3-16 A typical clutch and band application chart.

Automatic Transmission Problem Diagnosis

clutch or band, or it may be due to problems within the transmission's hydraulic system such as an internal leak, worn pump, sticky valve, or an inoperative modulator system.

A pressure test then will indicate any major hydraulic system malfunction and thereby eliminates the system as the cause of a particular problem. A hydraulic pressure test eliminates guess work as to the exact cause of a transmission problem, and this test should be part of your diagnosis procedure before tearing down the transmission for visual inspection and replacement of worn parts.

typical procedure

In order to perform a hydraulic pressure test, you must have a hydraulic pressure gauge set, a vacuum pump and gauge (refer to Section 1), and an accurate tachometer. The vacuum pump and gauge will not be necessary if the transmission does not have a vacuum modulator system. In addition, you must have access to the hydraulic pressure charts for the particular transmission you are testing.

preliminary steps

To prepare to pressure test a transmission still in the vehicle, perform the following steps:

1. Attach a tachometer to the engine, following the manufacturer's instructions.
2. Connect the vacuum pump and gauge assembly to the vacuum diaphragm unit (Fig. 3-17), and plug the vacuum hose.

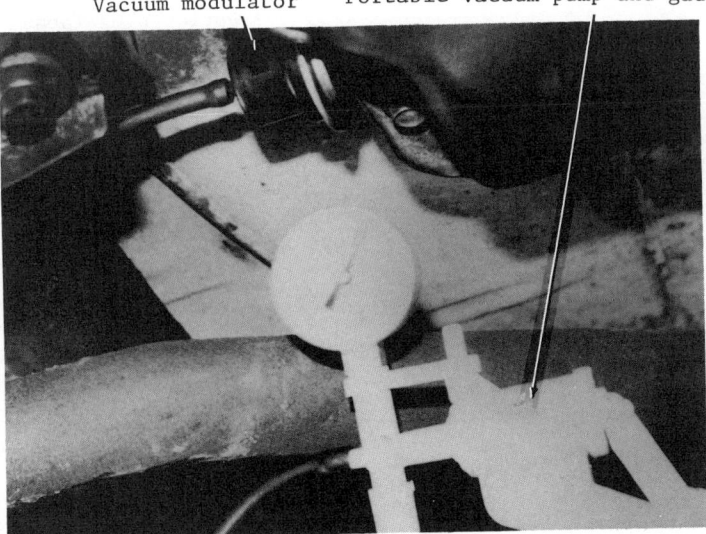

FIG. 3-17 Testing a vacuum-modulator assembly with a vacuum pump and gauge.

3. With the pressure hose and fittings, attach a pressure gauge to the control-pressure outlet port on the transmission (Fig. 3-18).

4. Have an assistant available to start the engine and operate the transmission in its various operating ranges while applying the vehicle's service brakes.

5. Adjust the engine's idle speed to the specified rpm in Drive.
   <u>Note</u>: If engine idle speed cannot be brought to within limits by adjustment at the carburetor's idle-speed adjustment screw, check the throttle or downshift linkage for binding. If the linkage is satisfactory, check for vacuum leaks in the vacuum-diaphragm unit or its connecting tubes and hoses. Check as necessary for vacuum leaks in other vacuum operated units such as the power brakes or distributor vacuum advance.

<u>test one</u>

To perform hydraulic test number one, do the following:

1. Apply 18 inches of vacuum to the vacuum diaphragm unit by means of the vacuum pump.

2. Start the engine and allow it to warm up.

3. Operate the transmission in <u>all</u> its operating ranges. Note and record the pressure reading during each phase of transmission operation.

<u>test two</u>

To perform test two, do the following:

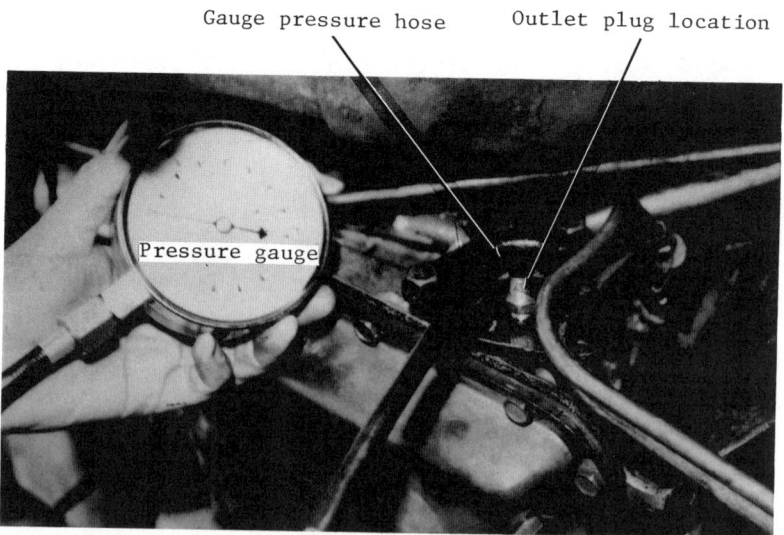

FIG. 3-18 Installing a hydraulic-pressure gauge assembly on the transmission's pressure outlet port.

Automatic Transmission Problem Diagnosis

1. Apply 13 to 16 inches of vacuum to the vacuum modulator unit.
2. Operate the transmission again in <u>all</u> its operating ranges. Note and record the pressure reading during each phase of transmission operation. The readings should be higher than they were during test one.

<u>test three</u>

To perform test three, do the following:

1. Apply no more than 1.5 inches of vacuum to the vacuum-modulator unit.
2. Operate the transmission for the final time in <u>all</u> its operating ranges. Note and record the pressure reading during each phase of transmission operation. <u>Note</u>: During this test it may be necessary to operate the engine between 1800 and 2000 rpm in order for the hydraulic pump to produce the specified pressure.

<u>test one--results and indications</u>

If hydraulic pressure at engine idle is below factory specifications in <u>all</u> selector positions during test one, this is an indication of problems other than the vacuum diaphragm unit. Check for excessive leakage in the front hydraulic pump, transmission case, and valve body.

If hydraulic pressure at engine idle is above factory specifications in all selector lever positions, the problem may be the vacuum diaphragm unit. To check these components with the engine stopped, perform the following steps:

1. Remove the diaphragm unit and diaphragm unit push rod.
2. Inspect the push rod for a bent condition and for corrosion.
3. Install the diaphragm unit into the transmission case to prevent fluid loss, but leave the push rod out. <u>Note</u>: With the push rod removed, the diaphragm unit cannot affect hydraulic pressure.
4. Start the engine and check hydraulic pressure again in all selector lever positions. If hydraulic pressure is still too high, the trouble is usually in the hydraulic pressure control system. If, on the other hand, the hydraulic pressure is now within specifications, the diaphragm unit was not operating properly and should be checked.

To check the unit for diaphragm leakage and the condition of its spring:

a. Remove the vacuum unit from the transmission.
b. Connect the vacuum hose from the vacuum pump and gauge (Fig. 3-19) to the unit's vacuum inlet port.
c. With the vacuum pump, apply 18 inches of vacuum to the vacuum unit. The vacuum unit should hold 18 inches of vacuum for at

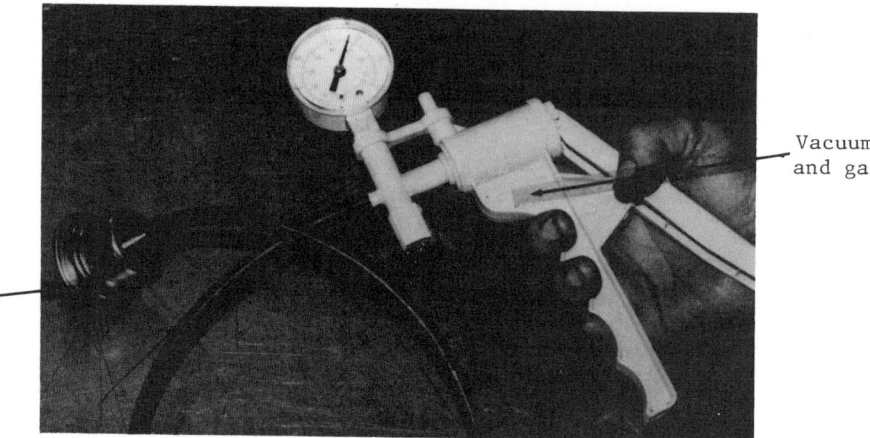

FIG. 3-19 Bench testing a vacuum-modulator assembly with a vacuum pump and gauge assembly.

least 30 seconds. If it does not, the diaphragm inside the unit is leaking.

d. Install the push rod into the vacuum unit.

e. With 18 inches of vacuum showing on the gauge, remove the hose from the vacuum unit while holding your finger over the end of the control rod. With the hose off, the internal spring within the vacuum unit should push the control rod outward. If the rod does not move outward, the spring is broken, or there is internal binding of components within the unit.

<u>test two--results and indications</u>

If hydraulic pressure did not rise in the forward driving ranges with the vacuum setting of 13 to 16 inches, again check the transmission's pressure rise capacity by operating the transmission in Reverse. If pressure rise is normal in Reverse, check the diaphragm unit as outlined under "test one results and indications". If the vacuum unit is serviceable, the pressure problem is in the hydraulic circuits to the clutches or servos.

<u>test three--results and indications</u>

If idle pressures are normal in both tests one and two, but test three pressure increases are not to specifications in all operating ranges, this indicates excessive hydraulic circuit leakage, low pump capacity, or restricted fluid screen or filter.

If pressures are not within specifications for specific driving ranges only, this indicates excessive hydraulic leakage in the clutch or servo circuits used in those ranges.

If control pressure is extremely erratic, check the diaphragm unit and push rod as mentioned earlier. Replace the unit and push rod as necessary and retest the transmission. If the pressure is still extremely erratic, clean and inspect the pressure regulator valve train and other valve body components.

adjusting the control pressure

Hydraulic pressure is adjustable on some automatic transmissions by means of an adjustable modulator or selective push rods. These devices are sometimes necessary to correct a soft or harsh shift condition; <u>the manufacturer does not produce these components to raise low hydraulic pressure caused by worn or leaking components.</u>

The adjustable vacuum diaphragm (modulator) replaces the factory installed nonadjustable unit. This adjustable unit has an adjustable screw in the vacuum inlet port (Fig. 3-20). By turning the screw in or out, the control pressure increases or decreases to correct the shift condition.

The use of selective push rods with a nonadjustable modulator will have the same effect. When installed, a longer than standard push rod raises hydraulic pressure. A shorter rod decreases hydraulic pressure.

To adjust the hydraulic pressure, perform the following steps:

1. Before installing an adjustable diaphragm (modulator) or selective pushrod, perform pressure tests one through three with the original nonadjustable diaphragm unit installed on the transmission. This action insures that pressures are all within specifications and that the shift problem is not due to other items within the transmission, or vacuum diaphragm unit.

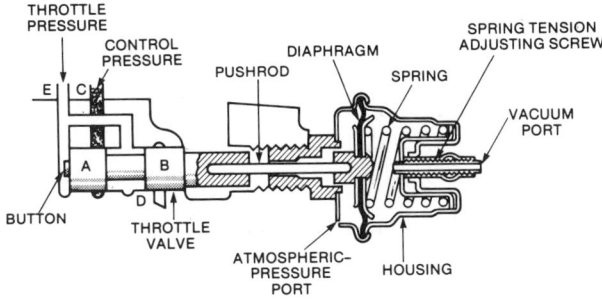

FIG. 3-20 Typical adjustable modulator assembly with an adjustment screw, located in the inlet port.

2. If pressures are within specifications, install the adjustable diaphragm unit, and operate transmission with 10 inches of vacuum applied to diaphragm. <u>Note</u>: It may be necessary to make an initial adjustment on this unit to provide the specified pressure at 10 inches of vacuum. Once this initial adjustment is complete, make any further adjustments on the unit to overcome the shift feel problem.

3. If shifts are harsh, reduce hydraulic pressure by turning the adjusting screw counterclockwise. <u>Note</u>: Installing a shorter selective pushrod with the original diaphragm will have the same effect.

4. If shifts are soft, increase control pressure by turning the adjusting screw clockwise. <u>Note</u>: Installing a longer selective pushrod with the original diaphragm unit will have the same effect.

5. Perform hydraulic pressure tests one through three again. The pressure obtained should still be within those specified in the service manual.

DIAGNOSIS GUIDES

All vehicle manufacturers provide diagnosis guides for their automatic transmissions (Fig. 3-21). These guides list the most common trouble symptoms and provide the items a technician should check to find the cause of the malfunction. The guides arrange these check items in a logical checking sequence which the serviceman should follow for quickest results. These guides are very useful tools to use during a stall or road test to accelerate the diagnosis procedure.

Let's now examine one of these guides and see how to use it in diagnosing a particular problem. The particular chart, shown in Fig. 3-21, begins at the top center of the page with the problem "no upshift". Moving straight down the guide from the symptom, you will see the first item to check--fluid level and condition. If the fluid level and condition are satisfactory, you continue to move down the list of check items, one at a time, until you locate the cause of the problem.

Notice that the check list is split into two groups beginning after the check for fluid level and condition. The check list on the left consists of mechanical items, linkage, and band adjustments. The list to the right includes hydraulic system components.

Note the horizontal line near the bottom of the page with the arrow on the left end pointing upward, and the one on the right pointing downward. This is the dividing line between the checks that are performable with the transmission installed in the vehicle and those that require transmission removal. All check items above the line are accessible with the transmission still in the vehicle; those items below the line are not.

Automatic Transmission Problem Diagnosis 117

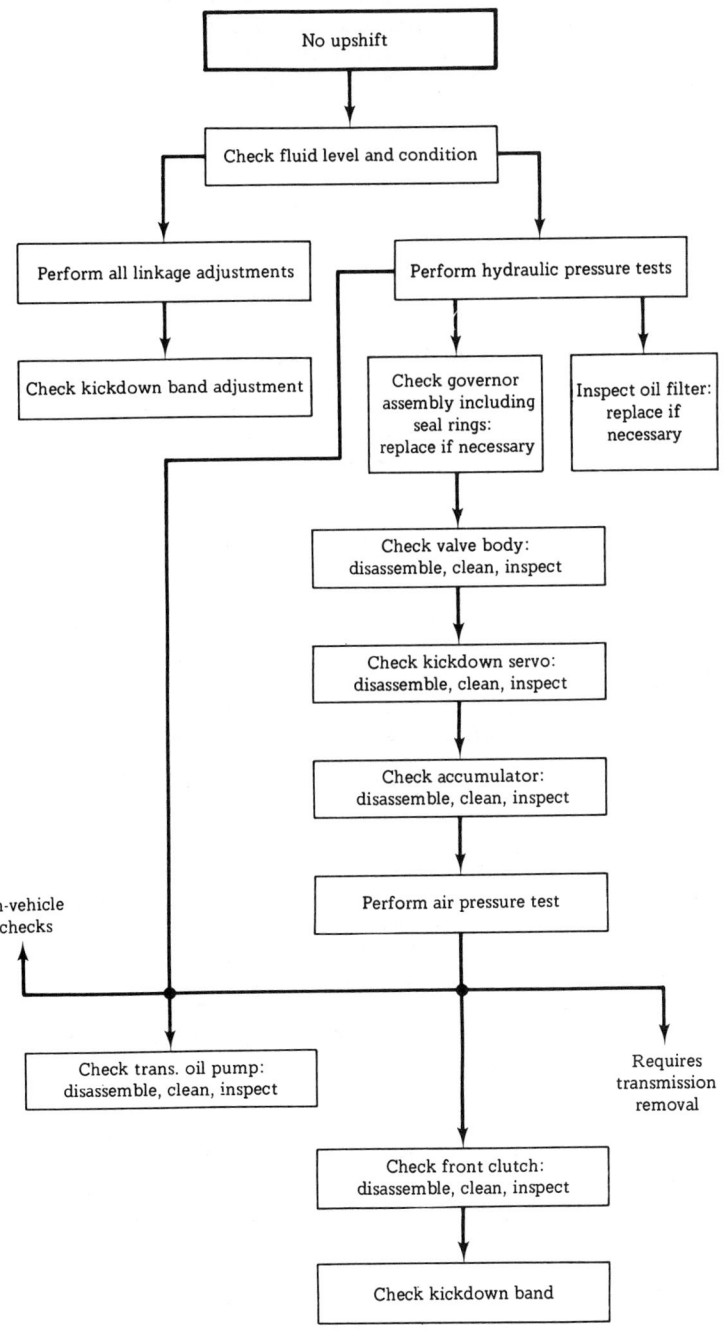

FIG. 3-21 Typical automatic transmission diagnosis guide.

There are a few things to remember in reference to these guides. There may not be a guide for the exact transmission malfunction you are attempting to diagnose. Remember that several transmission problems can cause a combination of several different symptoms or malfunctions, and the guide itself is only as good as how well you follow it.

ADDITIONAL DIAGNOSTIC TESTS

Air Checks

One of the check items on the diagnosis guide just discussed is an air pressure test. The air test, as its name implies, is the checking of certain components using compressed air. In transmission work, the air is a substitute for fluid pressure, and when directed into given hydraulic ports, it will indicate if a clutch, servo, or governor is functioning properly.

The air pressure test basically indicates if certain hydraulic components are leaking excessively. It is very possible in certain cases, to have an internal hydraulic leak inside the transmission that will not be detectable using the fluid pressure tests mentioned earlier in this section. The reason for this is that the transmission's hydraulic pump has a potential of producing high pressure and volume when in operation. Therefore, the fluid pressure test may not detect a small, internal hydraulic leak because the pump has the ability to maintain reasonable system pressure even when the system has a leaking hydraulic component. The air pressure test, on the other hand, can pinpoint the location of small and large internal leaks.

The air pressure test is performable with the transmission still in the vehicle or on the workbench. If it is done with the transmission in the vehicle, the test will be part of a diagnostic procedure used to supplement a stall, road, or hydraulic pressure test. However, it is more common to air check a transmission during the overhaul procedure either to verify a suspected leak before disassembling the transmission or to test clutches and servos to ensure that they function properly when assembling the unit.

<u>procedure</u>

Not all automatic transmission designs are such that you can air test them. Therefore, before attempting to air check a unit, check the appropriate transmission or service manual to see if the transmission is testable using air pressure, and what the approved procedure for performing the test is.

To perform a typical air test on a transmission still installed in a vehicle, complete the following steps:

1. Raise the vehicle on an overhead hoist or with a jack to a suitable

working height. **If raised with a jack, lower the vehicle on safety stands before working under the vehicle.**

2. Place a drain container under the transmission, and drain the fluid from the converter and transmission as outlined under Section 4.

3. Remove the fluid pan, gasket, and screen or filter as outlined under Section 4.

4. Loosen all the valve body attaching bolts, and carefully remove the valve body. Place the valve body on a clean work bench.

5. Check the transmission or vehicle service manual for the appropriate passage identification chart (Fig. 3-22).

6. Apply 25 to 30 psi air pressure to all the appropriate passages (Fig. 3-23). Hold the air nozzle on each port opening for several seconds and listen for a hissing noise, indicating a leak.

7. Listen also for a dull thud or clunk as the clutches and servos apply. Watch for the operation of the bands by observing them tighten and loosen around their respective drums as you apply and release the air pressure.

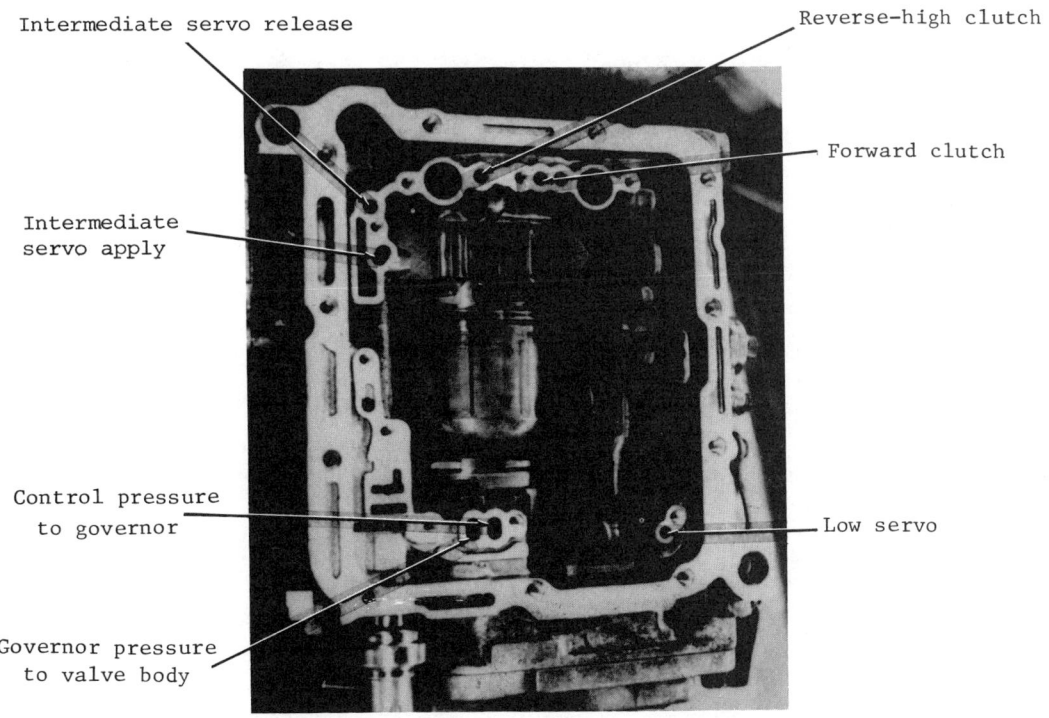

FIG. 3-22 Typical automatic transmission fluid-passage identification chart.

FIG. 3-23 Applying air pressure to a fluid passage to check a particular transmission hydraulic circuit.

8. After air checking the unit, reinstall the valve body and torque its attaching bolts to specifications.
9. Reinstall the filter or screen.
10. Reinstall the fluid pan, using a new gasket. Torque all of its attaching bolts to specifications.
11. Lower the vehicle and refill the transmission with fluid using the procedure outlined in Section 4.

Note: When air testing a transmission on the workbench, you will need to **follow** only steps 5 through 10.

<u>results and indications</u>

If during the air test procedures the following problems are detectable, remove and disassemble the transmission and its components for inspection and service:

1. A hissing sound. This indicates a leak usually caused by defective rubber or metal sealing rings.
2. No indication of clutch or band operation. This is also an indication of possible defective sealing rings, a binding clutch or servo piston, or broken piston return spring or springs.
3. No whistling, clicking, or buzzing from the governor assembly. This is a definite sign of a seized governor valve.

Automatic Transmission Problem Diagnosis

Noise Detection

Noise can be one of the hardest problems to diagnose because noise can travel from one area of the vehicle to another through its metal parts. In addition, a vibration often accompanies a noise, and when one or both are present, the problem can be in the tires, suspension, engine and its accessories, drive line, and the transmission. There are, of course, certain things you can do to locate the exact cause of the noise.

The first thing to do is to isolate the area where the noise originates from. To do this, road test the vehicle and look and listen for the following:

1. The speed at which the noise or vibration occurs.
2. The engine rpm, measured with a tachometer, at which the noise and or vibration occurs.
3. Noise levels that change with different selector lever positions or gear ratios.
4. Noise levels that increase or decrease with engine temperature.
5. Harshness of vehicle ride. In this case, check tire inflation.
6. Consistency of the noise. If the noise changes intensity several times a minute, there could be more than one problem.
7. Changes in the noise level during acceleration and deceleration.
8. Presence of noise when the vehicle is coasting, or standing stationary with the engine idling.

Hoist Tests

After the road test is complete, further diagnosis is performable with the vehicle on a hoist. With the wheels free to revolve, operate the vehicle at the road speed where the noise is noticeable. Use a stethoscope (**Fig. 3-24**) to assist in pinpointing the source of the noise by placing the tip of its probe as necessary on the converter housing, transmission case, extension housing, differential housing, axle housings, and the various engine components.

results and indications

The following conditions are examples of noise or vibration problems and the sources of the problem.

1. A vibration at 1,000 to about 2,000 rpm after an engine or transmission overhaul is a good indication of a crankshaft, flywheel, or torque converter unbalance.
2. A high-pitched whine coming from the converter housing area can result from either a low fluid level, clogged filter, or a worn or defective pump. Note: If a similar type of noise is detectable only

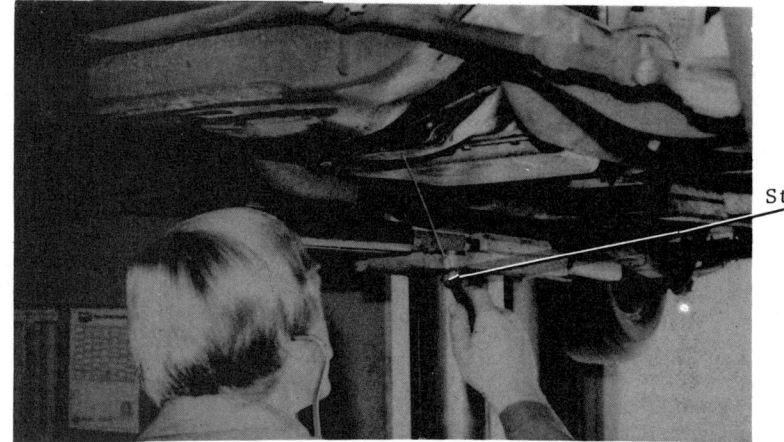

FIG. 3-24 Using a stethoscope to pinpoint the source of a noise.

during a stall test or on vehicle acceleration, the stator's one-way clutch is probably defective, which necessitates converter replacement.

3. A grinding noise in Neutral only is usually due to a very worn planetary gear train.

4. A noise in the lower gear ratios but not in third or direct drive is a good indication of worn planetary gear train thrust washers or needle bearings.

5. Noise in high gear only. This usually indicates a problem other than the transmission, like a worn driveshaft center-support bearing, differential bearing, carrier bearing, or axle bearing.

6. Vibration and possible noise on vehicle acceleration up to 45 mph are usually the result of a bent drive line or worn universal joints.

7. Noise or a howl when the vehicle is stationary with the engine running can indicate defective bearings in the air conditioner compressor, alternator, air pump, or power steering pump.

Dynamometer Tests

Many transmission repair shops now use a special transmission dynamometer to test re-manufactured units or those that have malfunctions. The dynamometer itself simulates the operating conditions the transmission encounters while in service in a motor vehicle without the need of the unit being in the vehicle itself. The machine bench checks the rebuilt or malfunctioning transmission for various problems with the unit operating under various load conditions, and the technician can spot and repair defects in the transmission before installing it in the vehicle.

Automatic Transmission Problem Diagnosis

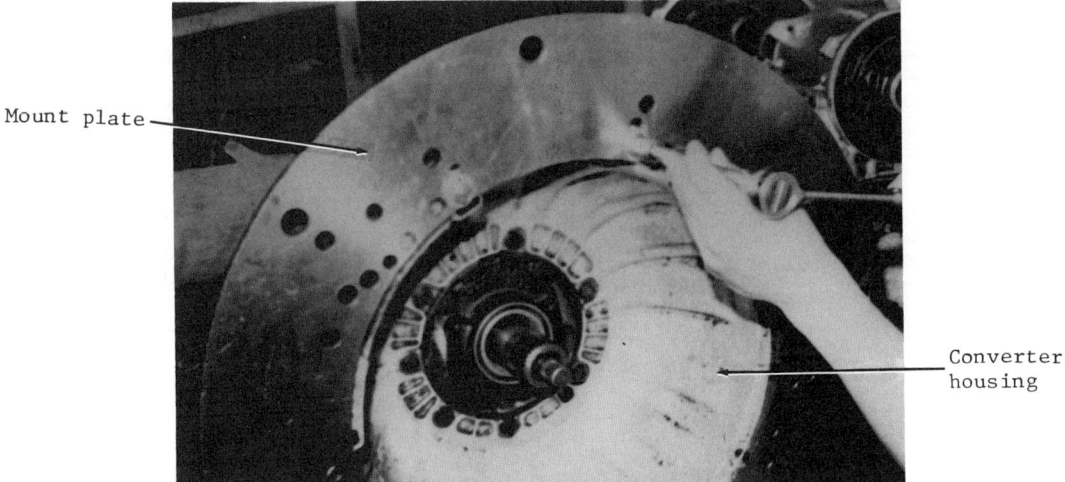

FIG. 3-25 Bolting the mount plate onto the transmission's converter housing.

set-up procedure

To set up and test a typical transmission on the dynamometer mentioned in Section 1, perform the following steps:

1. On the bench, bolt the appropriate mount plate to the transmission converter housing, using at least 4 bolts (Fig. 3-25). The converter housing dowel pins or those supplied with the machine will assist in locating the mount plate onto the housing.

2. Remove all the test pipe plugs from the transmission case, and install the quick-disconnect fitting into their tapped holes (Fig. 3-26).

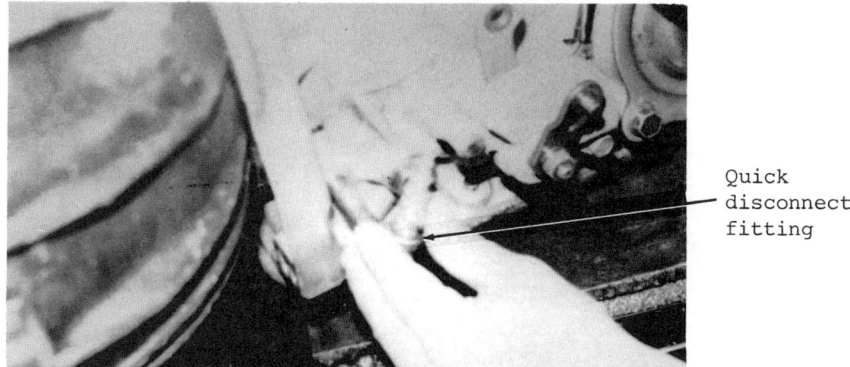

FIG. 3-26 Installing a quick-disconnect fitting into a pressure test port.

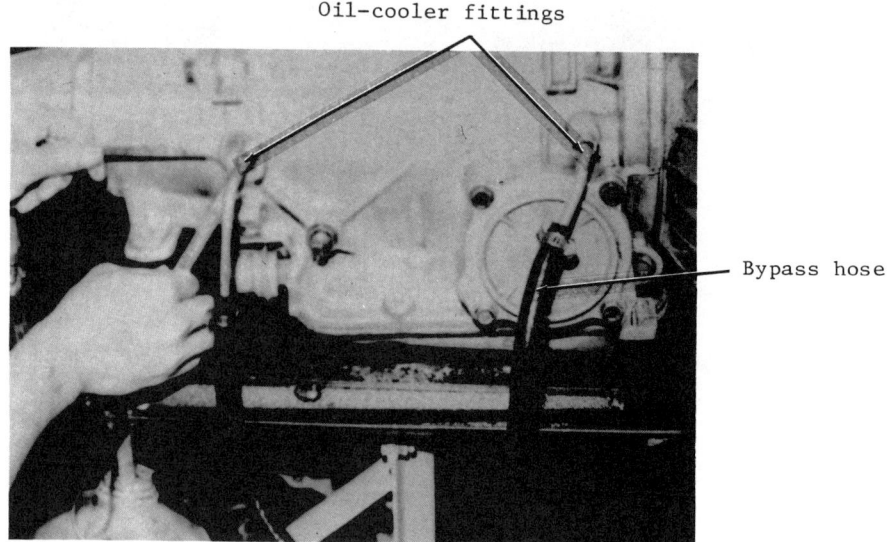

FIG. 3-27 Installing a cooler bypass hose to the transmission's two cooler fittings.

3. Attach a bypass hose between the two oil cooler fittings (Fig. 3-27). Note: On air cooled transmissions, this step is not necessary.

4. Install two impact risers over the converter drive studs 180 degrees apart (Fig. 3-28).

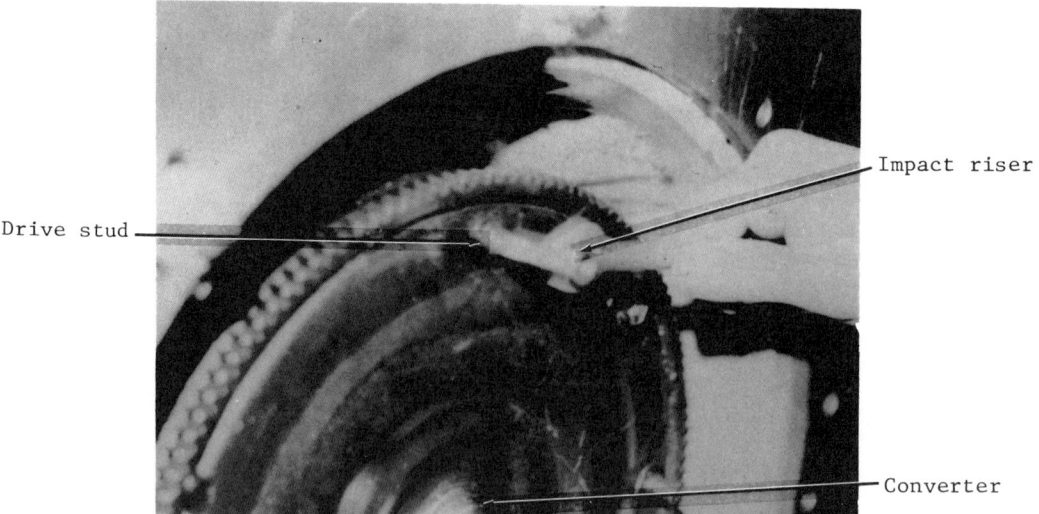

FIG. 3-28 Installing the impact risers to two converter-drive studs.

Automatic Transmission Problem Diagnosis

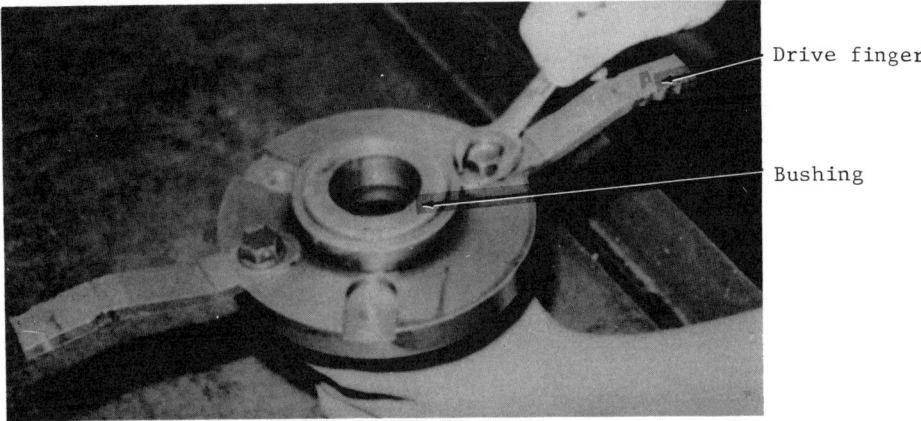

FIG. 3-29 Installing the converter bushing into and the drive fingers onto the converter driver.

5. Using an overhead crane and chain fall, transfer the transmission from the workbench to the machine.

6. Install the two centering-dowel bolts, and one top attaching bolt, and tighten these bolts securely.

7. Install the appropriate drive fingers on the converter driver, and insert the correct size bushing into the center of the driver (Fig. 3-29). Position the completed converter drive assembly into the input-shaft socket.

8. Release the cradle mount slide lock and slide the cradle mount forward to fully engage the converter with the converter drive assembly. Make certain that the drive fingers make

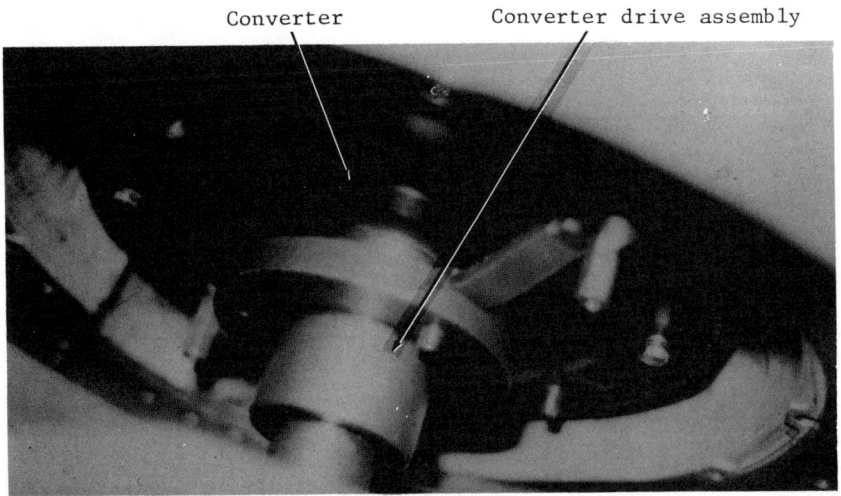

FIG. 3-30 Engaging the converter with the converter-drive assembly.

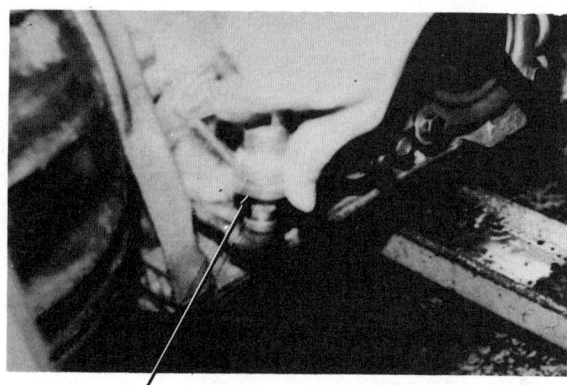

Pressure-gauge hose fitting

FIG. 3-31 Connecting the machine's pressure hose to the quick-disconnect fitting.

contact with the impact risers on their driving sides (Fig. 3-30). Lock the cradle mount in this position.

9. Connect the appropriate number of pressure gauge hoses to the quick-disconnect fittings (Fig. 3-31).
10. Slide the correct splined sleeve onto the transmission's output shaft (Fig. 3-32), and install the front U-joint into this splined sleeve.
11. Release the lock on the load section and the dynamometer's output-shaft set screw (Fig. 3-33). Bring both units forward until the

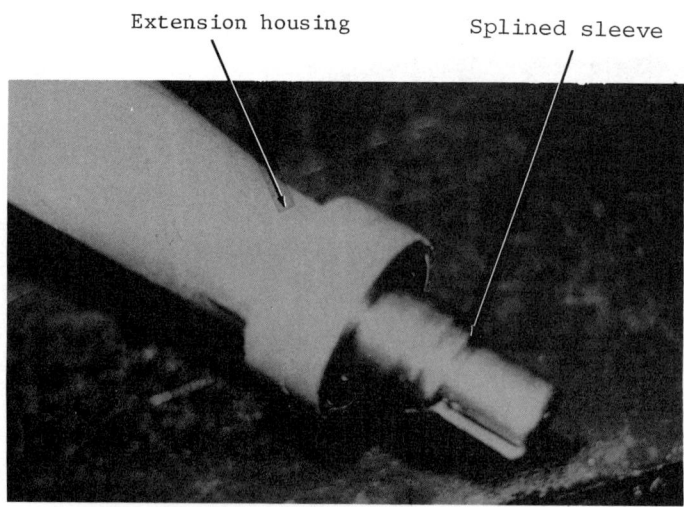

FIG. 3-32 Installing the splined sleeve over the output shaft.

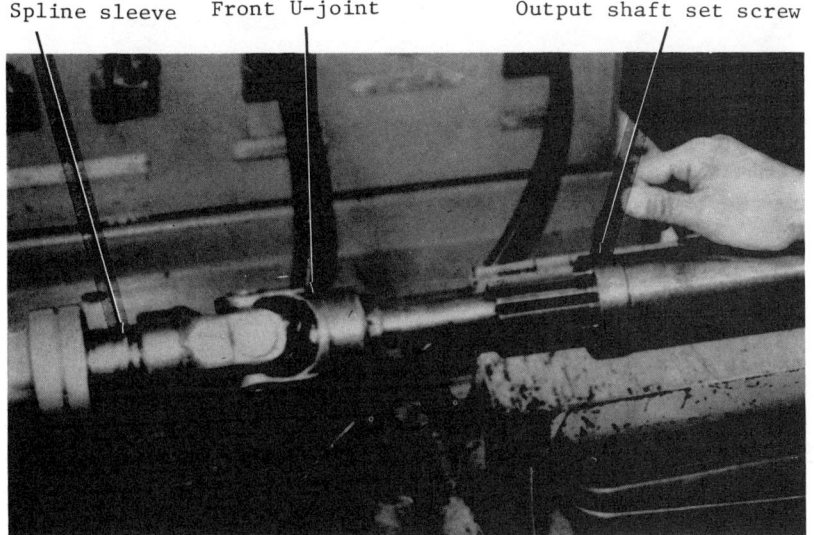

FIG. 3-33 Loosening the output shaft set screw.

drive shaft engages into the U-joint. Move the load section back until about 2 inches of splines are showing on the output shaft. Tighten the drive shaft set screw and load-section lock securely (Fig. 3-34).

12. Position and lock the drive shaft safety shield in place (Fig. 3-35).
13. Position the torque converter safety shield into place (Fig. 3-36).
14. Connect the machine's 12-volt test leads to the transmission electrical kickdown system if so equipped.
15. Install the machine's vacuum gauge hose over the transmission modulator's inlet port if so equipped (Fig. 3-37).

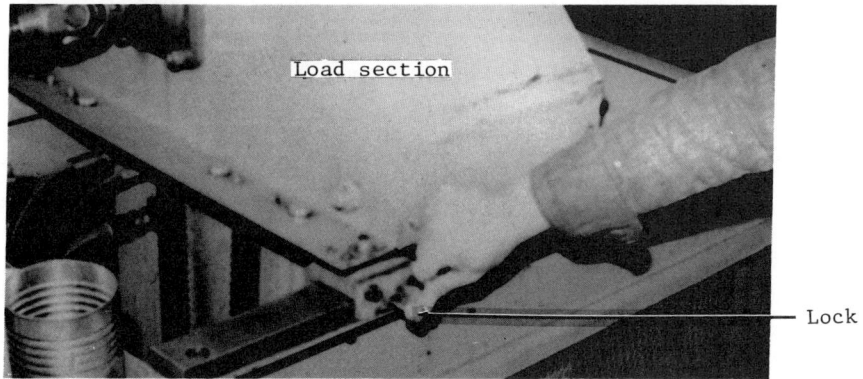

FIG. 3-34 Tightening the load-section lock.

FIG. 3-35 Locking the drive-shaft safety shield in position.

16. Connect the oil filter tube to the transmission.

test procedure

The following items require your observation during the upcoming test procedure:

1. Pressure fluctuations or improper pressure (may be due to low fluid level in transmission).
2. Loud or unusual noise from transmission (use a stethoscope to locate source of noise).
3. Unusual vibration in the transmission.
4. External leakage.

FIG. 3-36 Positioning the torque converter safety shield into place.

Automatic Transmission Problem Diagnosis

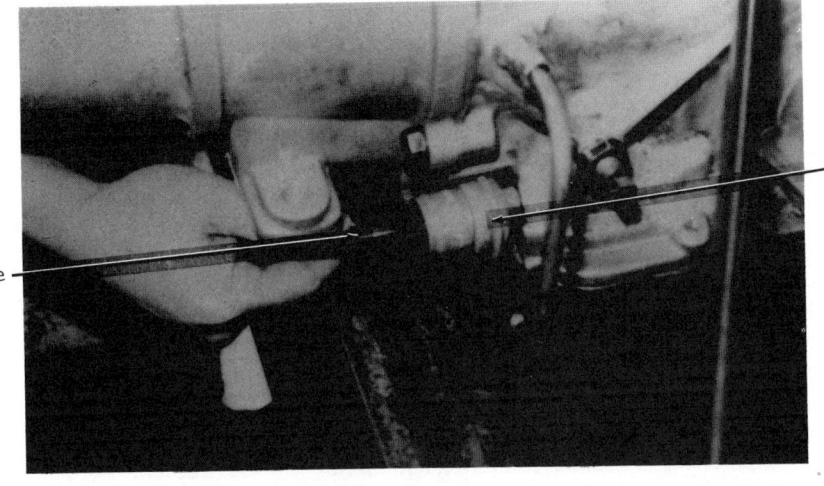

FIG. 3-37 Connecting the vacuum line to the modulator.

5. Missing of upshifts and downshifts.
6. Signs of slippage.

To carry out the dynamometer test, perform the following steps:

1. Service the transmission, with the fluid-supply hose, with 6 quarts of fluid (Fig. 3-38). Start the machine and allow the torque converter to fill. Continue to service the transmission with fluid

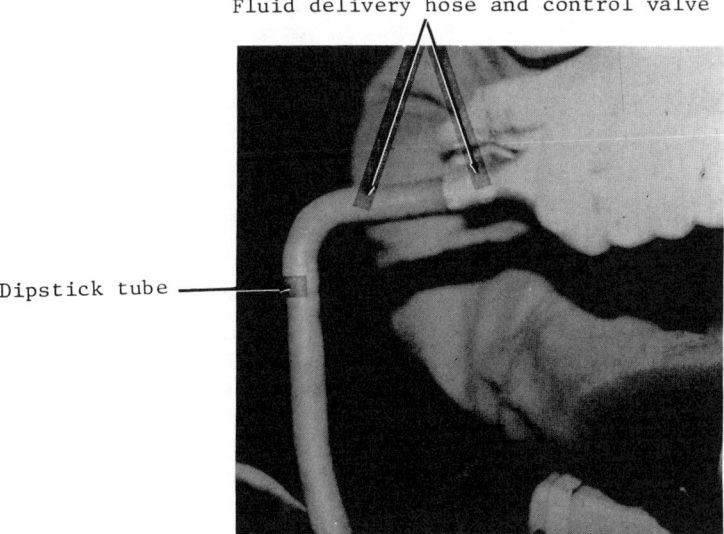

FIG. 3-38 Servicing the transmission with fluid from the fluid-supply house.

until the level on the dipstick reads full. At the conclusion of this servicing procedure, transmission fluid pressure should be stable and to specifications with the transmission operating in Neutral.

2. With the transmission turning at idle speed, shift the manual valve lever through the Drive, Intermediate, Low, and Reverse ranges, noting the control pressure of the transmission during each operating range. The pressure readings should be to specifications.

3. Return the manual valve lever to the Drive position. With the speed control handle, slowly increase input shaft speed while observing both the input and output tachometers. The upshift from low to intermediate should occur at about 900 rpm, and the shift from second to high should occur at about 1400 rpm (input shaft speed).

4. With the input speed set at 1600 rpm, remove the vacuum hose from the modulator. Control pressure should now increase to about 150 psi, and the transmission should downshift to second.

5. Reconnect the modulator hose. The transmission should upshift again to third, and the hydraulic pressure should drop back to the normal reading, about 60 psi.

6. Apply a dynamometer load of about 1200 psi to the transmission, and check the zero vacuum downshift (step 4) again. The downshift should now occur much more abruptly. Reconnect the vacuum hose as in step 5. The transmission should again upshift to third, but the shift should be sharper and more distinct than it was during step 5.

7. With dynamometer load still applied, reduce input-shaft speed to idle, observing the downshift of the transmission into second and finally to Low as input speed decreases. Increase the input speed again, observing the transmission's upshift pattern as input speed increases. The upshifts should occur at about the same points as they did with no load. Reduce input speed to idle and cut the dynamometer load on the transmission.

8. Set the disc brake lever to stall the drive shaft. Increase the input speed to about 750 rpm, allowing the transmission to operate at this speed for several minutes. This action will cause the internal temperature of the transmission to rise.

9. Increase the input speed to about 1100 to 1200 rpm. At this speed, the driveshaft should attempt to break loose from the disc brake, and the input motor will be laboring. If these characteristics are not present, try to increase input shaft speed. It should not be possible to reach 1300 rpm with the driveshaft locked unless internal slipping is present within the transmission itself. Reduce input speed to idle, release the disc brake, and allow the transmission to idle in Drive range for about 1 minute.

10. Apply a full dynamometer load (1600 psi) to the transmission, and adjust vacuum at modulator to 10 inches. Increase the input speed rapidly; the transmission should upshift to third at about 2200 rpm.

Automatic Transmission Problem Diagnosis   131

11. With the transmission operating in third at 2200 rpm, move the kickdown lever to the detent position; the transmission should downshift to second. Release the kickdown lever; the transmission should upshift again to third. Finally, reduce the dynamometer load and reduce input speed to idle.

12. When overall performance of the transmission satisfies all requirements, drain the torque converter and transmission fluid into the tray. Remove the transmission from the machine and return it to the work bench. Remove all the quick-disconnect fittings, replace them with plugs, and remove the mount plate and drive accessories.

post-test observations

When draining the transmission and torque converter, observe the fluid as it comes out:

1. Does the fluid contain particles of friction material or metal? This is a good indication of either internal failure or misfitting of components.

2. Is the fluid sharply discolored in relation to its color before the test? This indicates that the fluid overheated due to internal transmission slippage or to improper test procedures.

CHECK-UP QUESTIONS

The questions listed below will assist you in determining how well you remember the material contained in this section. Read each question carefully before adding the word or words necessary to complete the sentence. If you can't complete the sentence, review that portion of the section that covers the question.

1. To be successful in troubleshooting a transmission, the mechanic must follow a _____, _____, and _____ test procedure.
2. The first check to perform for any transmission malfunction is fluid _____ and _____.
3. Excess fluid can be pulled out of an automatic transmission using a _____ _____.
4. Fluid being burned inside the engine is a common cause of fluid loss in a transmission, using a _____ _____.
5. To help locate the source of a leak, place a piece of white _____ or _____ under the vehicle.
6. In order to determine the source of a leak from the converter housing,

visually inspect the housing area, or use a _____ _____ to determine if the fluid is coming from the transmission.

7. Fluid leakage into the engine cooling system usually results from a crack or hole in the _____ _____ inside the radiator.

8. There is always a mechanical link between the gearshift selector lever and the _____ _____ in the valve body.

9. If a transmission has a vacuum modulator system instead of a mechanically operated throttle-valve system, the modulator will then vary _____ _____ and _____ _____ to match engine load.

10. To test a modulator diaphragm, connect a _____ _____ and _____ _____ to the modulator.

11. The stall test finds the highest engine speed at _____ _____ with the wheel locked and the transmission in gear.

12. To verify an owner's complaint, _____ _____ the vehicle.

13. The initial engagement part of the road test determines how smooth a _____ or _____ engages.

14. A handy reference for analyzing the results of a road test is a _____ and _____ chart.

15. A _____ test indicates any major hydraulic system malfunction.

16. During pressure test number three, apply no more than _____ inches of vacuum to the modulator.

17. If a vacuum modulator has a ruptured diaphragm, hydraulic pressure readings will be _____ during hydraulic tests one and two.

18. Hydraulic pressure is adjustable on some automatic transmissions by means of an adjustable _____ or _____ push rods.

19. _____ guides list the most common trouble symptoms and provide the items a technician should check to find the cause of the malfunction.

20. An _____ test will determine if a clutch, servo, or governor is functioning properly.

21. It is more common to perform an air check during transmission _____.

22. An internal hydraulic leak is usually due to a defective _____ or _____ sealing ring.

23. A _____ is a handy tool in locating the source of a noise.

24. Many transmission shops use a _____ to test rebuilt or malfunctioning transmissions.

25. The dynamometer's 12-volt system connects to a transmission's electrical _____ system.

SECTION

4

Changing the Transmission Fluid and Filter

REFERENCES: Automatic Transmission Fundamentals, Chapter 3.
Automatic Transmission Service, Sections 1, 2, and 3.
Vehicle or transmission service manual.

Changing fluid is one type of in-vehicle service work done on a transmission. Mechanics usually change the fluid within the automatic transmission for one of two reasons. The first and most pressing reason is that the fluid has become oxidized or has lost part or all of its lubricating properties. Section 3 of this manual explains how to check the fluid for this condition. The second practical reason is for preventive maintenance, which includes changing the fluid and filter to comply with factory recommendations or specifications.

The majority of automatic transmission manufacturers recommend fluid changes in their respective units at given intervals, but these interval periods differ from one manufacturer to another. A given manufacturer may specify the interval period in chronological time while another will use vehicle mileage. In either case, the owner's manual or service manual for the vehicle contains the recommended fluid change intervals.

A typical mileage interval may be as low as 20,000 miles or as high as 60,000 miles. Under normal driving conditions, the owner should adhere to these specifications. If the owner does not follow these specifications and change the fluid at the interval period, the fluid can lose its ability to function as a lubricant; and the transmission will wear out prematurely.

In addition, under certain driving situations, the owner may need to have the transmission fluid changed more often. The driving conditions that reduce the useful life of the fluid are the constant operation of a vehicle in hot and or dusty climates, the frequent use of the vehicle to haul heavy loads, and the constant use of the vehicle in stop-and-go driving such as in heavy traffic. These conditions can reduce the life of the fluid considerably and require the owner to replace it at 6000-mile intervals or less.

A total fluid change of an automatic transmission requires a lot more than just draining and refilling the reservoir or pan. For example, every fluid change should include the removal of the pan and the cleaning or replacement of the filter screen. Furthermore, the mechanic should drain or pump the dirty fluid from the torque converter.

DRAINING THE FLUID FROM THE TRANSMISSION

To change the fluid in a typical automatic transmission, proceed as follows:

1. Raise and level the vehicle using either a hydraulic lift or suitable jacks so that the fluid pan and converter inspection plate are accessible. When using jacks for this purpose, always insert safety stands under the frame at appropriate locations. These stands prevent the vehicle from accidentally falling on the mechanic as he works underneath the transmission, and they should remain in place until he completes the repair job.

2. Inspect the transmission pan for a drain plug (Fig. 4-1). Most of the current model transmissions will not have a drain plug. Transmission manufacturers for the most part have deliberately eliminated this plug to assure a more complete fluid change and filter service. Consequently, the mechanic must remove the pan to change the fluid in the transmission.

3. Inspect the transmission for a pan mounted dipstick. The mechanic can also drain some transmissions by removing the dipstick (fluid filler) tube from the side of the pan (Fig. 4-2). Tube fittings secure these types of filler tubes to the side of the pan.

4. Place a drain pan under the transmission. If the transmission pan has a drain plug or detachable filler tube, remove either one and permit the dirty fluid to drain into the container.

5. To inspect, clean, or change the filter, remove the transmission pan's attaching bolts and lower the pan. Be careful: the pan may still be very hot.

6. If the pan has no drain plug or detachable filler tube, remove the pan using the following procedure: Install a drain pan under the transmission. From the rear end of the pan, remove all the attaching

Changing the Transmission Fluid and Filter

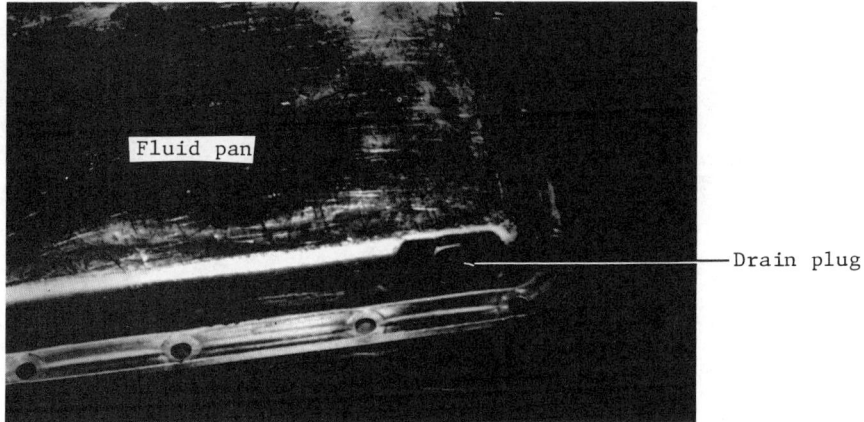

FIG. 4-1   Typical transmission pan drain plug.

bolts (Fig. 4-3). Then, one at a time, begin to remove a bolt from first one side and then the other. This procedure permits the rear end of the fluid pan to slowly tip downward for easy draining.

If the mechanic does not follow this procedure, the hot oil will gush out. This makes it very difficult to catch all the dirty fluid in the drain pan and usually causes a mess on the floor. In addition, the hot oil rushing out of the pan can burn the mechanic's hands as he works under the unit.

7. As soon as the draining fluid has slowed down to a trickle, remove the remaining attaching bolts from the pan and place it on a work bench.

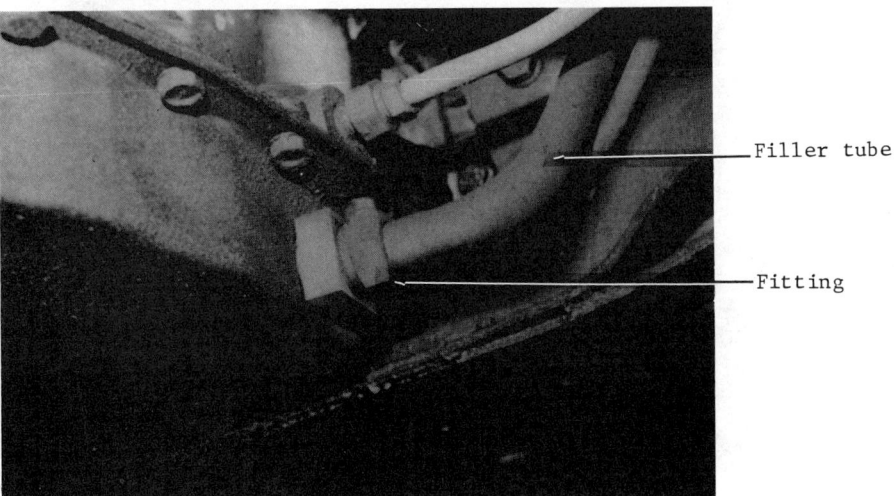

FIG. 4-2   Dipstick located on the side of the pan.

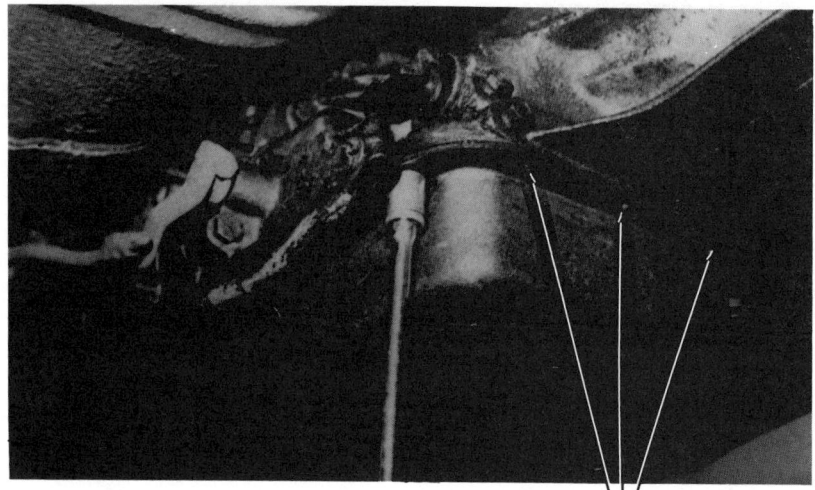

Rear attaching bolt holes

FIG. 4-3 Mechanic removing the bolts from the rear section of the pan.

INSPECTING AND SERVICING THE PAN

1. Carefully inspect the inside of the pan for the presence of metal or friction particles (Fig. 4-4). The presence of a few particles of these materials is normal, but large quantities indicate a transmission with

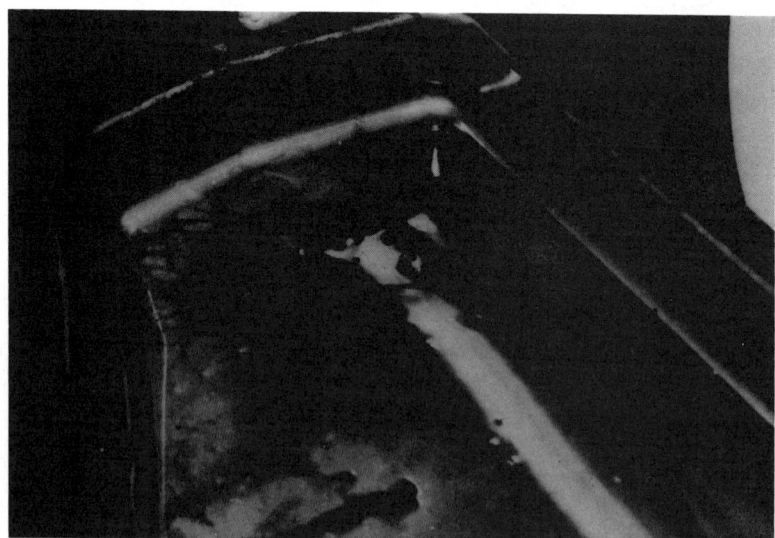

FIG. 4-4 Inspection of the inside of the pan for the presence of metal and friction particles.

Changing the Transmission Fluid and Filter

excessive wear. In this situation, the mechanic should tear the transmission apart for further inspection and repair.

Inside some late model Ford pans, the mechanic may find a small, round, plastic plug with a short stem. <u>This plug did not fall out of the transmission.</u> The transmission manufacturer inserts this device into the dipstick hole after assembling the unit to prevent the entrance of foreign material into the transmission.

When the assembler mounts the transmission onto the engine and before he installs the dipstick, he is supposed to remove the plug. If he does not remove the plug, inserting the dipstick forces the plug into the pan. The plug will not harm any of the transmission components, but it no longer serves a useful purpose. Therefore, if you find one, just throw it away.

2. Inspect the pan and its mating surfaces for distortion on a flat surface. Check the bolt holes on the gasket side of the pan to see if prior torquing of the attaching bolts has raised (dimpled) the metal surface (Fig. 4-5).

    To remove these dimples, place a flat metal bar or plate under each hole, on the bolt side of the pan, and carefully flatten the dimples, using a ball-peen hammer (Fig. 4-6). <u>Be careful not to distort the pan by striking the surface with excessive force.</u> If this process does not successfully flatten the dimples, or the pan is distorted, replace it.

3. After removing all traces of the old gasket, wash the pan in clean solvent, using a brush (Fig. 4-7). Then, using low-pressure compressed air, blow the pan dry. <u>Never use a cloth to dry the pan because any rag lint left behind may enter the hydraulic system and cause valves to stick.</u>

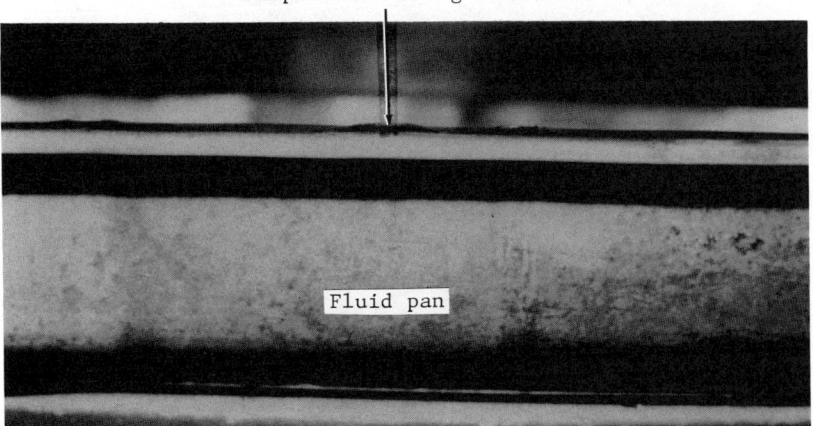

FIG. 4-5 Pan with dimpled bolt holes.

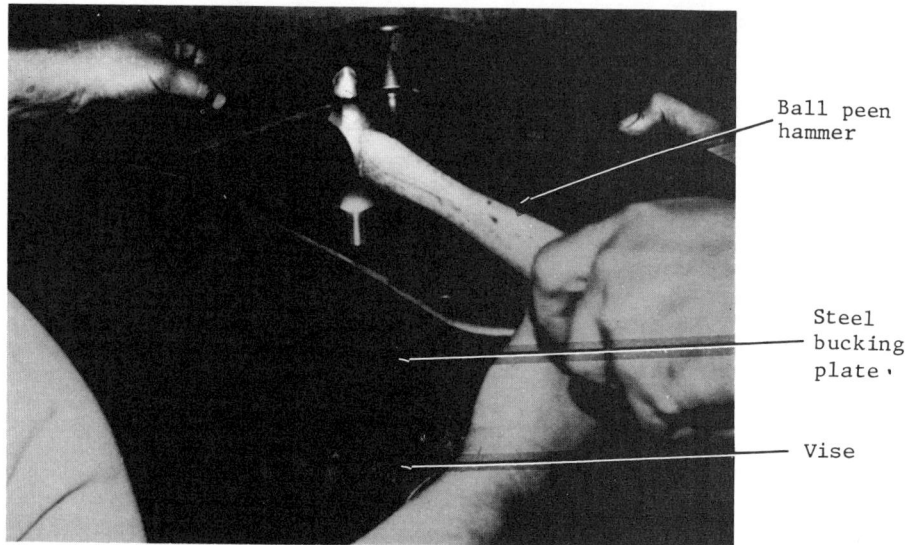

FIG. 4-6 Mechanic flattening a dimpled pan with a ball-peen hammer.

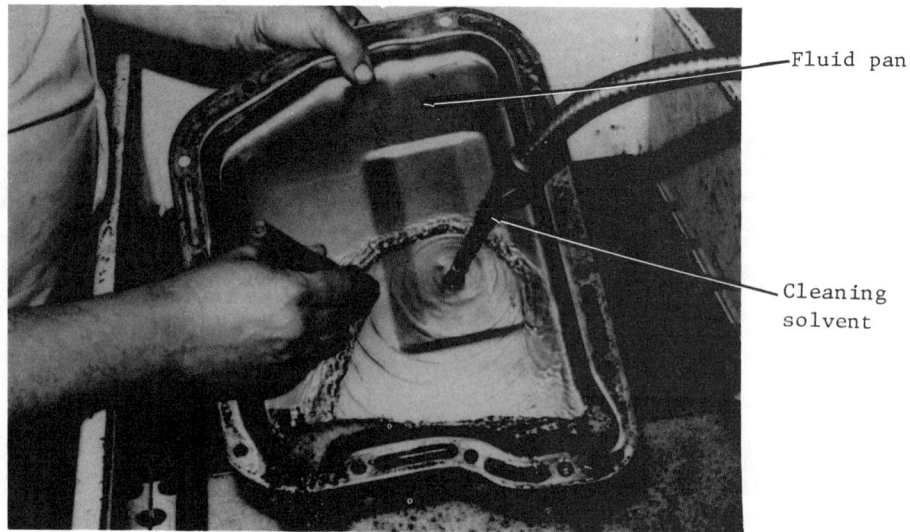

FIG. 4-7 Washing a pan in clean solvent.

Changing the Transmission Fluid and Filter

FILTER OR SCREEN SERVICE

1. While the pan is off, clean or replace the filtering device, located under the valve body (Fig. 4-8). The mechanic can often clean and reuse a metallic-type filter screen as long as it is in good condition. <u>But the mechanic must replace a paper or fabric filter</u>.

2. To actually remove a filter, remove the screws, bolts, or spring clips that secure the filter to the valve body and discard the gasket when used. Be careful when removing the screen from 1970 and some later C-4 transmissions, because a tab section of the screen's housing holds the throttle-pressure limit valve into the valve body (Fig. 4-9). When working on this type of transmission, therefore, carefully lower the screen so that you can reach over the tab and grab the valve and spring before they fall out.

3. Wash the wire screen thoroughly with clean solvent and a brush (Fig. 4-10). If solvent will not dissolve the varnish, built up on the surface of the screen, soak the filter in a good grade of carburetor cleaner for 15 to 30 minutes and rinse it out thoroughly with water.

4. Blow the filter screen dry with low pressure compressed air. Never dry a filter screen with any type of rag. A piece of lint from the rag, trapped in the screen, can work its way into the valve body and cause a valve to stick in its bore.

5. Inspect the screen after cleaning. If the cleaning job does not remove all the varnish, replace the assembly. Varnish build-up will reduce the fluid flow through the screen and cause problems within the

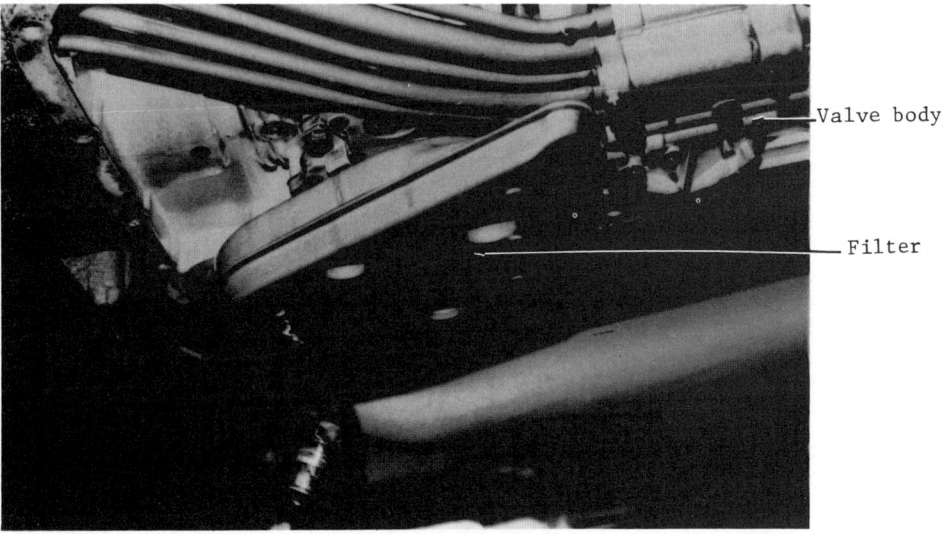

FIG. 4-8 Typical filter location under the valve body.

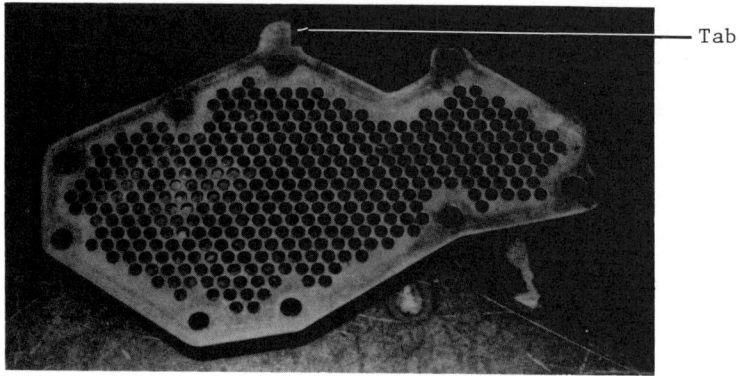

FIG. 4-9 Tab section of a C-4 filter screen assembly.

transmission because of fluid starvation. Finally, if the screen is bent excessively or torn open, replace it.

Installation

1. Replacing a filter or reinstalling a screen is usually not a hard job because they fit onto the valve body only one way. There are several reasons for this. First, the filter or screen manufacturer forms the unit a certain shape, either square or rectangular, and its attaching screw holes are machined so that the filter **must** fasten to the valve body a given way (Fig. 4-11). Second, to make it easy for the mechanic to quickly locate the filter on the valve body, some manufacturers stamp the word "front" on the filter itself. This indicates to the mechanic the proper position of the filter relative to the valve body.

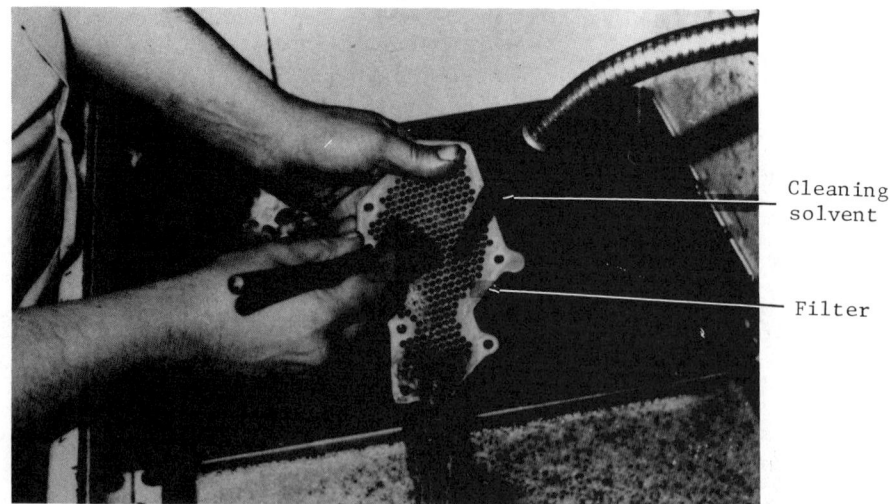

FIG. 4-10 Washing a screen-type filter in cleaning solvent.

Changing the Transmission Fluid and Filter

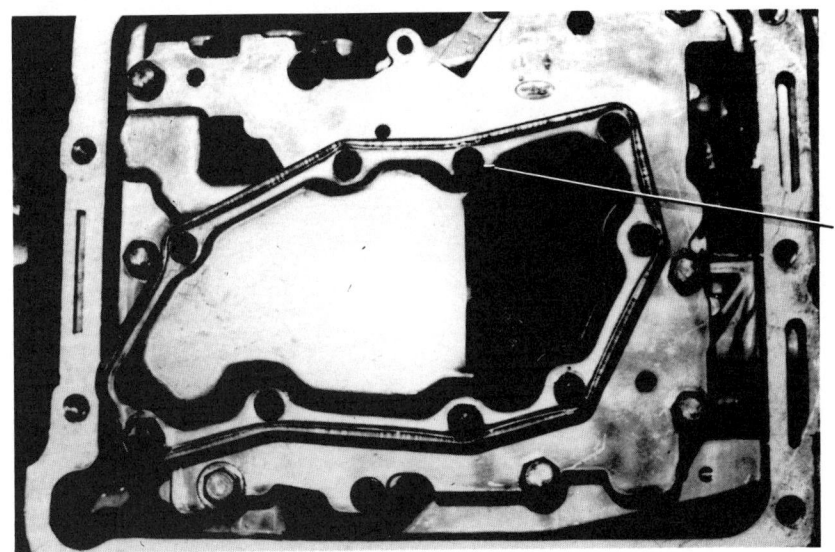

FIG. 4-11 The attaching screw pattern of a C-4 filter screen.

2. When installing the filter or screen, always install a new sealing device, seal or gasket, between the filter and the valve body. If the screen or filter has a mounting tube and it has an O-ring seal, always replace it. Then push the screen assembly straight up into position until the tube seats in its bore in the transmission case.

3. Install any screws, bolts, or attaching straps that secure the filter to the valve body. Tighten all attaching screws or bolts to factory specifications.

PAN INSTALLATION

1. Reinstall the fluid pan, using a new gasket. Place the new gasket on the pan's mounting flange (Fig. 4-12). To hold the gasket in place, coat the flange with petroleum jelly or grease. <u>Do not use gasket cement or sealer</u>.

2. Reinstall the pan onto the transmission case, making sure the gasket does not slip out of alignment or become pinched. Install all the pan's attaching bolts and washers if used. Tighten all bolts finger tight (Fig. 4-13).

3. With a suitable torque wrench, tighten the pan bolts to factory specifications, using a crisscross pattern. It is sometimes advisable to set the initial tightening torque to one-half of normal factory specifications to compress the gasket slowly. Then retorque all bolts to the full amount (Fig. 4-14). The crisscross pattern and this

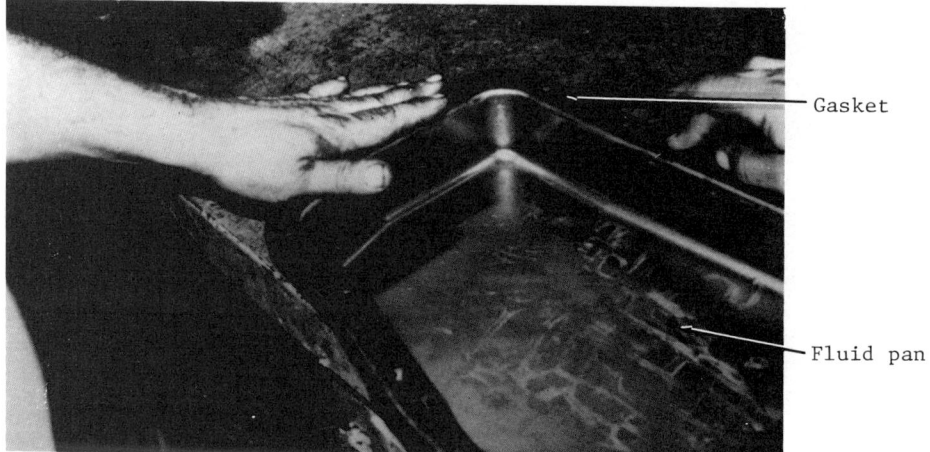

FIG. 4-12 Installing the gasket on the pan.

torquing sequence prevent the pinching, breaking, or distorting of the gasket out of shape as well as dimpling of the pan's bolt holes.

4. If the transmission has a filter tube that attaches directly to the pan, connect the tube and tighten its fitting finger tight. Using a torque wrench, tighten the fitting to factory specifications.

5. If the pan has a drain plug, install a new gasket over it and thread it into the pan finger tight. Then torque the plug to factory specifications.

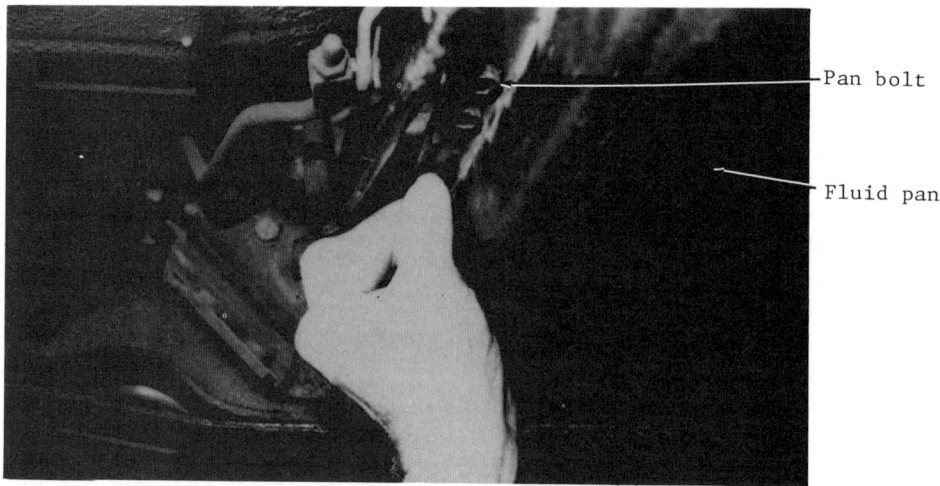

FIG. 4-13 Installing the pan attaching bolts finger-tight.

Changing the Transmission Fluid and Filter   143

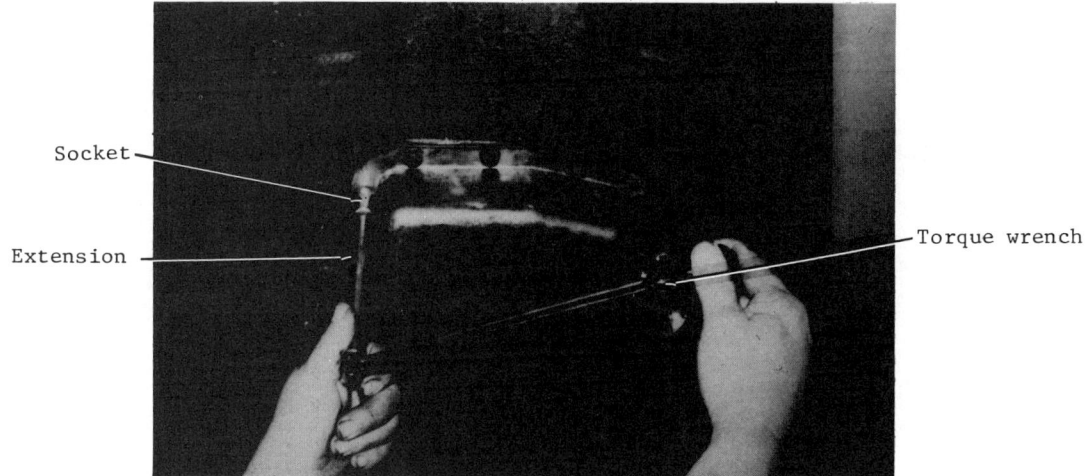

FIG. 4-14 Torquing the pan's attaching bolts.

DRAINING THE CONVERTER

Draining the dirty fluid from a torque converter can be a problem because some converters do not have drain plugs. For instance, in 1978, Chrysler Corporation eliminated the drain plugs from their converters. General Motors has not used plugs in any of their converters since the mid-1960s. Consequently, the only automotive converters that the mechanic can actually "drain" are the Ford, pre-1978 Chrysler, and some early G.M. converters.

To service those units that have a drain plug perform the following steps:

1. Remove the converter inspection or access panel from the bell housing (Fig. 4-15).

2. Using a remote starter button or the starter switch, turn the engine over until the converter drain plug is in its 6 o'clock position (Fig. 4-16). <u>Never pry on the starter ring gear or flexplate with a screwdriver or use a wrench on the converter mount bolt in order to turn the engine. This action could distort the ring gear or flex plate, or damage the converter bolts.</u>

3. Remove the plug and allow the dirty fluid to drain into the fluid container (Fig. 4-17).

4. After the old fluid has drained out of the converter, replace and tighten the plug.

5. Install the access panel back onto the bellhousing, torquing all attaching bolts to specifications.

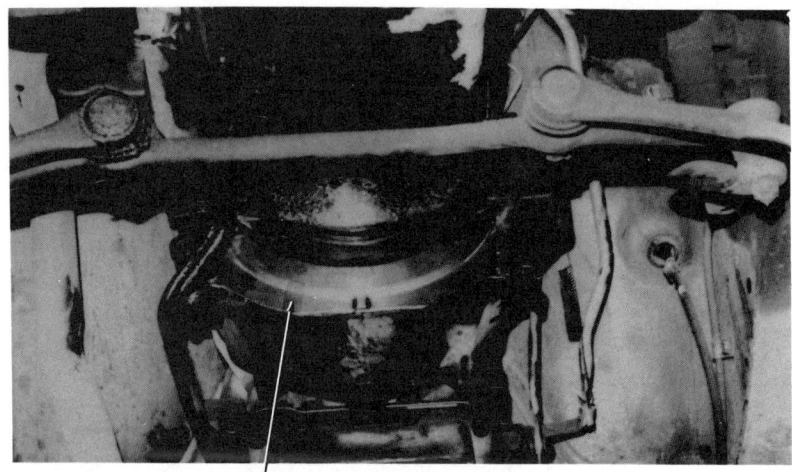

Converter access panel

FIG. 4-15 Typical converter access panel.

PUMPING THE FLUID FROM THE CONVERTER

If the torque converter has no drain plug, a mechanic can pump the majority of the dirty fluid out of the unit by following this technique.

1. At the radiator, disconnect the inlet-pressure line fitting at the transmission cooler and remove the tube (Fig. 4-18).

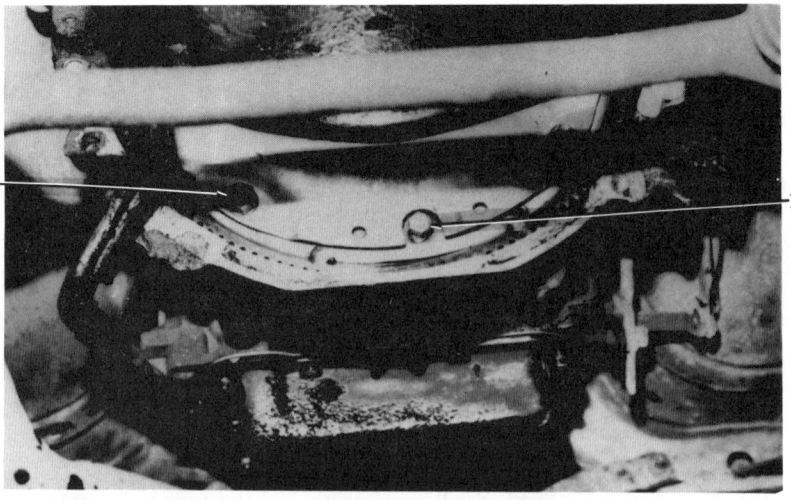

FIG. 4-16 Converter drain plug, positioned for removal.

Changing the Transmission Fluid and Filter 145

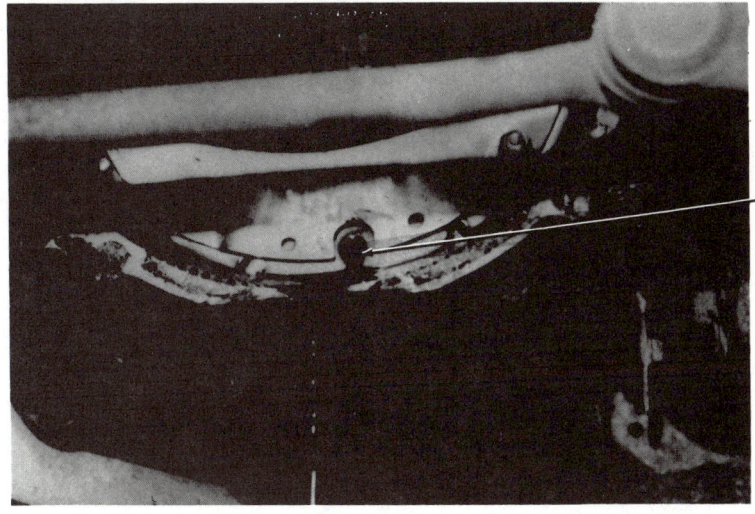

FIG. 4-17 Fluid drains easily from a converter with the plug removed.

2. Slip a suitable snug-fitting drain hose over the end of the tube and place the free end of the hose in a drain pan (Fig. 4-19). If in doubt as to which line is the inlet, disconnect both lines and attach drain hoses to both.

3. After cleaning the pan and servicing or replacing the filter, add 4 quarts of the correct type of ATF to the transmission through the filler tube.

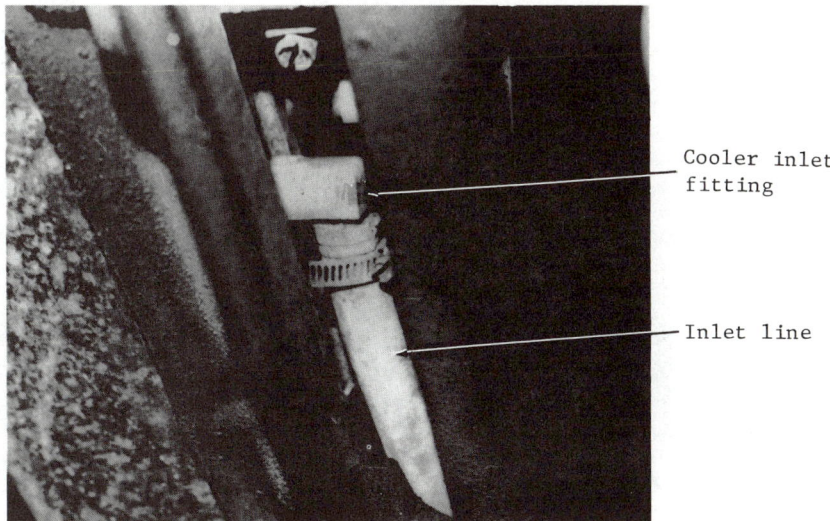

FIG. 4-18 Typical cooler inlet-pressure line fitting.

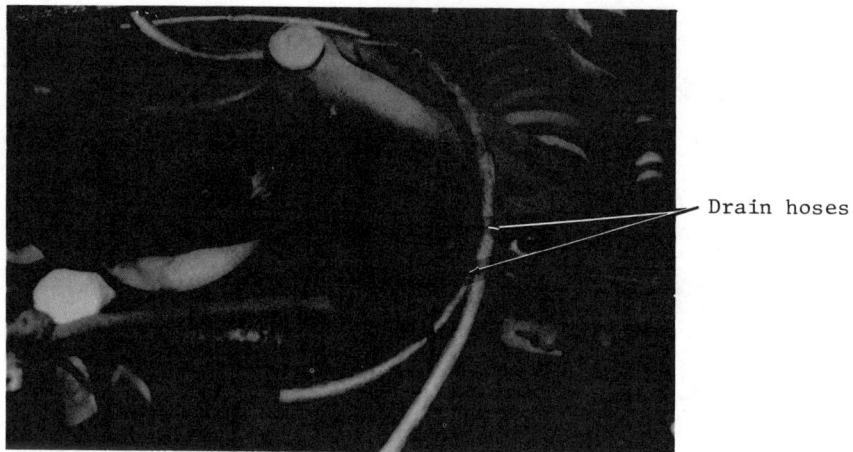

FIG. 4-19 Two drain hoses from the cooler lines leading down to a drain pan.

4. While observing the hoses in the drain pan, have an assistant start the engine. Dirty fluid should begin to flow out of the inlet hose.

5. Keep the engine running at idle rpm until the fluid changes color, indicating new fluid is now in the converter and cooler line, or until air bubbles appear (Fig. 4-20). If air bubbles appear before the fluid changes color, stop the engine and add 2 additional quarts of fluid to the transmission. Restart the engine and continue the process until all the dirty fluid is out of the converter.

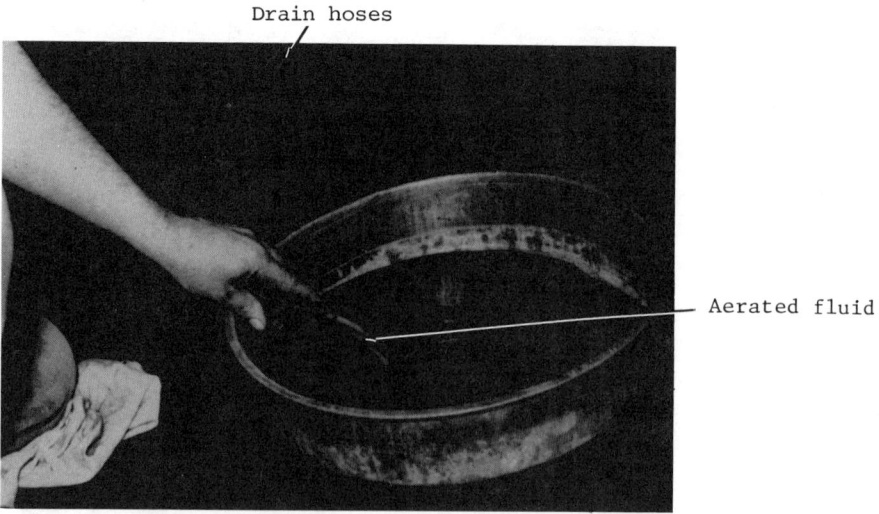

FIG. 4-20 Air bubbles coming out of the drain hose indicates the pump is sucking air because the pan is empty.

Changing the Transmission Fluid and Filter   147

> Note: Always stop the engine immediately when air bubbles appear, because this indicates the pan is empty of fluid and the pump is sucking air.

6. When clean fluid begins to pour from the inlet line, shut the engine off and reconnect the cooler lines to the radiator. Torque the fittings to factory specifications.

FILLING THE TRANSMISSION WITH NEW FLUID

1. Look up the total fluid capacity of the transmission and converter in the shop manual. The actual amount of fluid to initially add and the amount necessary to completely fill the unit will depend on two factors: (1) the total capacity of the converter and transmission, and (2) whether the mechanic drained or pumped the dirty fluid out of the converter. If the transmission and converter hold 10 quarts and you drained the pan and the torque converter, add 4 quarts of ATF to the transmission before starting the engine (Fig. 4-21). If on the other hand, you pumped the fluid out of the converter until it changed color, add only 3 quarts to the transmission before restarting the engine. Finally, if you drained the pan by itself, add 1 quart <u>less</u> than pan capacity before starting the engine.
2. With the gearshift selector lever in Park, start the engine and allow it to operate at idle rpm. <u>If you drained the converter via the plug, immediately add 4 quarts of fluid to the transmission.</u>
3. With the parking brake set and the brake pedal held down, operate the transmission in all its operating ranges, stopping in each range until the transmission fully engages.

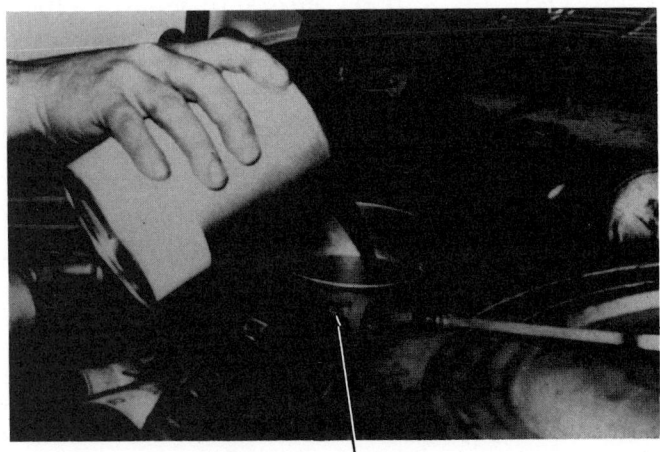

Funnel-type transmission filler

FIG. 4-21 Servicing the transmission with fluid.

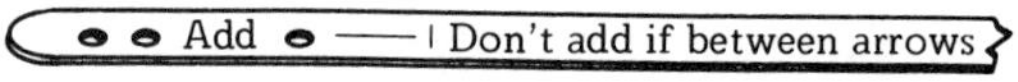

FIG. 4-22 Add mark on a typical dipstick.

4. Return the gearshift selector lever to Neutral or Park as specified by the manufacturer and check the fluid level.
5. Add more fluid as required to raise the level up, but never higher than the <u>add</u> mark on the dipstick (Fig. 4-22).
6. Permit the transmission and engine to operate until they reach normal operating temperature and then recheck the fluid level. Now, bring the fluid level up to the <u>full</u> mark on the dipstick by adding fluid carefully. <u>Never overfill the transmission</u>.
7. Inspect the cooler fittings, drain plug, pan gasket, and filler tube installations for external leakage. Perform any repair necessary to stop any observed leakage.

CHECK-UP QUESTIONS

The questions listed below will assist you in determining how well you remember the material contained in this section. Read each question carefully before adding the word or words necessary to complete the sentence. If you can't complete the sentence, review that portion of the section that covers the question.

1. If the transmission fluid becomes _____, it should be changed immediately.
2. Fluid change intervals for a given automatic transmission can be found in the _____ _____ or the _____ _____.
3. A total fluid change of an automatic transmission includes draining the _____, _____, and servicing the filter.
4. Some fluid pans are drainable by removing the _____ attached to its side.
5. To remove a pan that has no drain plug, remove the _____ attaching bolts first.
6. A large amount of metal or friction particles found in the fluid pan is a good indication that the transmission has excessive _____.
7. Excessive bolt torque can cause the pan's bolt holes to _____.
8. The mechanic should never use rags to dry a pan or filter because _____ may get into the hydraulic system.
9. The mechanic should never attempt to clean a _____ type filter.

Changing the Transmission Fluid and Filter

10. To hold the gasket in place on the pan, coat the mounting flange with _____ _____ or _____.

11. The mechanic should tighten the pan's attaching bolts with a _____ _____ to _____ _____.

12. To remove the converter's drain plug, rotate the engine until the plug is in its _____ _____ position.

13. If the converter has no drain plug, the mechanic can _____ the majority of the fluid out of it.

14. If air bubbles appear in the fluid flowing from the converter drain hose, _____ the engine immediately.

15. The mechanic can look up the fluid capacity of a given automatic transmission in the _____ _____ for the unit.

SECTION 5

Band and Linkage Adjustments

REFERENCES: Automatic Transmission Fundamentals, Chapters 6, 10, and 11.
Automatic Transmission Service, Sections 1, 2, and 3.
Vehicle or transmission service manual.

Along with fluid and filter changes, other important types of service work that a mechanic can perform while a transmission is still in a vehicle are band and linkage adjustments. The adjustment of these components must be correct (to factory specifications), or the transmission can not operate correctly. In other words, many transmission malfunctions can result from improper band or linkage adjustments. However, these adjustments obviously will not correct all transmission problems as many people believe; the material contained in Section Three of this manual pointed this out rather clearly.

BAND ADJUSTMENTS

Effects Of Improper Band Adjustments

In order for a band to perform its task of holding a planetary member stationary, the mechanic must follow a certain adjustment procedure that establishes the correct amount of open or operating clearance between the band's lining and the drum. If, for example, a band is too loose or there

Band and Linkage Adjustments

is <u>excessive clearance</u> between this lining and the drum it must stop, the band's servo piston will not be able to firmly lock the band around the drum before the piston bottoms in its bore. As a result, the band permits the drum and its attached planetary member to slip, which causes the engine to overspeed. In other words, the transmission will slip in any driving range where this excessively loose band applies.

If, on the other hand, the band is too tight or there is <u>insufficient clearance</u> between the lining and the drum, the band may drag or not release its hold on the drum at all. This will cause a total band failure or cause the planetary gear train to seize up, depending on how tight the band is. If the band adjustment is such that the lining constantly drags on the drum, the band lining will prematurely burn out from excessive friction caused by a lack of or reduction in fluid lubrication between the lining and drum.

If the band adjustment is so tight that the lining will not permit the drum to move at all, the planetary gear train will, most likely, seize as it attempts to change from one ratio to another. This planetary seizure results from the inability of the band-controlled planetary member to free wheel when it should. In other words, the planetary gear train would, technically speaking, be trying to operate in two different gear ratios at the same time; this is what causes the unit to seize, or bind up.

As previously stated, many people still think that a band adjustment will correct most transmission problems; this is not really the case. While a band adjustment does play an important part in the overall operation of the transmission, it may not always be the cause of a slipping condition. For example, low system pressure on the servo piston that applies the band or a defective friction or overrunning clutch can also cause an automatic transmission to slip. Consequently, the adjustment of the band, in an attempt to correct a slipping condition caused by any of the other conditions, would be a waste of time and effort.

Frequency of Band Adjustments

Each transmission manufacturer recommends a given interval period between band adjustments. This interval may be in vehicle mileage such as at every 20,000 miles, or it may be at every fluid change, or only during overhaul. In any case, the manufacturer makes these recommendations for a reason, and the vehicle owner and mechanic should adhere to them.

The main purpose behind adjusting a band at a given mileage interval or fluid change is to compensate for lining wear. The manufacturer does expect a given amount of lining wear over a certain period of time. Consequently, the owner has to have someone perform a transmission band adjustment to compensate for this wear, or the band will begin to slip after a while.

Manufacturers recommend band adjustments during unit overhaul for several reasons. First, the band adjustment procedure, if followed correctly, centers the band properly around the drum it will hold. Second, the adjustment procedure inserts the specified lubrication clearance between the band lining and the drum, which prevents premature lining burn out.

Types of Band Adjustments

Basically, there are two ways of making band adjustments--externally or internally. If the transmission design is such that the mechanic can adjust a band externally, the transmission will have a threaded adjuster that screws into the side of the transmission case (Fig. 5-1). Since the head of this adjuster is on the outside of the case, it provides the means by which the technician can perform an external adjustment.

If the transmission design is such that the adjuster is not accessible from the outside of the case, the mechanic must first remove the fluid pan in order to make an internal adjustment. Usually when a transmission has internal adjusters (Fig. 5-2), the manufacturer recommends a band adjustment during a fluid change only. The reason for this, of course, is to reduce the number of times the fluid pan has to be pulled off for maintenance.

In some transmissions, the manufacturer does not provide a means to adjust the bands. For example, the General Motors T-400 transmission has two bands, which are not adjustable by the normal method. In a situation like this one, the band-to-drum clearance is a factor determined by the diameter of the drum, the thickness of the band and lining, and the stroke of the servo piston. With this arrangement, when the band-to-drum

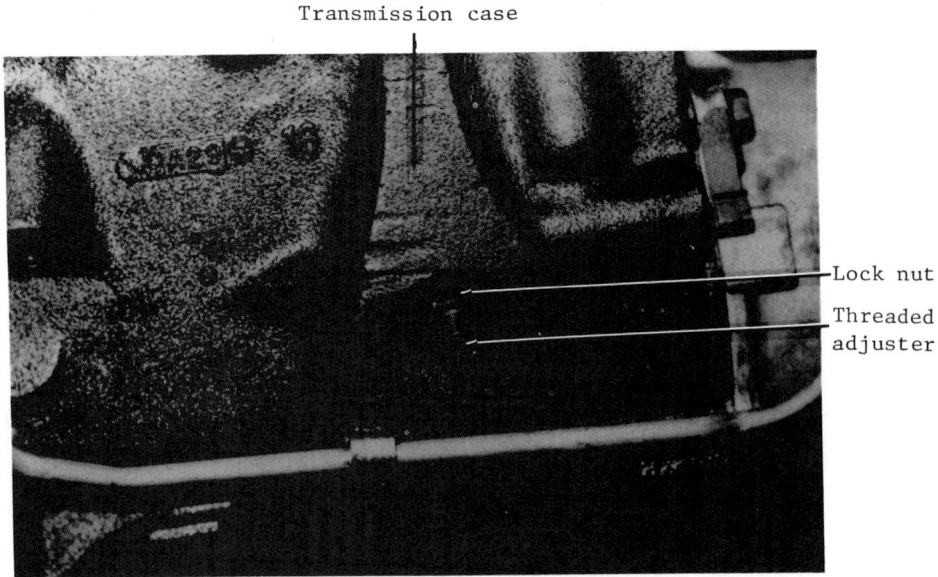

FIG. 5-1 Typical external-type band adjuster.

Band and Linkage Adjustments

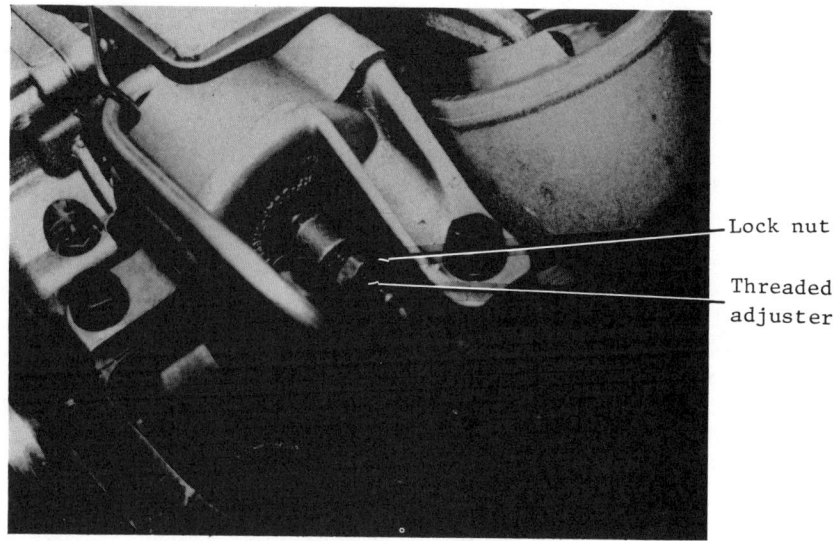

FIG. 5-2 Typical internal-type band adjuster.

clearance becomes excessive enough to cause slippage, the mechanic has no alternative but to replace the band.

If and when a mechanic overhauls a transmission with nonadjustable bands, he must check the band-to-drum clearance using a special gauge (Fig. 5-3). If the band is in good condition but the clearance is not to specifications, the technician must change the selective servo apply pin. In other words,

FIG. 5-3 Checking the rear-servo apply-pin length of a T-400 transmission using a special gauge.

this transmission manufacturer provides (selective) servo apply pins of various lengths in order to adjust band clearance during the overhaul process.

Common Band Adjustment Service Operations

Although band adjustment procedures vary from one transmission type to another, there are certain service operations common to all: torquing the adjuster and then either tightening or loosening the adjuster a specified number of turns. The first important service operation in band adjustment is, of course, the torquing of the adjuster to a given specification. This operation actually loads (tightens) the band around the drum, and if the mechanic does not use an accurate torque wrench or follow the torque specifications, the second operation or part of the adjustment procedure will be inaccurate.

The second, and equally important operation, is the actual establishment of the correct band-to-drum clearance. The mechanic accomplishes this by turning the adjuster either in or out a specified number of turns. The proper clearance, in other words, will exist between the band and drum if the mechanic first torques the adjuster and then turns the adjuster in or out the number of turns specified by the manufacturer.

Equipment and Tools Necessary To Adjust Bands

As previously mentioned, some transmissions require the use of a special piece of equipment to check and/or set band-to-drum clearance (Fig. 5-3). Figure 5-4 shows another form of equipment -- a special gauge block and torque wrench -- used to adjust the front (intermediate) band on an FMX transmission. The 1/4-gauge block fits between the intermediate, servo-piston stem, and the adjuster screw. A torque wrench with a given setting is built into the handle of the tool; this handle breaks or slips when the adjuster screw reaches a torque of 10 pounds-inch.

The tools also necessary to adjust the most common types of bands usually include such items as an Allen or 8-point socket and an accurate torque wrench. The Allen or 8-point socket is necessary because the manufacturers use these head designs for the adjustment screws to differentiate them from other transmission hardware (Fig. 5-5). The torque wrench, remember, is necessary to tighten the adjustment screw a given amount, either in inch- or foot-pounds.

Band Adjustment Specifications

Before the mechanic performs a band adjustment on any transmission, it is important that he look up the specifications as to the initial torque on the adjustment screw and the number of turns it must be turned out or in (Fig. 5-6). When looking up these specifications in a manual, the

Band and Linkage Adjustments   155

FIG. 5-4 Special gauge block and wrench used to adjust the front (intermediate) band on an FMX transmission.

manufacturer may require some specific information from the transmission itself. For example, it may be necessary for the mechanic to check the data tag or transmission case for the transmission's serial number.

This transmission serial number now, in most cases, not only identifies the vehicle the unit fits into but also the engine type and displacement. The knowledge of the engine type and displacement is necessary because manufacturers install the same transmission model with several engine styles. When the manufacturer does this, the internal components of the transmission are usually somewhat different and so are the adjustments. Therefore, a mechanic should always check the serial number on the transmission before looking up the specifications in the manual.

FIG. 5-5 Head designs of typical band adjustment screws.

Torqueflite transmission
band adjusting chart

Kickdown band

Transmission model	Engine size	Adjuster torque	Number of turns
A-904	225 and 318	72 inch pounds	2
A-904 LA	360 (1)	72 inch pounds	2
A-904 LA	225, 318, 360 (2)	72 inch pounds	2-1/2
A-727	360, 400, 440	72 inch pounds	2-1/2

Low and reverse band

A-904	225	41 inch pounds	7
A-904	318	72 inch pounds	4
A-904 LA	360 (1)	72 inch pounds	4
A-904 LA	225, 318, 360 (2)	72 inch pounds	2
A-727	360, 400, 440	72 inch pounds	2

(1) Transmissions that have five clutch discs in the reverse-high clutch pack.

(2) Transmissions that have three clutch discs in the reverse-high clutch pack.

FIG. 5-6   Typical chart listing band adjustment specifications.

Typical Internal Band Adjustment Procedures

FMX transmission

To perform a routine intermediate band adjustment on certain FMX transmissions, proceed as follows:

1. Drain the fluid from the transmission pan, and remove the pan and fluid screen as outlined under the section on fluid changes.

2. Loosen the intermediate (front) servo adjustment screw locknut two full turns (Fig. 5-7). Check the adjusting screw for free rotation in the actuating lever after loosening the locknut, and free the screw if necessary. Note: The adjusting screw will not receive the proper torque value if it does not rotate freely.

3. Pull the adjusting screw end of the actuating lever away from the servo body. Insert the adjusting-tool's 1/4-inch gauge block between the servo-piston stem and the adjusting screw (Fig. 5-8).

4. Tighten the adjusting screw with the adjusting-tool wrench handle until the wrench overruns the screw. If this special wrench is not available, torque the screw to 10 pounds-inch with a common torque wrench.

5. Remove the gauge block and either tighten or loosen the adjustment screw the specified number of turns.

Band and Linkage Adjustments

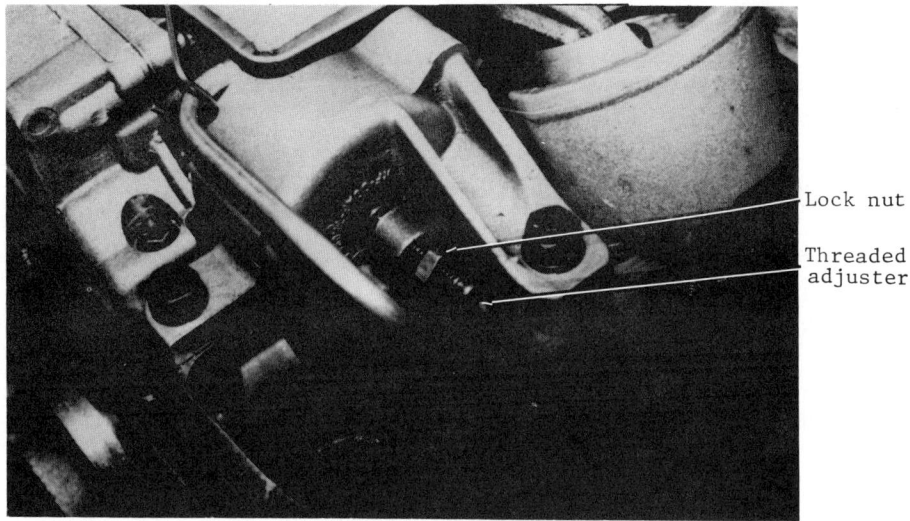

FIG. 5-7 FMX intermediate band adjustment screw.

6. Hold the adjusting screw stationary and torque the locknut to 20 to 25 pounds-foot.
7. Reinstall the filter screw and fluid pan as outlined in the section on fluid changes.
8. Refill the transmission to the correct level with Type F Fluid.

FIG. 5-8 Adjusting the FMX intermediate band using the special gauge block and wrench.

Chrysler

The adjustment of the low reverse band of a Torqueflite transmission does not require the use of a special gauge. To adjust this band, proceed as follows:

1. Raise the vehicle, drain the transmission fluid, and remove the pan as outlined under the section on fluid changes.

2. Loosen the adjusting-screw locknut and back it off approximately five turns (Fig. 5-9). Check the adjusting screw itself for free turning in the lever.

3. Torque the adjusting screw the specified number of turns (Fig. 5-10). Then, while holding the adjusting screw, tighten the locknut to specifications.

4. Back off the adjusting screw the specified number of turns. Then, while holding the adjusting screw, tighten the locknut to specifications.

5. Reinstall the fluid pan and refill the transmission with Dexron fluid as outlined under the section on fluid changes.

Typical External Band Adjustment Procedure--C-4 Transmissions

To adjust the bands on a typical C-4 transmission proceed as follows:

1. Clean all dirt away from the intermediate band adjusting screw area (Fig. 5-11

2. While holding the adjusting screw, loosen, remove, and discard the locknut. This type of locknut has a seal bonded to its inside surface, so it will no longer seal properly if reused.

3. Install a new locknut loosely onto the adjusting screw.

4. Torque the adjusting screw to 10 foot-pounds (Fig. 5-12).

FIG. 5-9 Torqueflite low reverse band adjuster.

Band and Linkage Adjustments

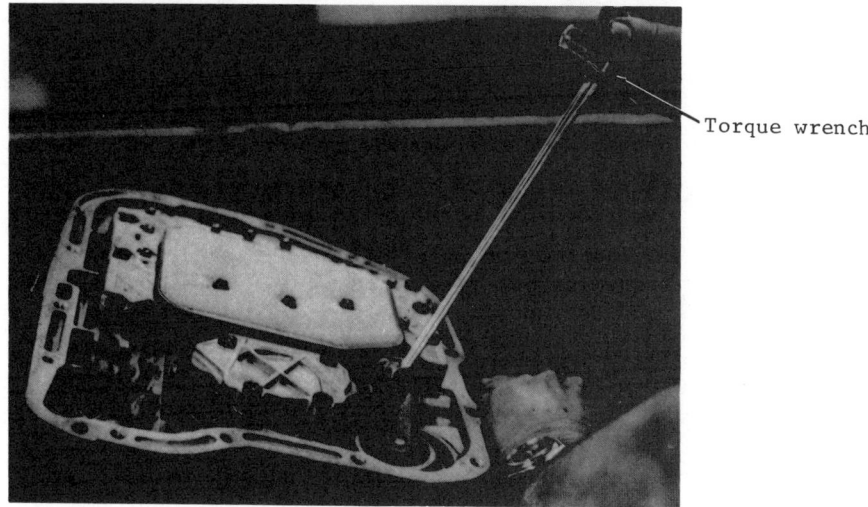

FIG. 5-10 Torquing the low reverse band adjuster.

5. Back off the adjusting screw the number of turns specified by the manufacturer.

6. While holding the adjusting screw, torque the locknut to 35 to 45 pounds-foot.

<u>low-reverse</u>

1. Clean away any accumulated dirt from around the band adjusting screw area (Fig. 5-13).

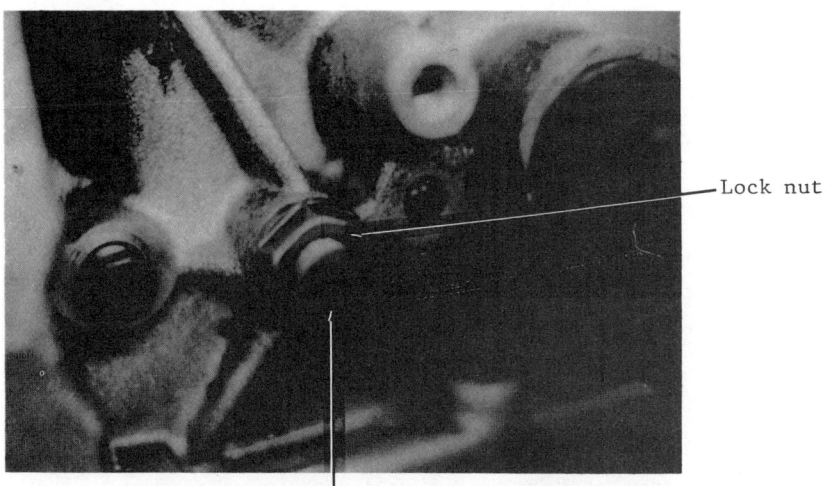

FIG. 5-11 C-4 intermediate band adjuster.

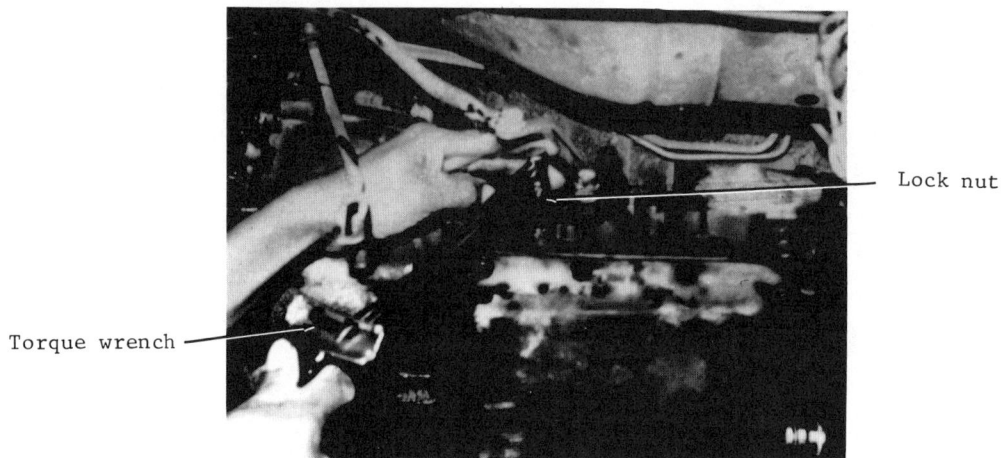

FIG. 5-12 Torquing the C-4 intermediate band adjuster.

2. While holding the adjusting screw, loosen, remove and discard the locknut. This locknut has a seal bonded to its inside surface, so it will no longer seal properly if reused.
3. Install a new locknut loosely onto the adjusting screw.
4. Torque the adjusting screw 10 pounds-foot (Fig. 5-14).
5. Back off the adjusting screw the number of turns specified by the manufacturer.
6. While holding the adjusting screw, tighten the locknut to 35 to 45 pounds-foot.

FIG. 5-13 C-4 low reverse band adjusting screw.

Band and Linkage Adjustments

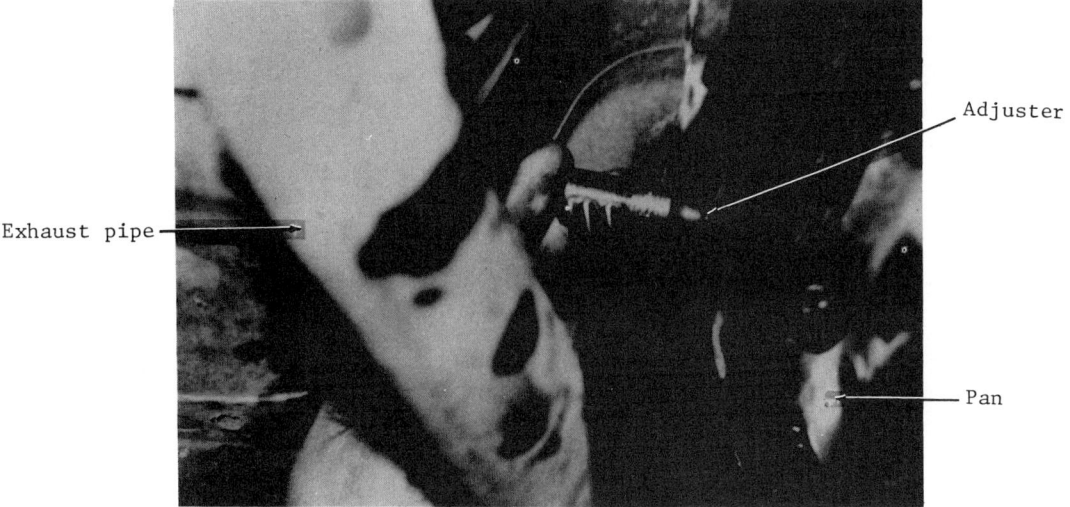

FIG. 5-14 Torquing the C-4 low reverse band adjusting screw.

LINKAGE ADJUSTMENTS

As mentioned earlier in this section, correct linkage adjustments play a very important role in transmission operation. The reason for this is fairly simple when you consider what the linkage really does. The linkage itself provides control input into the automatic transmission from various sources, the gearshift selector lever, and the accelerator pedal. In other words, with the gearshift selector lever and its linkage, the operator manually selects a given driving range within the transmission. Furthermore, the accelerator pedal, with its attached linkage(s) to the transmission, mechanically signals the transmission as to the load on the engine at any given time. With these facts in mind, it should be easy to figure out what could happen if one or more of these linkages are out of adjustment.

The four types of linkages that this section will discuss are the gearshift (manual valve) linkage, accelerator pedal, throttle valve, and downshift linkages. Not every vehicle has all four types of linkages; their usage, therefore, depends on the type of transmission and the motor vehicle containing the transmission. Consequently, when in doubt as to the types of linkages used in a given vehicle, check the appropriate service manual for the type, location, and adjustment.

Gearshift (Manual Valve) Linkage

The gearshift linkage moves the manual valve within its bore in the valve body. The manual valve, in turn, sets up the actual operating condition within the transmission by directing fluid to apply the clutches and/or bands. If this linkage is out of adjustment, the transmission can malfunction because the manual valve will not be in its correct position, thus permitting clutch-apply or band-apply fluid to leak past the valve and back to the sump.

Certain conditions can cause gearshift linkage to become out of adjustment. For instance, if the mechanic removes the transmission or the gearshift linkage to perform maintenance, he must check and readjust the linkage as required. In addition, linkage adjustment may be necessary because of wear caused by constant use, or because of loose linkage fasteners.

In practice, two components control the actual position of the manual valve within its bore in the valve body: a detent pawl and a shift gate. The detent pawl is part of the manual valve lever arrangement mounted on the valve body (Fig. 5-15). An arm on this lever fits into a groove on one end of the manual valve itself. With this design, whenever the lever moves due to the action of the gearshift selector lever the manual valve moves in its bore.

Another section of this very same lever forms the detent pawl that has notches for each gearshift selector position. A spring-loaded detent ball that fits into a special bore in the valve body engages into these notches, one at a time, as the manual valve moves from position to position. With this arrangement then, the detent ball and pawl hold the manual valve into one selected position at a time.

The shift gate is part of the gearshift selector lever mechanism, built either into the column or console. Figure 5-16 shows a typical column-shift

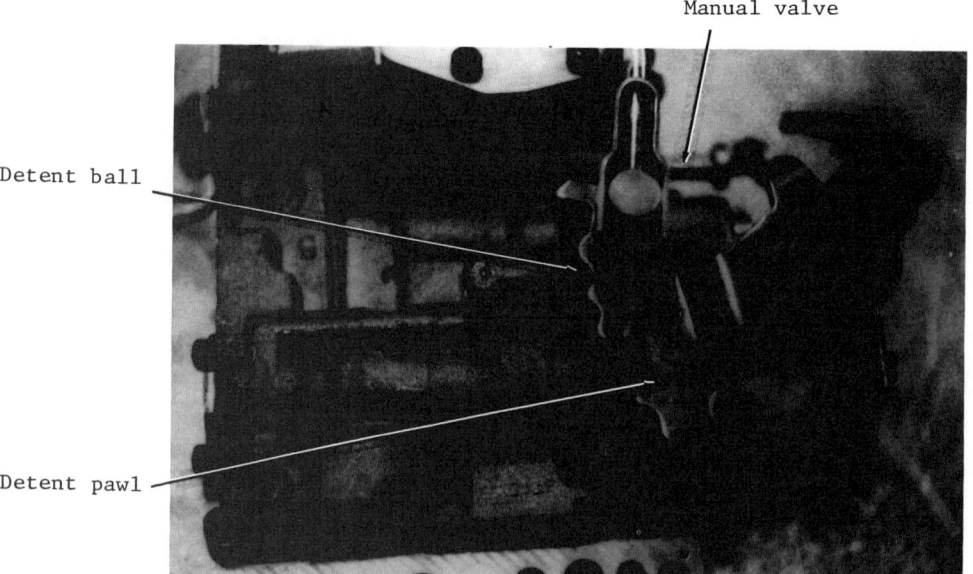

FIG. 5-15 Typical detent pawl mounted on a valve body.

Band and Linkage Adjustments

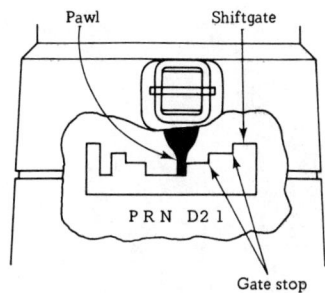

FIG. 5-16 Typical shift gate and pawl.

gate with pawl. The shift gate itself has a series of notches or stops that correspond to the notches in the pawl section of the manual valve lever. The gate pawl moves with the gearshift selector lever, and it engages into the various gate stops. When the gearshift selector pawl and gate-stop position synchronize with the pawl and notch position on the manual valve lever, the shift linkage correctly positions the manual valve.

<u>adjustment checks</u>

The following procedure for checking linkage adjustment is typical of those commonly used in the automotive industry. Because the actual procedure varies somewhat from one vehicle to another, always refer to the appropriate service manual before checking or altering linkage adjustment.

1. Position the gearshift selector lever into N (Neutral).
2. Lift upward on the gearshift selector lever and move it toward the D (Drive) position until you feel the manual-valve pawl detent drop into place in the Drive position (Fig. 5-16).
3. Lower the gearshift selector lever until the pawl stops against the shift gate.
4. Without raising the gearshift selector lever, attempt to shift the lever into the 2 position. You should feel little or no movement in the lever. If excessive movement exists, the linkage requires adjustment.

<u>adjustment procedure</u>

1. Place the gearshift selector lever into P (Park) position (Fig. 5-17).
2. Raise the vehicle on a hoist or with a jack to a suitable working height.
3. Locate and loosen the control-rod swivel-clamp screw a few turns (Fig. 5-18).
4. Move the transmission control lever all the way to the rear, the P (Park) position.
5. With the transmission control lever in P detent position and the gearshift selector lever in Park position, tighten the swivel clamp screw securely.

FIG. 5-17 Gearshift selector lever positioned in park (P).

6. Recheck all the gearshift selector lever detent positions, and then lower vehicle and road test.

Accelerator, Throttle, and Kickdown Valve Linkage

In order for the automatic transmission to respond to the varying loads placed on the engine, the linkages from the accelerator pedal to the carburetor and from the carburetor to the transmission have to be adjusted properly. This is especially true for vehicles which use such transmissions

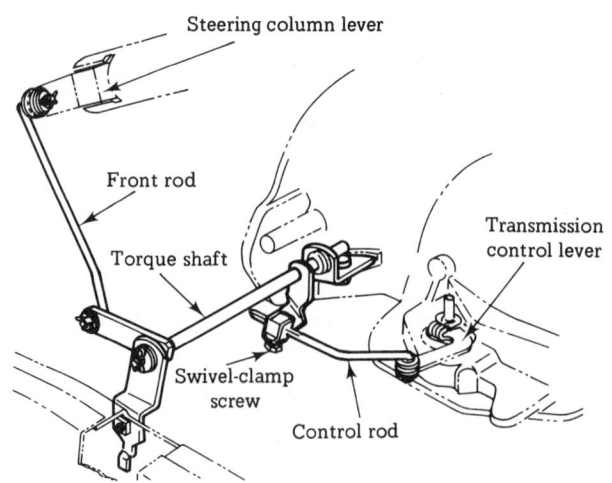

FIG. 5-18 Adjustment of the manual valve linkage.

Band and Linkage Adjustments

as the Powerglide, T-200, and Torqueflite because these units utilize mechanic linkage or a cable from the carburetor to operate both the throttle and kickdown valves. This is also true to a lesser degree for transmissions that use linkage to control only the kickdown valve.

Several transmission malfunctions are evident if the combination throttle and kickdown linkage or cable is out of adjustment. The transmission may upshift too soon or too late in relation to the engine load and vehicle speed. A forced downshift or kickdown may not occur when the driver depresses the accelerator pedal enough to move the linkage through the detent position, or the downshift may even occur before reaching the detent position. This premature or no kickdown condition can also occur in a transmission that has only a kickdown linkage if this linkage is out of adjustment.

Although the actual procedures for checking and adjusting accelerator, throttle, and kickdown linkage may differ between vehicle manufacturers, in practice, there are three important rules to always follow: First, refer to the linkage adjustment section of the service manual for the make and model of the vehicle you are repairing before checking or altering any adjustments. Two, make sure that all linkages are intact and serviceable. Three, make certain that when the accelerator pedal is all the way to the floor, the carburetor throttle plate is wide open with the throttle lever against the wide open throttle stop. And equally important is that as you release the accelerator pedal, the linkage returns freely and immediately to the idle position.

typical adjustment procedure

1. Disconnect the automatic choke at the carburetor or block the choke valve in the full-open position with a screwdriver (Fig. 5-19). Open the throttle slightly to release the fast idle cam. Return the carburetor linkage to the curb idle position.

2. Remove the spring, cotter or retaining pin, washer, and slotted throttle-rod adjuster from the bell crank lever pin.

3. By means of the transmission throttle rod, hold the transmission lever forward against its stop (rod or lever must not move vertically while you move the lever against its stop). Adjust the length of the transmission rod, with the threaded adjuster at its upper end. After adjustment, the rear end of the slot should just contact the bellcrank lever pin without exerting any additional forward force on the throttle rod.

4. Lengthen the throttle rod by turning the adjuster one full turn.

5. Assemble the slotted adjuster to bell crank lever pin and reinstall washer and retainer pin. Position the transmission linkage return spring in place.

6. Check the transmission throttle rod linkage for freedom of operation by moving the slotted adjuster to its full rearward position. Allow it to return slowly, making sure the adjuster returns to its full forward position.

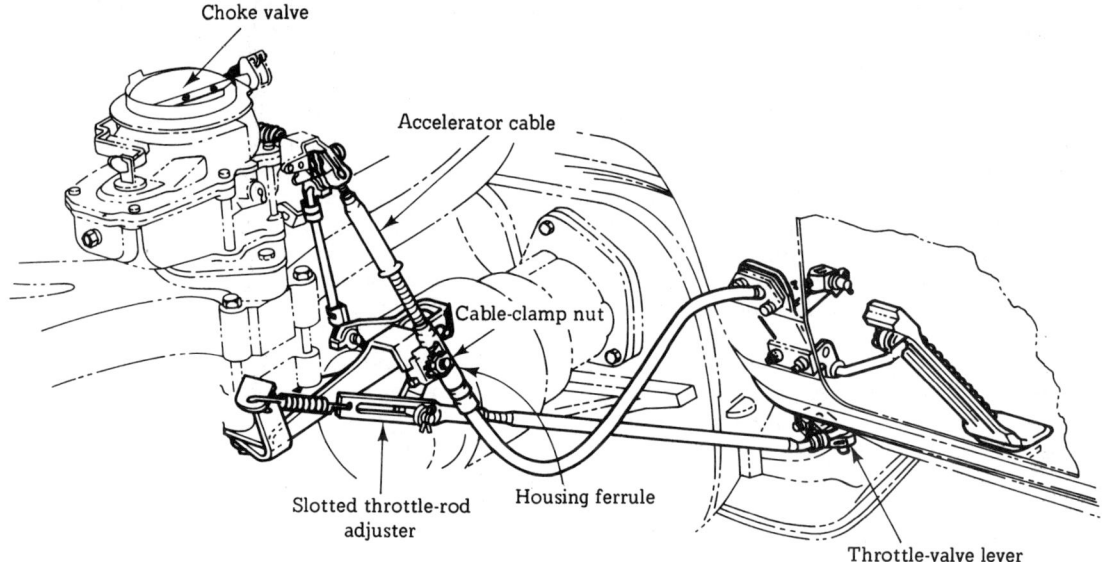

FIG. 5-19 Typical accelerator and throttle valve linkage adjustments.

7. To adjust the accelerator cable, loosen the cable clamp nut. Adjust the position of the cable housing ferrule in the clamp so that you remove all the slack from the cable with the carburetor at curb idle. To remove slack from the cable, move the ferrule, in the clamp, in a direction <u>away</u> from the carburetor lever.

8. Back off the ferrule 1/4 inch. This action provides 1/4-inch cable slack at curb idle. Finally, tighten the cable clamp nut securely.

9. Route the cable so that it does not interfere with the transmission throttle rod throughout its full range of travel.

10. Reconnect the automatic choke rod, or remove the screwdriver that blocks open the choke valve or plate.

11. Road test vehicle and check shift points and kickdown valve operation.

typical throttle valve cable assemblies

If a transmission has a throttle valve cable, it serves the same function as the throttle rod just described. In other words, the cable connects the carburetor to the valve body, and the cable transfers any carburetor throttle plate movement directly to throttle lever and bracket assembly, attached to the valve body. This assembly then operates the throttle valve within its valve body bore.

Band and Linkage Adjustments

cable and component inspection

A broken, sticking, or misadjusted cable will also interfere with normal transmission performance. To see if the cable and its attached components are serviceable, follow this procedure:

1. Check cable for attachment at the carburetor lever and the transmission rod (Fig. 5-20).
2. Remove any sharp bends as necessary by rerouting the cable.
3. Inspect the transmission rod between the end of the cable and the throttle valve lever for alignment. Straighten or replace the rod as necessary.
4. If necessary, remove the fluid pan and inspect the throttle valve lever and bracket assembly. If the lever, bracket, and retaining pin are worn or damaged, replace them as necessary.

cable adjustment

To adjust a throttle valve cable proceed as follows:

1. Unlock or disengage the snaplock on the cable (Fig. 5-20).
2. Turn the cable lever on the carburetor to its wide open position.
3. While holding this lever in its wide open position, push the snap lock downward until its top is flush with the rest of the cable. Engage the snap lock.
4. Slowly return the carburetor lever to return it to its closed position.

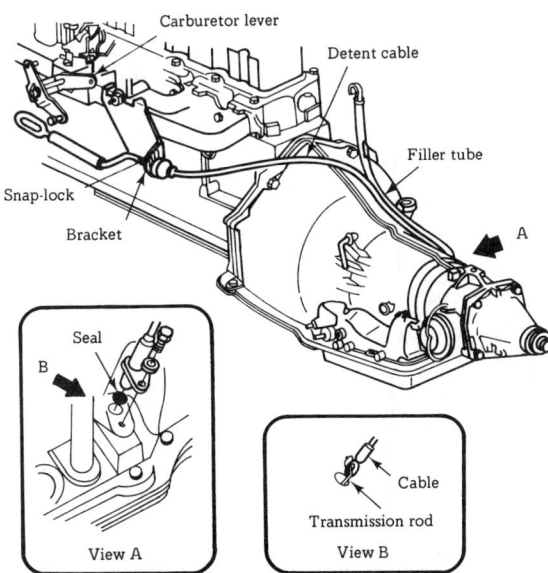

FIG. 5-20 Adjusting a common-type throttle valve cable.

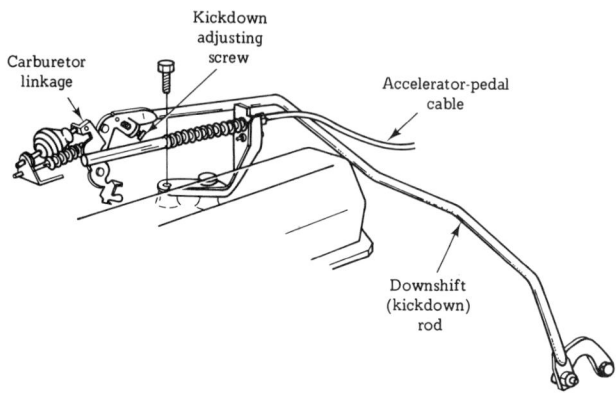

FIG. 5-21 Typical kickdown rod adjustment.

5. Road test vehicle to check upshift points and through-detent downshifts.

kickdown linkage

As previously mentioned, some transmissions require only a kickdown or downshift rod. This rod that operates by accelerator linkage causes a forced transmission downshift at wide-open throttle or as the driver depresses the accelerator pedal to the floor. To accomplish the forced downshift, the kickdown rod (Fig. 5-21) activates the downshift valve in the valve body; this valve, in turn, actually downshifts the transmission.

kickdown linkage adjustment

To check and adjust a typical kickdown linkage, follow these instructions:

1. Check and if necessary adjust the accelerator pedal and carburetor linkages.
2. Rotate and block the carburetor throttle valve open and against its wide open stop.
3. Depress the kickdown rod until the kickdown valve bottoms against its stop within the valve body.
4. While holding the kickdown rod in this position, check the clearance between the throttle lever and the end of the kickdown adjusting screw, located on the carburetor lever linkage. Adjust the screw as required to obtain the specified clearances (Fig. 5-21).
5. Release the kickdown rod and allow it to return to its normal position. Unblock and allow the throttle valve to return to idle position.
6. Road test the vehicle. A detent downshift should now occur at the appropriate road speed.

Band and Linkage Adjustments

electrical kickdown switches

Other transmissions, like the T-300 and 400 units, have an electrical kickdown switch (Fig. 5-22). Wide open travel of the accelerator pedal operates the switch, and the switch, in turn, directs an electrical signal to a downshift solenoid inside the transmission. This solenoid then causes a hydraulic detent and kickdown valve to operate with the kickdown valve itself actually responsible for downshifting the transmission.

typical switch adjustment

When kickdown shift points are not within specifications, the kickdown switch position should be adjusted as follows:

1. Check and adjust accelerator pedal and carburetor linkages as necessary.
2. Loosen mounting bolts (Fig. 5-22).
3. With automatic choke and accelerator linkage blocked in the wide open position, depress the detent switch plunger until it bottoms in the switch.
4. While holding the plunger in, move the switch toward the throttle lever paddle until you obtain a clearance of .23 inch between paddle and plunger.
5. Tighten the mount bolts, and unblock the choke and accelerator linkage.
6. Road test the vehicle to check the detent downshift. If the downshift still does not occur properly, the solenoid, detent, or downshift valves may be at fault and should be checked one at a time.

CHECK-UP QUESTIONS

The questions listed below will assist you in determining how well you remember the material contained in this section. Read each question carefully before adding the word or words necessary to complete the sentence If you can't complete the sentence, review that portion of the section that covers the question.

1. The task of a band is to hold a _____ _____ stationary.

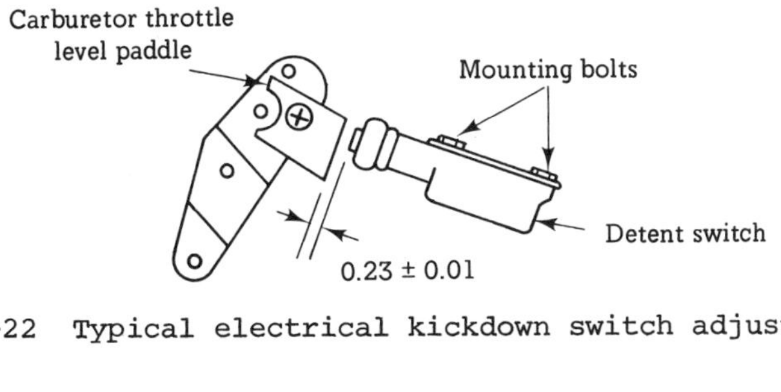

FIG. 5-22 Typical electrical kickdown switch adjustment.

2. A band will burn out if its adjustment is too _____.
3. The main reason why a band is adjusted at a given mileage interval is to compensate for _____ wear.
4. If the head of the band adjuster is on the outside of the transmission case, the mechanic can perform an _____ adjustment.
5. In transmissions with nonadjustable bands, the manufacturer sometimes supplies _____ servo apply pins of various lengths to adjust band clearance during an overhaul.
6. When the mechanic torques a band adjuster, he _____ the band around the drum.
7. To tighten most band adjusters, the mechanic must use either a _____ or _____ socket.
8. Before adjusting any band, look up the _____ in the service manual.
9. In order to make an internal band adjustment, you must first drain the _____ and _____ the fluid _____.
10. The linkages provide _____ _____ into the transmission.
11. The gearshift linkage moves the _____ _____ within the valve body.
12. The _____ _____ and the _____ _____ control the position of the manual valve in the valve body.
13. On some transmissions, linkage from the carburetor to the transmission controls the operation of both the _____ and the _____ valves.
14. Before making any linkage adjustments, look up the procedures in the _____ _____ for the make and model of vehicle you are repairing.
15. Instead of a throttle rod, some transmissions use a _____.

SECTION

6

Transmission and Seal Removal and Installation

REFERENCES: <u>Automatic Transmission Service</u>, Chapters 1, 2, and 4.
Vehicle or transmission service manual.

It may be necessary for a mechanic to remove the automatic transmission for different reasons. Transmission removal is necessary to replace the front pump seal and bushing if a leak develops around the converter hub. The mechanic may even have to pull the transmission to perform repair work on the engine. The most common reason to remove the transmission is, of course, to rebuild the unit.

There is no one single, detailed procedure that a mechanic can use to remove and reinstall all automatic transmissions. Each procedure is different because of the many model years, body styles, and engine and transmission configurations now in current use. All these things determine the location of exhaust system components, cross members, underbody parts, and linkages, which effect the manner or procedure by which the mechanic will remove and install the transmission. Consequently, the mechanic will find it necessary to follow the specific instructions found in the vehicle service manual. However, the following general instructions and procedures apply to most all makes and models of domestic vehicles.

TRANSMISSION REMOVAL

General Instructions

1. Open the hood and visually inspect the radiator hoses, fan, and other parts located at the front of the engine that might sustain damage during the removal process.
2. Visually inspect all components at the rear of the engine for working clearance between them and the firewall. This clearance is necessary to permit the engine to tilt downward at the back during transmission removal. If the vehicle has a V-8 engine with a rear-mounted distributor, remove the cap and rotor before removing the transmission.
3. In most cases, it is mandatory that the mechanic remove the transmission and torque converter as an assembly; otherwise, damage will occur to the converter drive plate, pump bushing, and oil seal. Furthermore, the drive plate itself will not support a load; therefore, the mechanic cannot allow any of the weight of the transmission to rest on this plate during removal.
4. There are three approved methods of turning the engine over to either drain the converter or remove its attaching bolts or nuts. The first method is to rotate the engine by using a wrench on the crankshaft front pulley bolt. The second and equally safe method is to turn the starter ring gear on the converter or flex plate with a flywheel wrench. The third method which is not as safe as the other two but which is necessary on some vehicle configurations, employs the use of a remote starter button and the starter to crank the engine over. Many manufacturers do not recommend this method, however, because of damage to the torque converter or flex plate that may occur due to starter motor torque after the mechanic loosens the converter mount bolts. Never use a screw driver on the ring gear teeth, the flex plate, or the converter itself; and never use a wrench on the converter drive bolts to turn the engine over. This action can damage either the ring gear teeth, the flex plate, or the converter mount drive bolts.

Procedure

1. If vehicle configuration is such that you must use the starter to rotate the engine, connect a remote control starter switch to the starter solenoid; then position the switch so that you can operate the starter from under the vehicle.
2. Disconnect the high tension wire from the distributor cap; this prevents the engine from accidentally starting as you turn the engine over with the remote switch.
3. With the hood still open, raise the vehicle on a hoist to a suitable working height. If you must use jacks instead of a hoist, locate the proper support points underneath the vehicle, and then lower it onto

Transmission and Seal Removal and Installation

safety stands. Note: Leaving the hood open, at this time, prevents damage to the hood by the air cleaner or other engine parts which can strike the hood as you lower or raise the engine during the transmission removal and installation process.

4. Remove the cover plate from the front of the converter to provide access to the converter drain plug and attaching bolts (Fig. 6-1).

5. Rotate the engine with the remote starter switch, large wrench, or fly wheel tool in order to bring the drain plug to the 6 o'clock position (Fig. 6-2). Place a large drain pan under the converter, remove the drain plug, and let the fluid drain from the converter into the pan.

6. Center punch or suitably mark the converter and drive plate to aid in reassembly. This is helpful because the crankshaft-flange bolt circle, the inner and outer circle of holes in the flex drive plate, and the front face of the converter, in many cases, will have one hole offset to assure the installation of these parts in a given position. This design maintains the balance of the engine and converter. These marks, then, quickly assist you in aligning the converter to the flex plate, and at the same time maintain engine and converter balance.

7. Replace the drain plug; then rotate the engine again in order to locate the two converter-to-drive plate attaching bolts at the 5 and 7 o'clock positions. Remove these two bolts, and turn the engine over once more to locate and remove the remaining two attaching bolts.

8. Place a drain pan beneath the fluid pan; remove the pan's attaching screws and pan (Fig. 6-3) in order to drain the fluid from the transmission. Note: Refer to Section 4 if necessary for further details of this procedure.

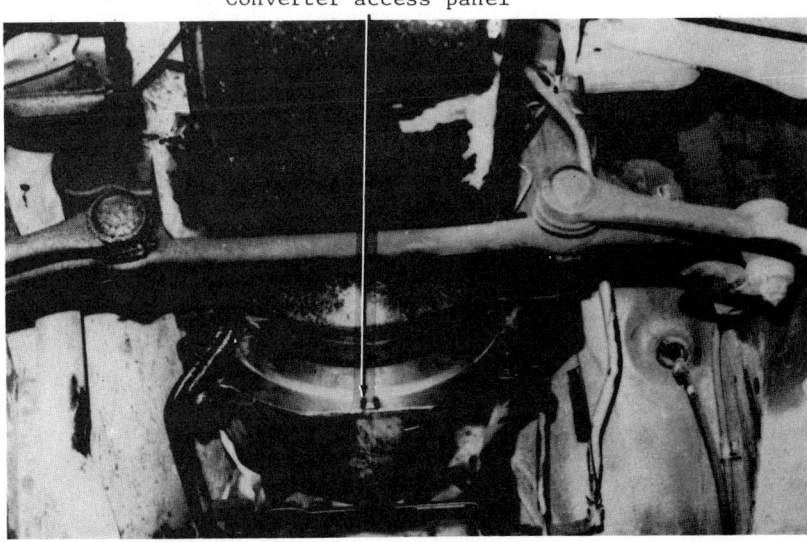

FIG. 6-1 Removing the torque converter access plate.

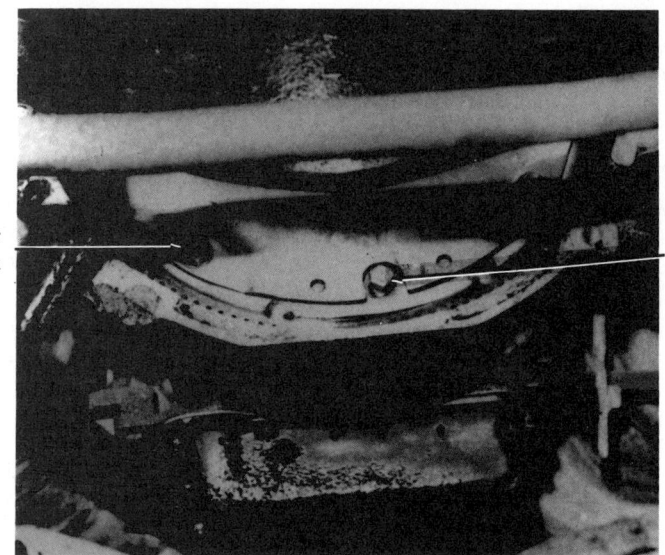

FIG. 6-2   Removing the torque converter drain plug and attaching hardware.

9. Reinstall the pan with several attaching bolts. The installed pan will provide a suitable lifting surface for the platform of the transmission jack.

10. Lower the vehicle if necessary and disconnect the negative (ground) cable from the battery (Fig. 6-4). This action prevents a short circuit that could occur if any electrical wires become grounded while removing the starter or other electrical components.

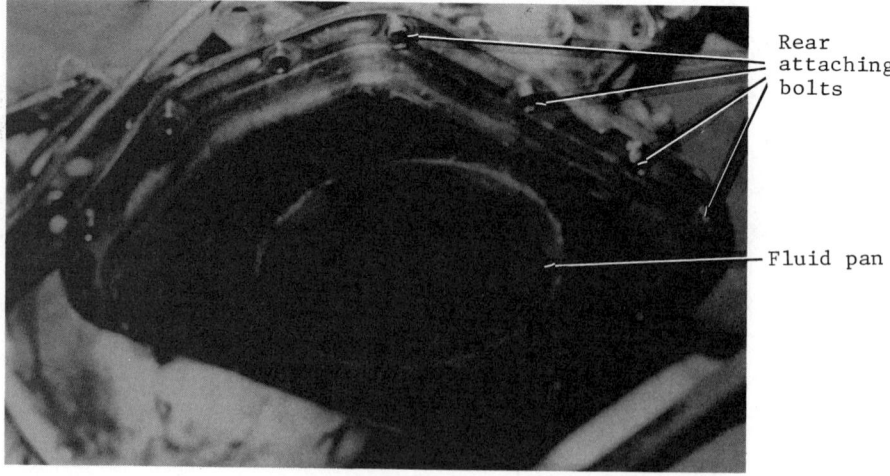

FIG. 6-3   Removing the fluid pan's attaching screws.

Transmission and Seal Removal and Installation

FIG. 6-4 Disconnecting the battery's negative ground cable.

11. Disconnect all the electrical leads from the starter solenoid, and unbolt and remove the starting motor (Fig. 6-5).
12. Disconnect the wires from the neutral safety switch (Fig. 6-6).
13. Disconnect the gearshift rod from the transmission lever. Remove the gearshift torque shaft from the transmission housing and the left-side rail (Fig. 6-7).
14. Disconnect the throttle rod from the throttle lever on the transmission (Fig. 6-8).

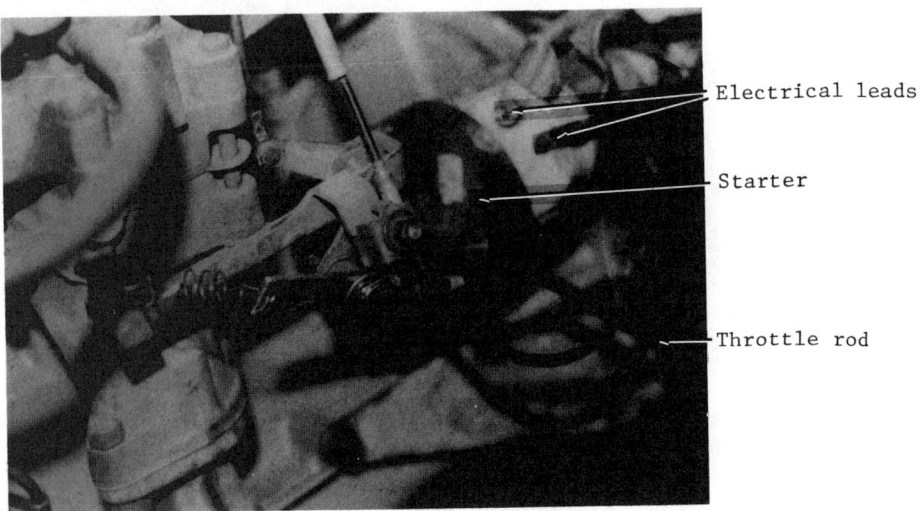

FIG. 6-5 Removing a starter assembly.

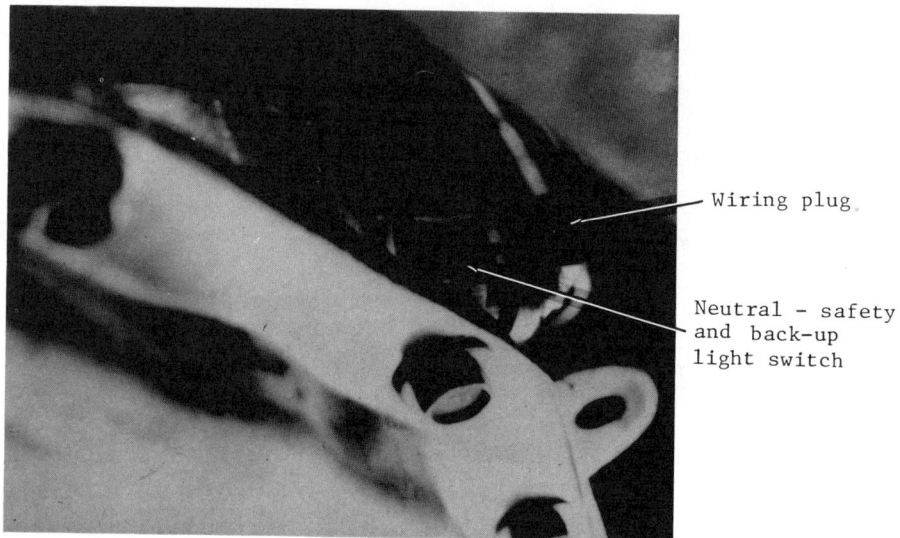

FIG. 6-6   Disconnecting the wire(s) from the transmission's neutral safety switch.

15. Remove any clamps or clips used to hold the transmission cooler lines in place on the engine or transmission.
16. Place a drain pan under the cooler line fittings, and with two wrenches, one placed on the line fitting and one on the transmission fitting,

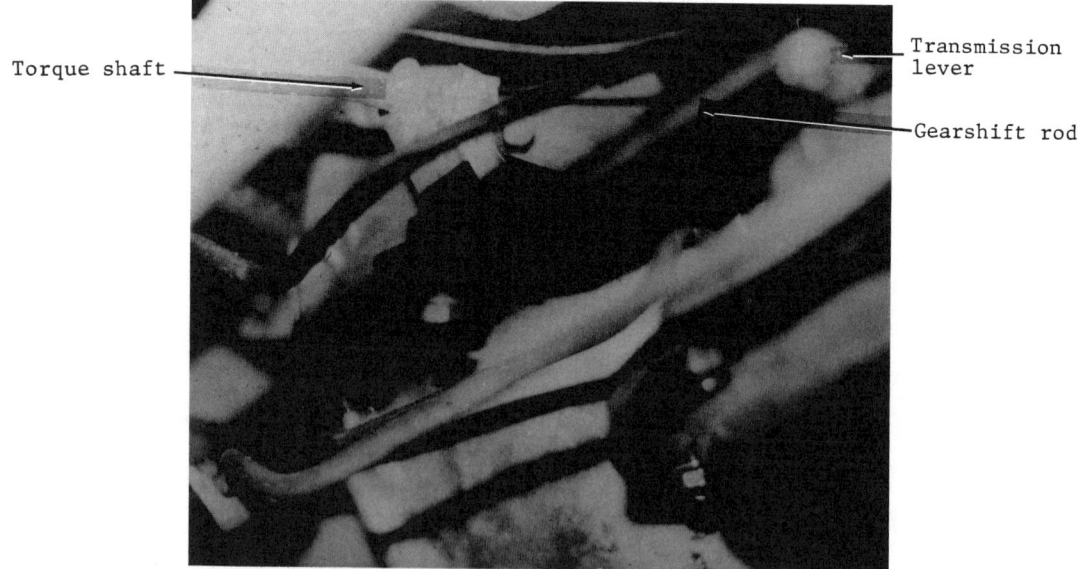

FIG. 6-7   Disconnecting the transmission's gearshift rod and torque shaft.

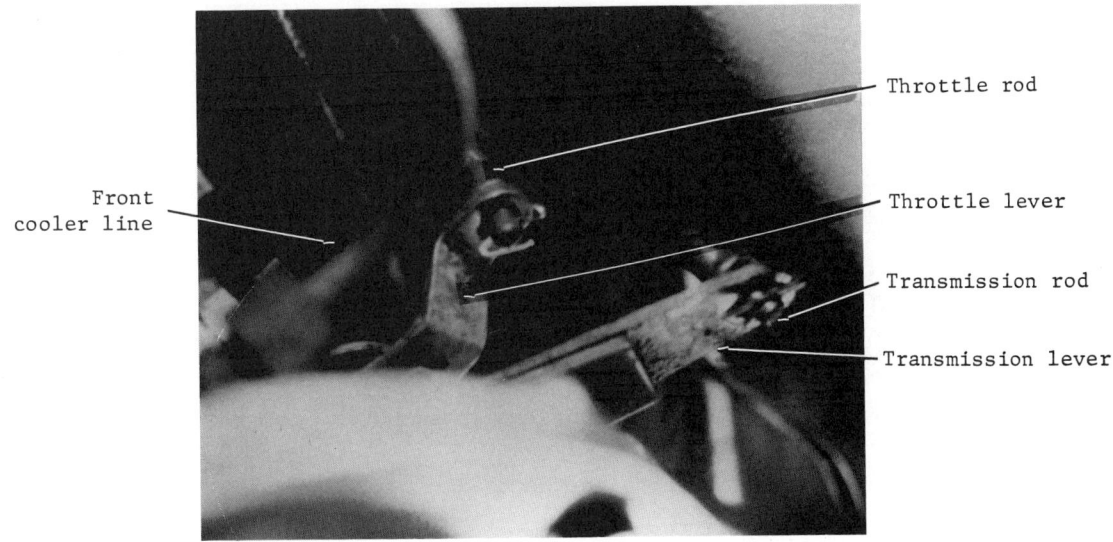

FIG. 6-8 Disconnecting the throttle rod from its lever on the transmission.

remove both cooler lines (Figs. 6-8 and 6-9). By holding the transmission fitting with one wrench while using the other to loosen the line fitting, you will prevent damage to the cooler line or the fittings. Finally, plug both the lines to keep dirt or moisture from entering.

17. Disconnect the speedometer cable from the transmission housing (Fig. 6-10) and move it out of the way.

18. Remove the dipstick attaching bolts, and remove the tube by pulling the

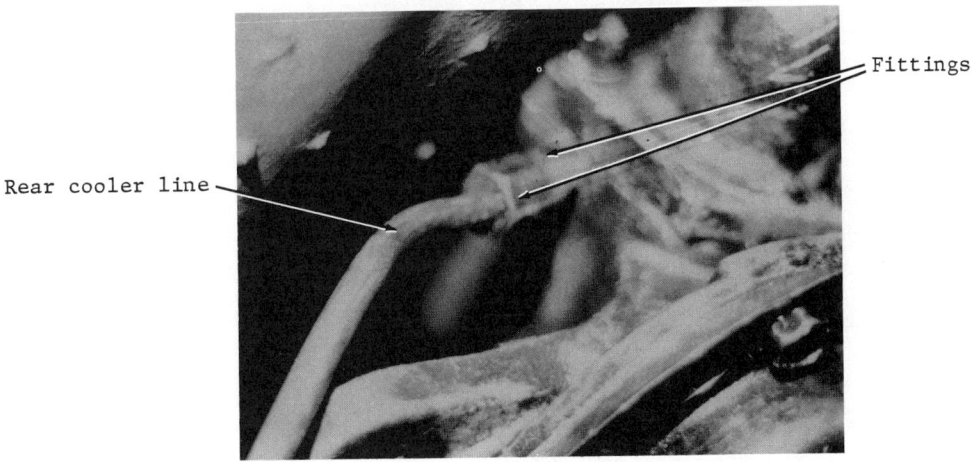

FIG. 6-9 Removing a cooler line.

FIG. 6-10 Disconnecting the speedometer cable.

tube straight out of its bore in the transmission case. <u>Be careful not to bend or twist the tube</u>.

19. Disconnect the drive shaft at the rear universal joint (Fig. 6-11). Mark the U-joint and companion flange to aid in reassembly. Remove the rear yoke or companion flange attaching hardware.

20. While holding the bearing caps and U-joint cross in position as you separate them from the rear yoke, secure both caps to the cross with wire or tape. Lower the rear end of the drive shaft downward a few inches.

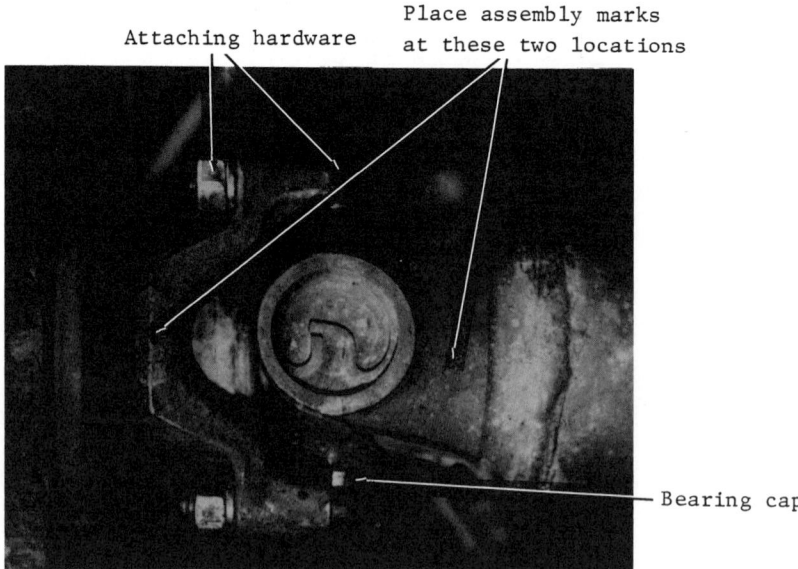

FIG. 6-11 Disconnecting the rear universal joint from the companion flange.

Transmission and Seal Removal and Installation   179

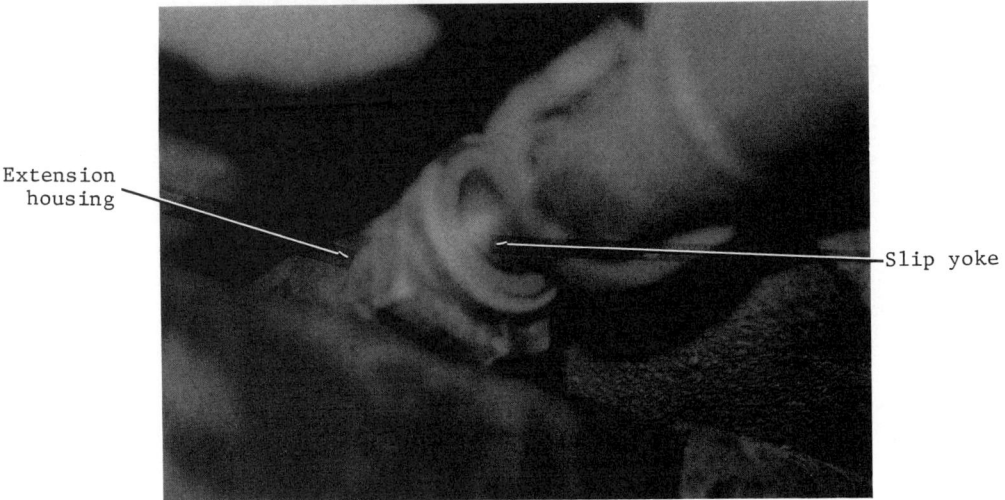

FIG. 6-12 Removing the drive shaft and slip yoke.

21. While supporting the drive shaft itself, carefully pull the drive shaft slip yoke from the extension housing (Fig. 6-12). Plug the extension housing with a seal driver, spare slip yoke, or a clean rag to prevent dirt from entering or fluid leaking from the housing.

22. Position a transmission jack squarely under the transmission, and adjust the height of its platform until it contacts the fluid pan (Fig. 6-13). If the jack has a tilting platform, adjust it to conform to pan angle and tighten the adjusting or locking screws. Move each corner bracket

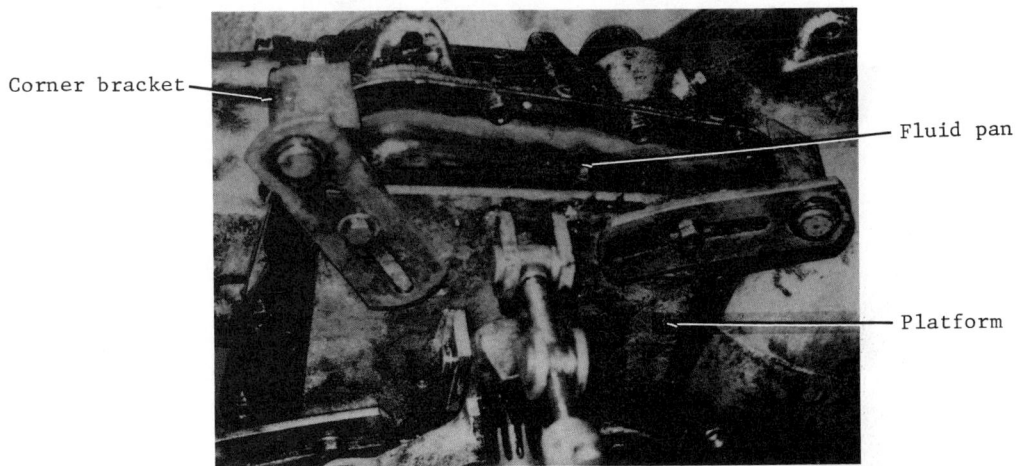

FIG. 6-13 Installation of the hydraulic tramsisssion jack under the fluid pan.

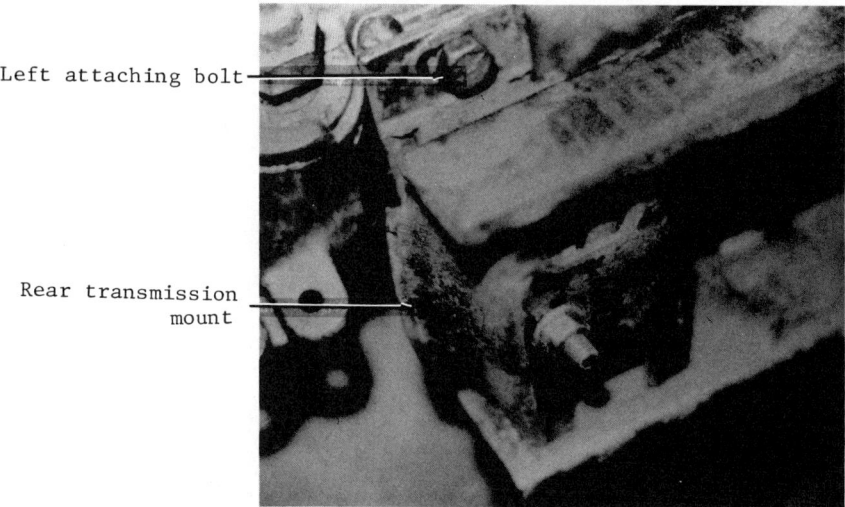

FIG. 6-14 Removing the attaching bolts from the rear mount.

of the platform into position around the corners of the pan and tighten their attaching bolts.

23. If the jack has a safety chain, install it around the transmission and secure the chain to the jack. Then slightly raise the jack to relieve the load on the rear transmission mount.

24. Remove the attaching bolts from the rear transmission mount (Fig. 6-14).

25. Remove the attaching bolts from the transmission crossmember and take out the crossmember (Fig. 6-15).

26. Remove the lower transmission bell housing-to-engine attaching bolts (Fig. 6-16).

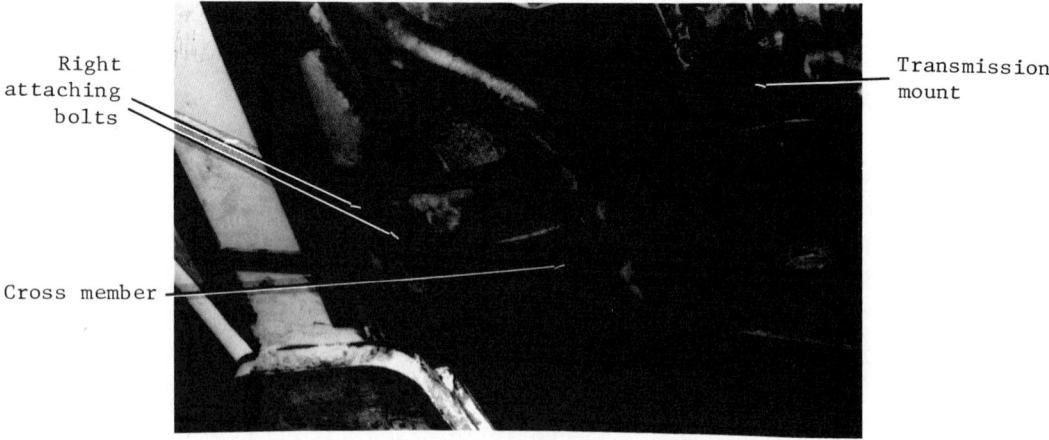

FIG. 6-15 Removing a typical crossmember.

Transmission and Seal Removal and Installation

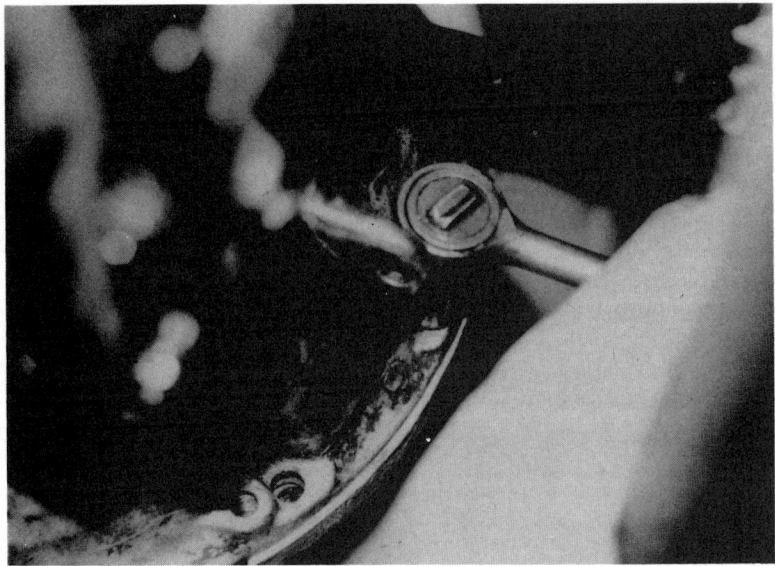

FIG. 6-16 Removing the transmission-to-engine attaching bolts.

27. If necessary, lower the transmission jack enough to provide access to the upper transmission bell housing-to-engine attaching bolts. Install an engine support bar at the rear of the engine and remove the upper attaching bolts.

28. Attach a small C-clamp to the edge of the bell housing to hold the torque converter in place during the removal of the transmission.

29. Carefully work the transmission rearward (off) the engine block dowels and disengage converter hub from the end of the crankshaft, making sure that no components are left to interfere with transmission removal.

30. Lower the transmission jack and remove the transmission and converter assembly away from the engine (Fig. 6-17).

31. Before removing the converter or the transmission from the jack, plug all transmission openings and steam clean the outside of the transmission case.

FRONT SEAL REPLACEMENT

1. Disconnect the safety chain from around the transmission, and place the transmission on a suitable work bench. Remove the C-clamp from the bell housing.

2. Carefully remove the torque converter from the front of the transmission (Fig. 6-18). Pull the converter straight out from the front pump; this

FIG. 6-17 Lowering the jack and transmission away from the engine.

will prevent scoring or damage to the front pump bushing and seal. Then place the converter on an unused portion of the workbench.

3. Using the seal puller shown in Fig. 6-19, remove the front pump seal from the pump housing.

4. With a flashlight, inspect the front pump bushing for scoring, wear, or damage. If the bushing is worn excessively, remove the pump and replace the bushing as outlined in Section 7. Note: <u>Excessive bushing wear can cause a leak at the front pump seal. If there is any doubt as to bushing condition, replace it, or a leak may develop.</u>

FIG. 6-18 Removing the torque converter.

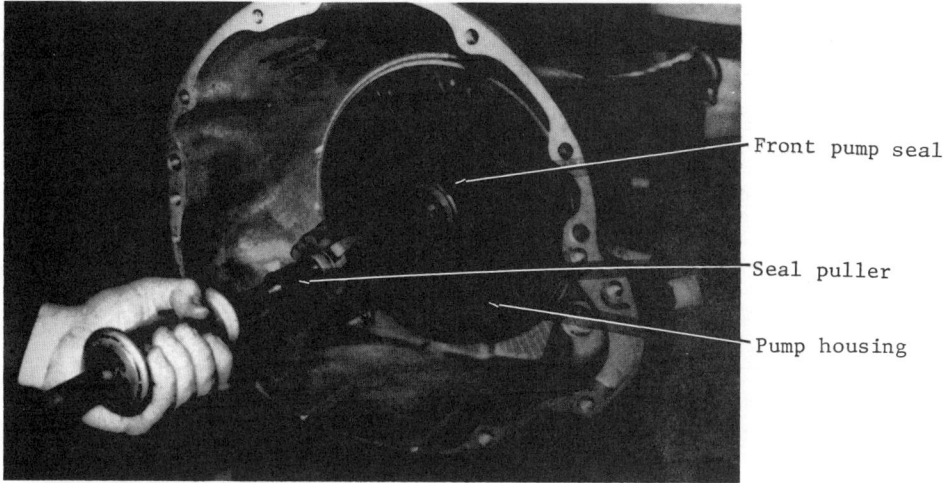

FIG. 6-19 Using a seal puller to remove the front-pump seal.

5. Before installing the new seal, coat the outer circumference of the seal backing with a nonhardening sealer. This procedure is unnecessary if the seal backing has a rubber or resin coating.

6. To actually install the new seal, place it in the opening of the pump housing with the lip side facing inward. With the tool shown in Fig. 6-20, drive the seal into its housing bore until it bottoms.

7. Lubricate the converter hub with transmission fluid, and carefully reinstall the converter into the front of the transmission (Fig. 6-21) by sliding the converter over the stator support and into the front pump. If necessary, rotate the converter until the converter hub

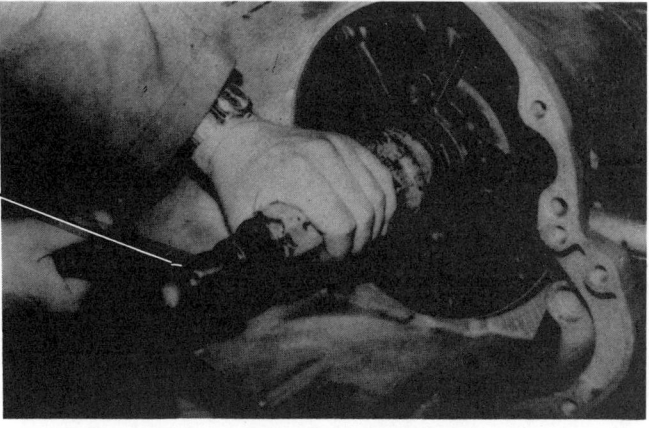

FIG. 6-20 Installation of the front-pump seal.

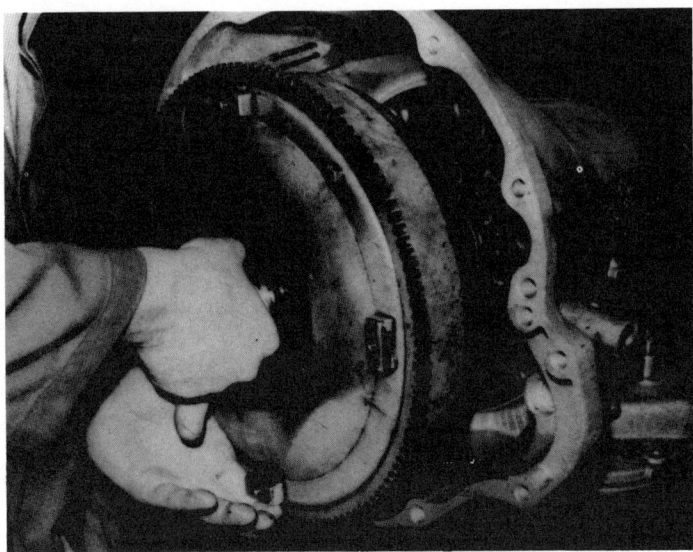

FIG. 6-21 Installation of the torque converter.

engages into the inner rotor or drive gear of the front pump. <u>Note: Severe damage can occur to the pump, converter, or flex plate if you attempt to install the transmission with the converter hub not engaged properly into the pump rotor or drive gear.</u>

8. Test for full engagement of the converter by placing a straight edge on the face of the bell housing. With the converter pushed in all the way, the distance from the edge of the straight edge to the drive lug should be to specifications, in this case 1/2 inch.

9. Reinstall the C-clamp to the edge of the bell housing to keep the converter from slipping out of place.

EXTENSION HOUSING SEAL REPLACEMENT

1. With a chisel or suitable puller, remove the extension housing seal (Fig. 6-22).

2. If the extension housing has a bushing in front of the seal, inspect it with a flashlight for scoring, wear, or damage. If the bushing is worn excessively, and the special bushing remover that can pull the bushing without the housing is not available, then remove the extension housing and replace the bushing as outlined in Sections 8 and 9. <u>Note: Excessive bushing wear can cause a leak at the extension housing seal. If there is any doubt as to bushing condition, replace it, or a leak may develop.</u>

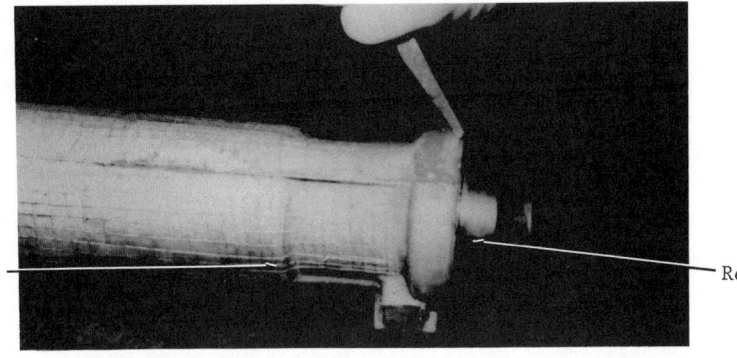

FIG. 6-22 Removal of the extension housing seal, using a chisel and hammer.

3. Before installing the new seal, coat the outer circumference of the seal with a nonhardening sealer. This procedure is unnecessary if the seal backing has a rubber or resin coating.

4. To actually install the new seal, place it in the opening of the extension housing with the lip side facing inward. With the tool shown in Fig. 6-23, drive the seal into its bore in the extension housing.

TRANSMISSION INSTALLATION

1. Before positioning the transmission back on the jack, replace the pan gasket and service the filter as outlined in Section 4 of this manual.

2. If the transmission has been rebuilt, flush out the cooler and its lines before installing the transmission. This procedure prevents friction and metal particles from getting back into the overhauled

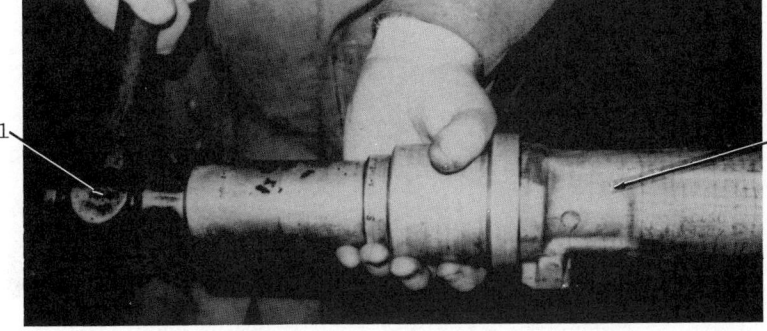

FIG. 6-23 Intallation of the extension housing seal.

transmission where it can cause the unit to fail again. Refer to Section 8 of this manual for details of this flushing operation.

3. If not accomplished previously, reinstall and torque the converter drain plug.

4. Position the transmission back onto the hydraulic jack; reconnect the safety chain around the transmission and secure it to the jack.

5. Inspect the converter drive flex plate for distortion or cracks and replace it if necessary. Coat the converter hub hole in the back of the crankshaft with wheel bearing grease.

6. Position the hydraulic jack and transmission under the vehicle for installation. Raise or tilt the platform as necessary until the transmission aligns with the engine (Fig. 6-24).

7. Rotate the converter so that the mark on the converter made during removal will align with the mark on the drive flex plate.

8. Carefully work the transmission assembly forward and over the engine block dowels; the converter hub should now enter the crankshaft opening.

9. Once the transmission is in position, install all the bell housing bolts (Fig. 6-25). Torque these attaching bolts to specifications.

10. Raise the transmission slightly and reinstall the crossmember and tighten its attaching bolts securely (Fig. 6-26).

11. Lower the transmission jack so that the extension housing aligns and rests on the rear mount. Install the rear mount attaching bolts and torque to specifications (Fig. 6-27).

12. Unfasten the safety chain from around the transmission and remove the jack from under the vehicle.

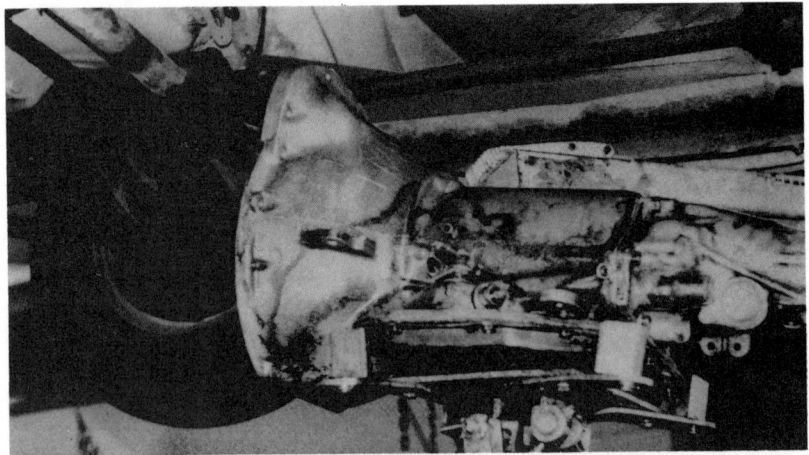

FIG. 6-24 Positioning the jack and transmission under the vehicle in preparation for the installation process.

Transmission and Seal Removal and Installation

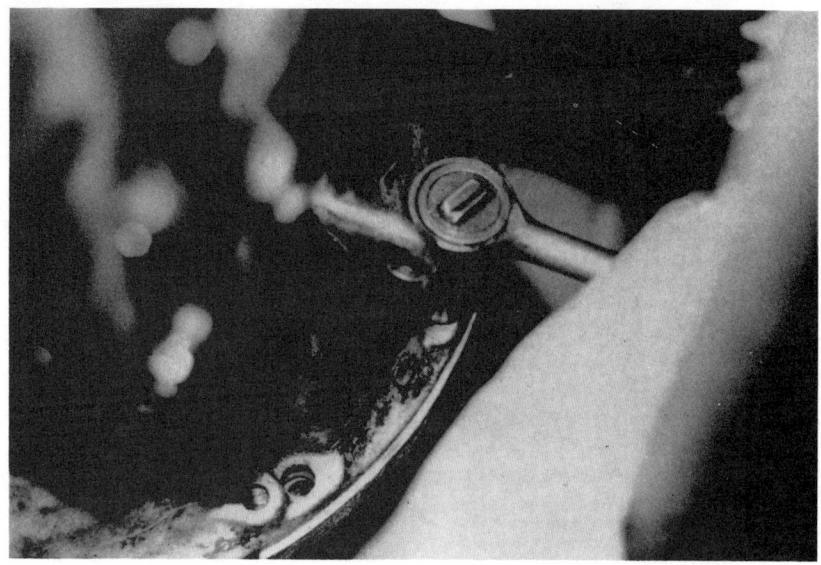

FIG. 6-25 Installation of bell housing bolts.

13. Install the two lower drive plate-to-converter attaching bolts and torque to specifications (Fig. 6-28).
14. Install the starter motor by reinstalling its attaching bolts and reconnecting all of its electrical leads (Fig. 6-29).

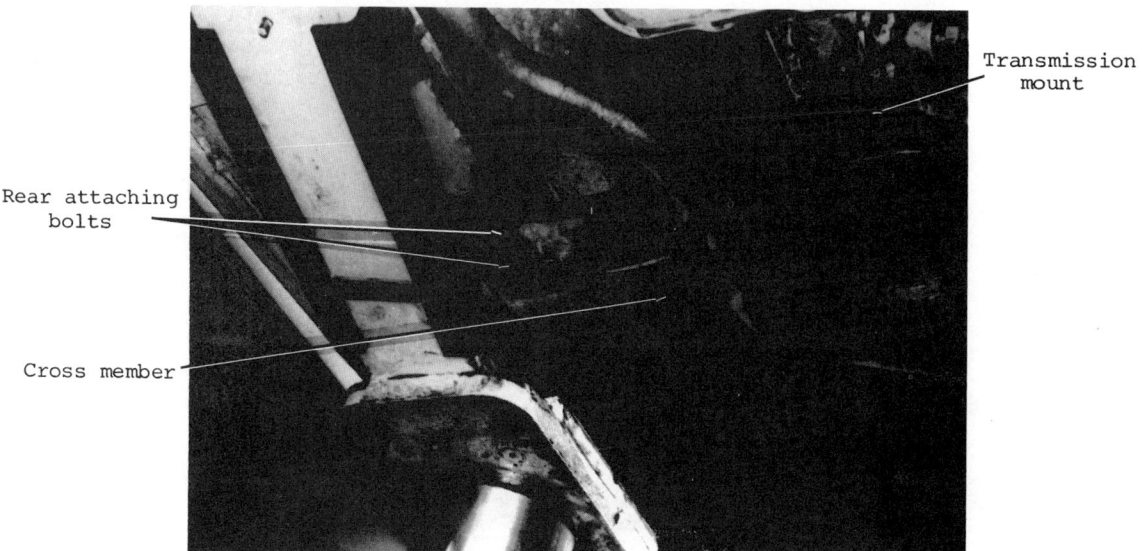

FIG. 6-26 Installing the cross member.

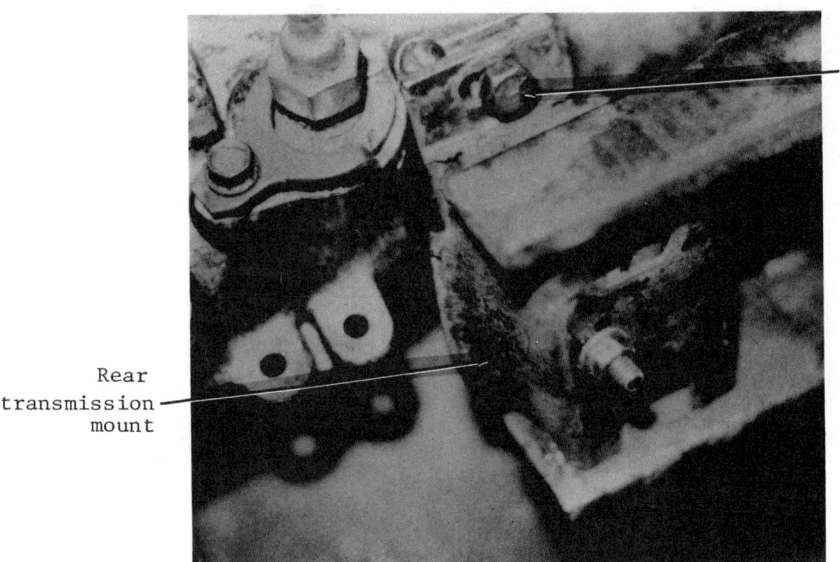

FIG. 6-27 Installing the rear-mount.

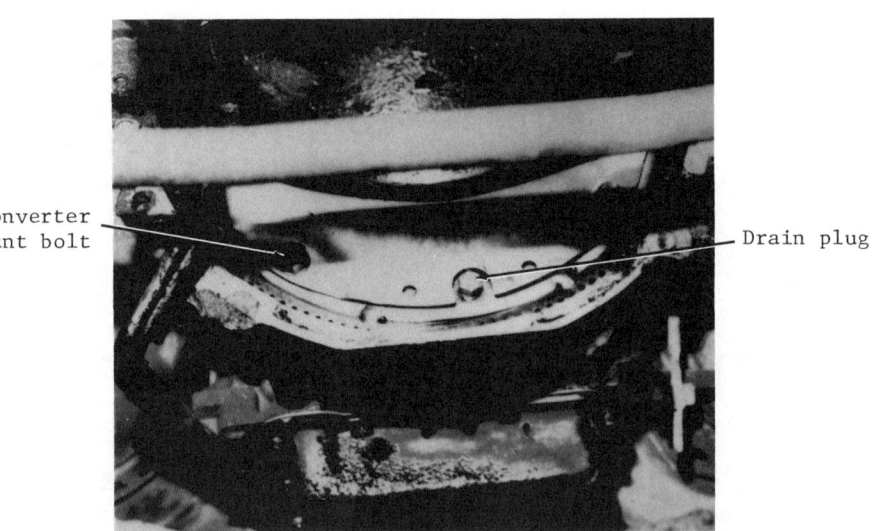

FIG. 6-28 Installing the torque-converter and flex-plate attaching bolts.

Transmission and Seal Removal and Installation

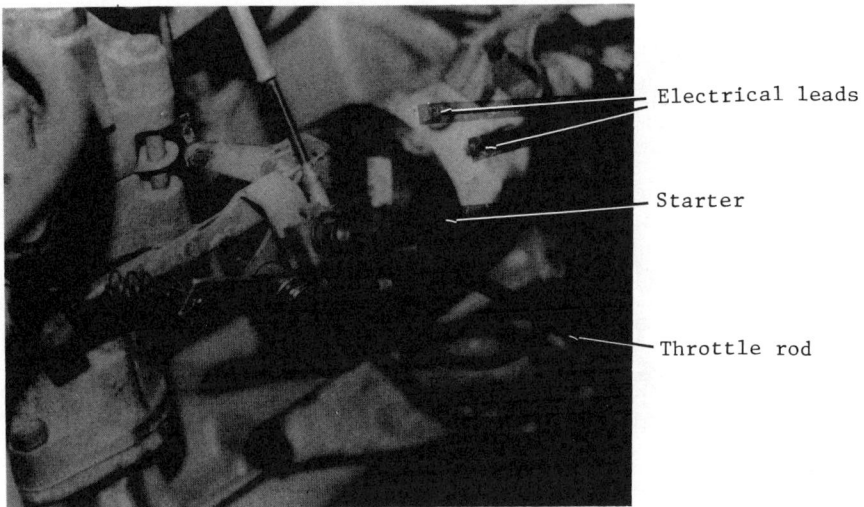

FIG. 6-29 Reinstalling the starter assembly.

15. Lower the vehicle if necessary and reconnect the negative (ground) cable to the battery (Fig. 6-30).
16. Rotate the engine with a remote starter switch, large wrench, or flywheel tool in order to bring the other two converter-to-drive plate attaching bolts to the 5 and 7 o'clock positions. Reinstall these bolts and torque to specifications.
17. Install the cover plate in front of the converter, torquing its attaching bolts to specifications (Fig. 6-31).

FIG. 6-30 Reconnecting the negative battery terminal.

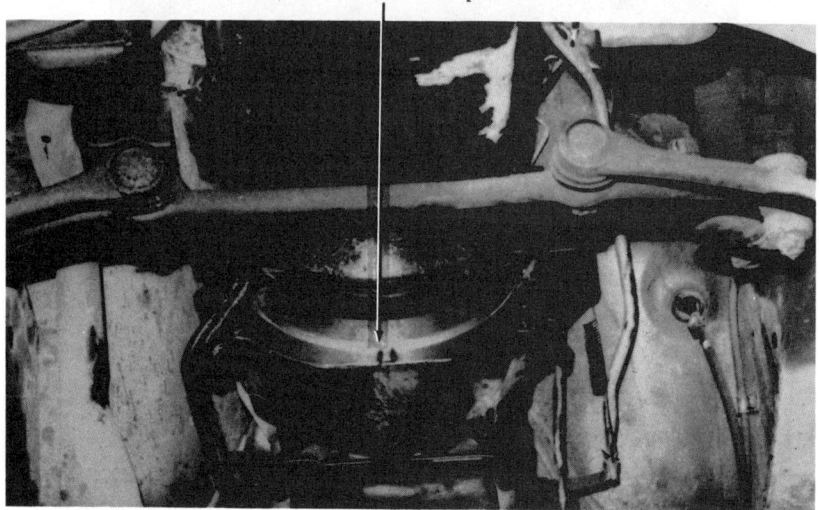

FIG. 6-31 Installing the converter cover.

18. Reinstall the dipstick filler tube carefully, replacing its seal as necessary. Seat the tube completely down into its bore and reinstall and torque the tube's attaching bolts.
19. Reconnect the wires to the neutral safety switch (Fig. 6-32).
20. Reinstall the throttle rod to the throttle lever on the transmission (Fig. 6-33).
21. Reinstall the gearshift rod to the transmission lever. Install the gearshift torque shaft to the transmission and to the left-side rail (Fig. 6-34).

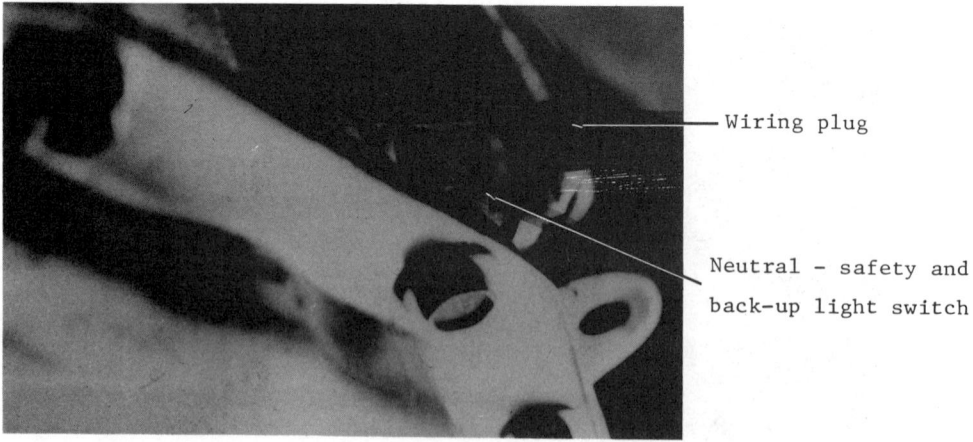

FIG. 6-32 Reinstalling the Neutral Safety Switch wire(s).

Transmission and Seal Removal and Installation

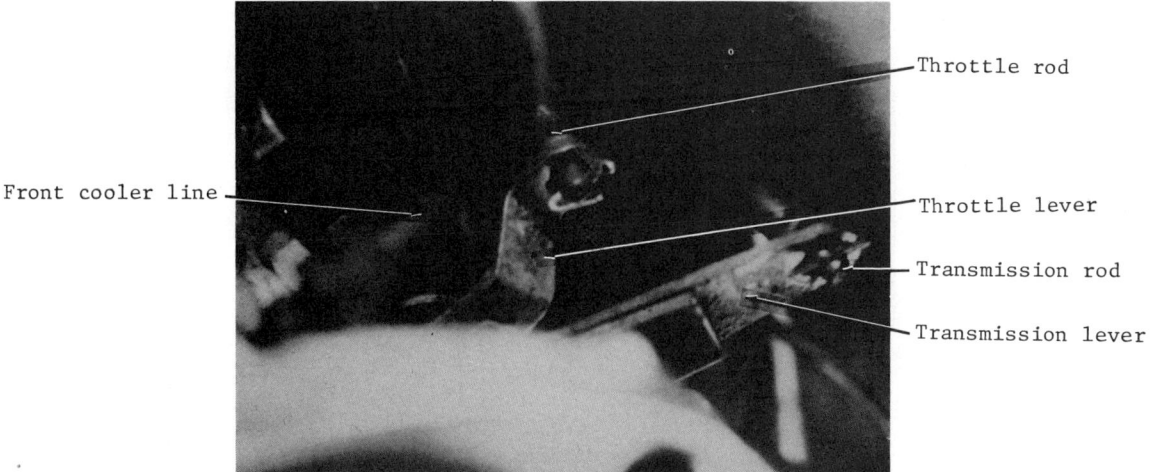

FIG. 6-33 Installing the throttle rod onto the throttle lever.

22. Reinstall any clamps or clips utilized to hold the transmission cooler lines in place on the engine or transmission.

23. Reconnect the cooler lines to the transmission case, using two wrenches (Fig. 6-35). Hold the transmission fitting with one wrench and tighten the tubing fitting with the other wrench.

24. Reconnect the speedometer cable to the transmission (Fig. 6-36).

25. Wipe the drive shaft slip yoke clean and lubricate with clean transmission fluid. Carefully guide this yoke into the extension housing and onto the output shaft splines (Fig. 6-37).

26. Align the U-joint yoke and companion flange marks made during disassembly (Fig. 6-38). If you do not reinstall the output shaft

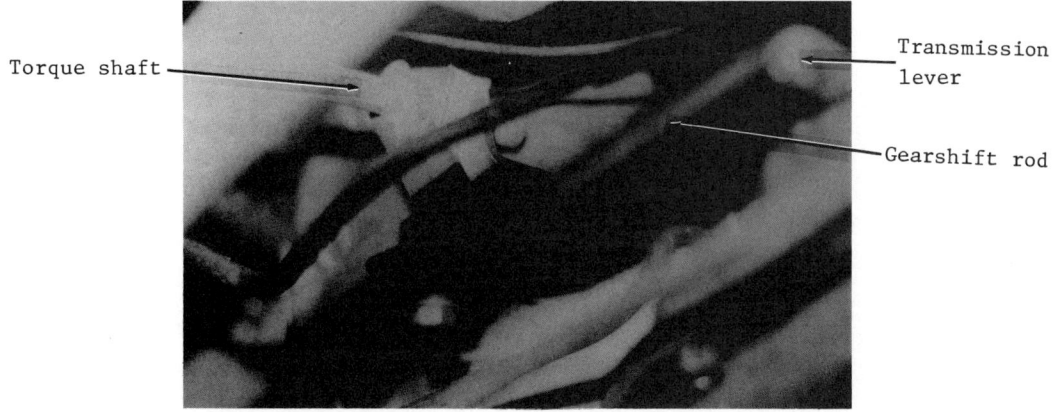

FIG. 6-34 Reinstalling the gearshift rod and torque shaft.

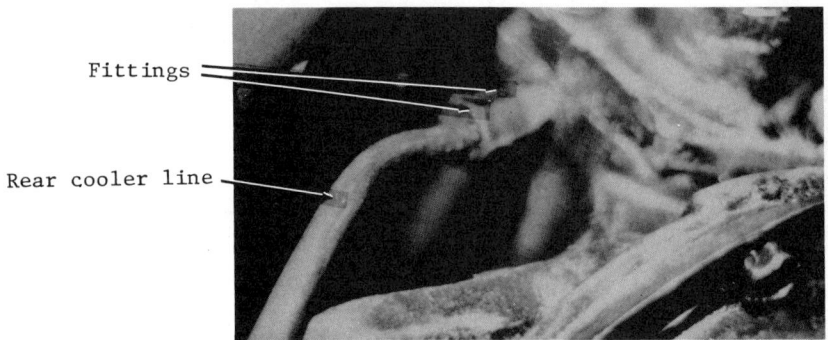

FIG. 6-35 Reinstalling a cooler line.

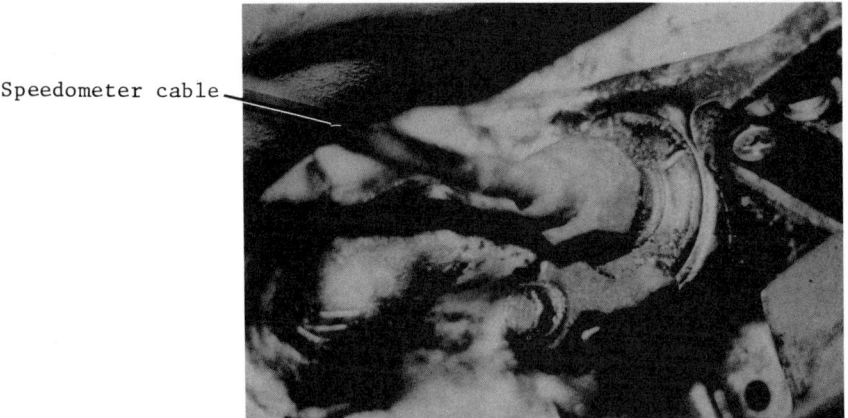

FIG. 6-36 Reconnecting the speedometer cable.

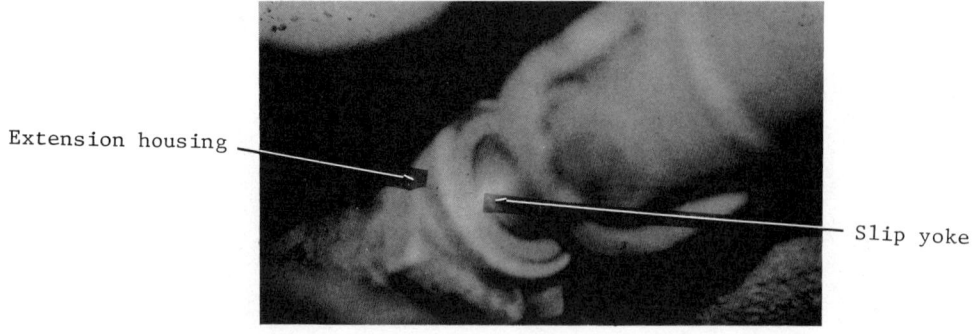

FIG. 6-37 Installing the slip yoke in the extension housing.

Transmission and Seal Removal and Installation

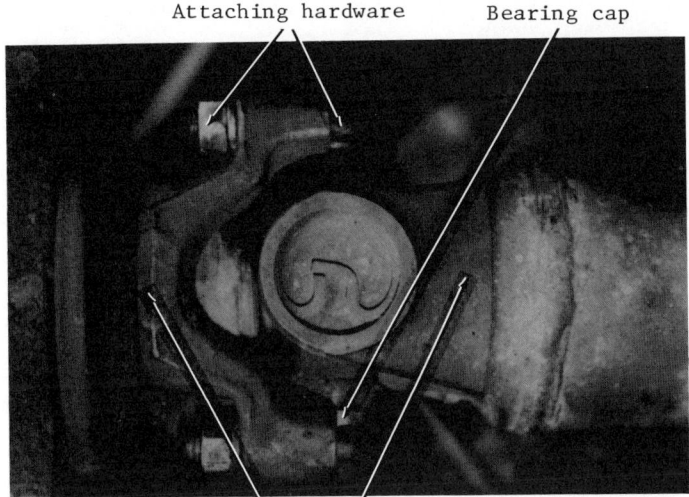

FIG. 6-38 Reconnecting the U-joint to the companion flange.

at this point exactly as you removed it, shaft runout will cause excessive vibration.

27. Remove the wire or tape holding the bearing cups to the cross, and reconnect the U-joint assembly to the companion flange (Fig. 6-39). Torque all the hardware to specifications.

28. Adjust all mechanic linkages to specifications. Refer to Section 5 for the general procedures.

29. Lower the vehicle and reconnect under the hood any parts removed or disconnected.

30. Fill the transmission to the correct level with the proper fluid. Refer to Section 4 for this procedure.

31. Visually inspect the transmission, converter, and cooling lines for leakage.

32. Road test the vehicle and check the operation of the transmission. Make any adjustments necessary to the linkage, and then recheck the fluid level.

CHECK-UP QUESTIONS

The questions listed below will assist you in determining how well you remember the material contained in this section. Read each question carefully before adding the word or words necessary to complete the sentence. If you can't complete the sentence, review that portion of the section that covers the question.

1. It is necessary to remove the transmission to replace the front-pump _____.

2. In order to remove most automatic transmissions, the engine has to tilt _____ in the back.

3. One of the safest methods of turning the engine over to remove the converter mount bolts is through the use of a _____ wrench on the starter ring gear.

4. Center punching or marking the converter and flex plate is an aid to _____ the two units.

5. Disconnect the battery ground wire to prevent a _____ _____.

6. In order to prevent the torque converter from sliding out, install a _____ to the edge of the bell housing.

7. Excessive _____ wear can cause a front pump seal to leak.

8. Install a front pump with its lip facing _____.

9. Severe damage can occur during transmission installation if the converter hub does not engage in the _____ _____ or _____ _____ of the front pump.

10. In most cases, it will be necessary to remove the extension housing to replace the _____.

11. Before installing the transmission, coat the converter hub hole with _____ _____ _____.

12. After installing the transmission, adjust all the _____ _____.

SECTION 7

Torque Converter and Hydraulic Pump Inspection, Testing, and Service

REFERENCES: Automatic Transmission Fundamentals, Chapters 4, 8, and 12.
Automatic Transmission Service, Sections 1, 2, 3, and 9.
Transmission or vehicle service manual.

The torque converter and hydraulic pump are two of the most expensive components of the automatic transmission. When the mechanic rebuilds or services the transmission, it makes good sense to do everything practical to save these components for re-use. With this in mind, let's examine the methods of checking whether the converter and pump are still usable, and the cleaning and repair operations necessary to restore them both to operating condition.

The converter is a sealed assembly. Consequently, the mechanic can perform only a limited number of service operations on it. The most important operation is to clean the converter thoroughly with a mechanically agitated cleaner whenever the mechanic rebuilds or services the transmission, and whenever the fluid has become contaminated. The special item of equipment, the torque converter flusher (Section 1), is necessary to perform this operation effectively and automatically. The mechanic should always use this machine to clean the converter, provided it passes a few simple inspection checks for serviceability.

196 SECTION 7

CONVERTER INSPECTION AND SERVICE

Serviceability Checks

The mechanic should perform the serviceability checks before he installs the converter on the cleaning machine since they take only a few minutes, and there is no point in cleaning a converter that is no longer serviceable. These serviceability checks include drive stud, lug, or flange inspection; converter hub inspection; stator-to-impeller interference test; stator-to-turbine interference test; stator one-way clutch test; turbine end-play test; and converter leakage test.

<u>drive stud, lug, or flange inspections</u>

Torque converters may have either drive studs, welded drive nuts (lugs), or a drilled drive flange (Fig. 7-1). These parts mate the converter to the flex plate, permit the engine to drive the converter, and in some cases, align the converter so that it will run true with the engine. To perform an inspection of these parts, place the converter on the work bench, hub facing down, and check the drive studs for the following conditions:

1. Tightness in the converter itself and broken support welds.

2. Damaged or worn threads.

3. On Ford-type converters, damaged or worn pilot shoulders (Fig. 7-2). <u>Note</u>: These shoulders pilot the converter to run true with the engine and transmission. Any converter misalignment caused by worn or damaged pilot shoulders causes the hub on the converter, which drives the pump, to operate eccentrically, and this would quickly destroy the

FIG. 7-1 Drive lugs of a typical torque converter.

Torque Converter and Hydraulic Pump Inspection, Testing, and Service

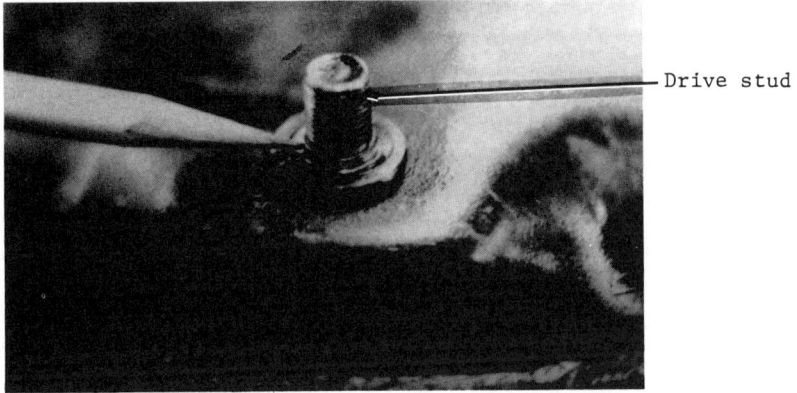

FIG. 7-2 Inspection of a drive stud with a pilot shoulder.

pump bushing or cause a possible vibration. If these shoulders are worn or damaged, therefore, replace the converter with a new or rebuilt unit.

Check a converter that has drive nuts or lugs for the following:

1. Loose lugs or cracked or broken support welds (Fig. 7-3).
2. Worn or damaged threads. Note: If there is no other converter damage, the mechanic can repair these threads as outlined in Sections 1 and 8.

Inspect the converter equipped with drilled drive flanges for the following:

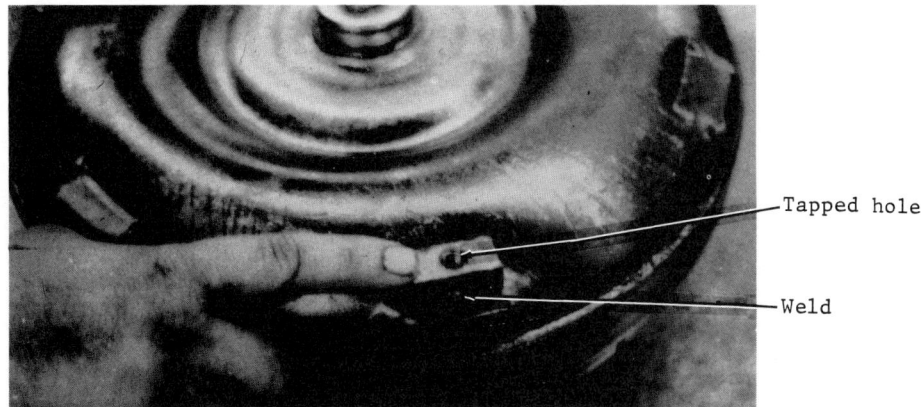

FIG. 7-3 Inspection of a typical converter with a threaded, drive lug.

1. Damaged or elongated holes that accommodate the drive bolts.
2. Cracked or broken drive flanges or mount welds. Note: The mechanic should replace the converter if the flanges or drive holes are worn or damaged.

converter hub inspections

Turn the converter over and take a close look at the outside of the hub that drives the pump (Fig. 7-4). There are opportunities for hub wear at two points: (1) where the hub operates inside the front-pump seal; and (2) where it operates inside the pump bushing. If the hub has score or wear marks, this indicates that the seal or bushing or both are also unserviceable, and each should be replaced as necessary, according to the location of the wear marks.

Frequently, score marks on the hub appear severe to the eye, but they may in fact be very shallow. To check whether the score marks are deep, run a fingernail across the score marks. If it catches in the score indentations, this is a good indication that the wear is excessive. Replace the converter with a new or rebuilt unit if scoring is excessive.

If the wear marks are not too deep, polish them out using this procedure:

1. Cover the hub opening to prevent particles from entering the converter itself during the polishing operation.
2. Wrap a piece of 600-grit crocus cloth around the hub and polish the marks away (Fig. 7-5).
3. Clean the hub thoroughly after polishing.

stator-to-impeller interference tests

This check, as its name implies, indicates whether the stator assembly

FIG. 7-4 Inspection of a typical converter hub.

Torque Converter and Hydraulic Pump Inspection, Testing, and Service

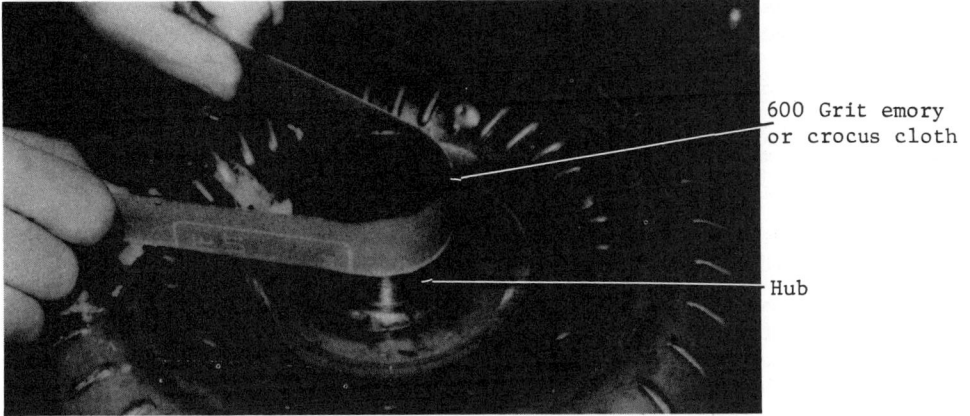

FIG. 7-5 Polishing the converter hub with emery cloth.

is touching or interfering with the normal rotation of the impeller. To perform such a test follow these steps:

1. Place a front pump assembly on a clean work bench with the splined end of the stator shaft pointed upward.

2. Install the torque converter over the pump assembly in such a manner that the splines on the stator one-way clutch inner race engage with the mating splines of the stator support, and the converter hub drive slots engage the pump drive gear (Fig. 7-6).

3. While holding the pump stationary, attempt to turn the converter counterclockwise. The converter should rotate freely without any signs of scraping or interference from within the converter assembly.

4. If there is an indication of scraping or binding, the tailing edges of the stator blades are probably interfering with the leading

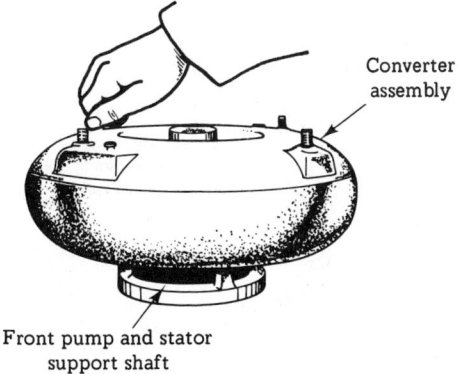

FIG. 7-6 Conducting a stator-to-impeller interference test.

edges of the impeller blades. If this occurs, replace the converter with a new or rebuilt unit.

stator-to-turbine interference tests

This check will indicate whether the stator is striking or interfering with the normal rotation of the turbine. To perform this test follow these steps:

1. Place the torque converter assembly on a clean work bench with its front side facing down.

2. Install a front pump assembly over the converter so that its stator shaft splines engage into the mating splines of the stator one-way clutch inner race, and its drive gear engages with the drive slots of the converter hub.

3. Install the input shaft and engage its splines into the mating splines within the turbine hub (Fig. 7-7).

4. While holding the pump stationary, try to turn the turbine with the input shaft. The turbine should turn freely in both directions without any audible signs of interference or scraping.

5. If interference does exist, the stator's front thrust washer is probably worn out, allowing the stator to hit the turbine. In this situation, replace the converter with a new or rebuilt unit.

stator one-way clutch tests

This test indicates whether or not the stator's one-way clutch will hold the stator against counterclockwise rotation. To perform this test accurately a special tool set, like the one shown in Fig. 7-8, is necessary; this set consists of a holding tool, gauge post, and a guide collar.

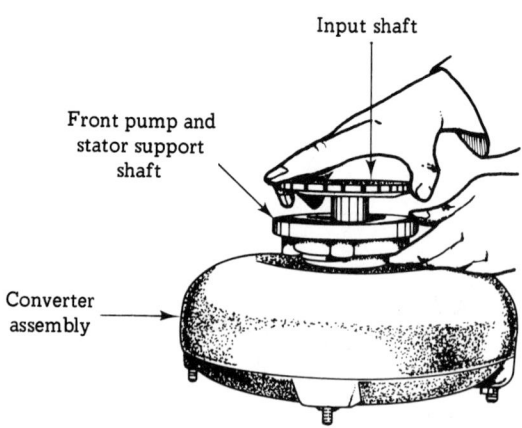

FIG. 7-7 Performing a typical stator-to-turbine interference test.

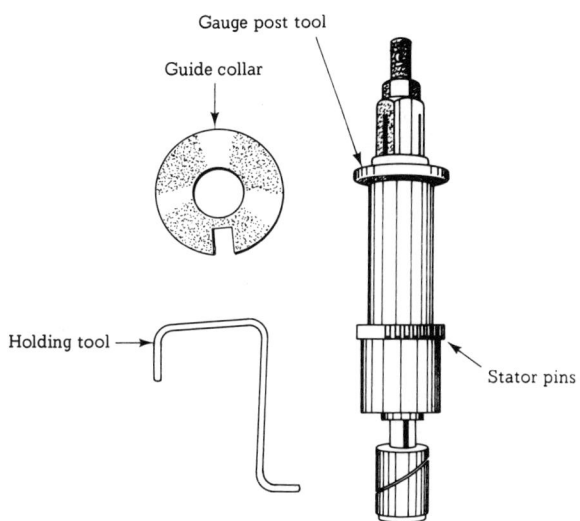

FIG. 7-8   Stator one-way clutch testing tool.

To use this tool set to perform the test, proceed as follows:

1. Install the end of the stator outer-race holding tool into one of the four holes provided in the stator itself. This tool must engage into one of these holes to prevent the stator from turning during the test (Fig. 7-9).

2. Insert the post tool so that its pins engage into the splines of the one-way clutch inner race.

3. Slide the guide collar over the post, hub, and holding tool.

4. While holding the guide collar and holding tool stationary, turn the post tool in both directions with a torque wrench (Fig. 7-9). The tool should freely turn clockwise, but it should hold at least a torque of 10 pounds-foot counterclockwise. Try the clutch for lockup and hold in at least five different locations around the converter.

5. If the one-way clutch fails to lock up and hold the 10 pounds-foot torque, replace the converter with a new or rebuilt unit.

   If the tool set (in Fig. 7-8) is not available and you suspect a total one-way clutch failure, check it using this simple method: Insert one finger into the splined, one-way clutch inner race and attempt to turn the race in both directions. The race should turn freely in the clockwise direction, but it should lock up when turned counterclockwise. This is not an accurate test of clutch reliability, of course, but it will indicate a total failure of the one-way clutch.

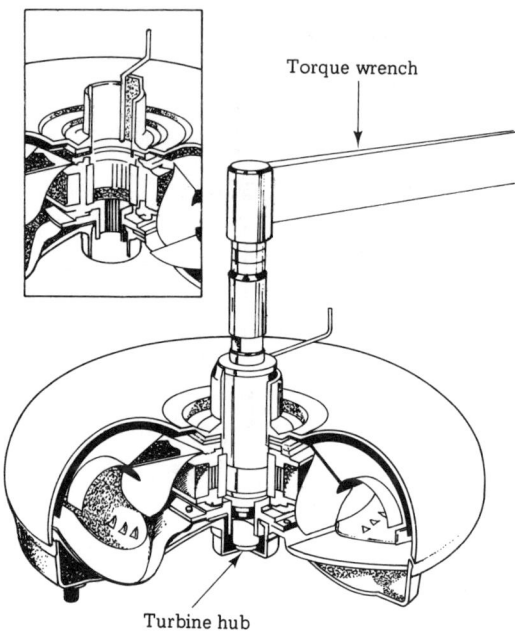

FIG. 7-9 Performing a torque-converter one-way clutch test.

turbine end-play tests

This check actually measures the amount of wear on the thrust washers or bearings separating the internal components within the torque converter. If excessive wear does exist, the converter turbine can lose some of its operating efficiency or produce varying levels of noise. In order to accurately check the turbine's end play, special equipment (refer to Section 1) is necessary.

To check any converter for turbine end play using the universal gauge, proceed as follows (Fig. 7-10):

1. Slide the gauge assembly into the slot in the bracket mounted on the converter flushing machine.

2. With the gauge installed into the mount bracket, set the converter hub down into the gauge cup.

3. Push up on the gauge handle until you feel the gauge tip make contact with the turbine splines inside the converter.

4. Adjust the indicator assembly up or down on the shaft until the dial indicator stem makes contact with the gauge cup. Lock the dial indicator in place with its large thumb screw.

5. Push up on the gauge handle lightly until you meet some resistance. Loosen the small thumb screw on the dial indicator face and zero the gauge.

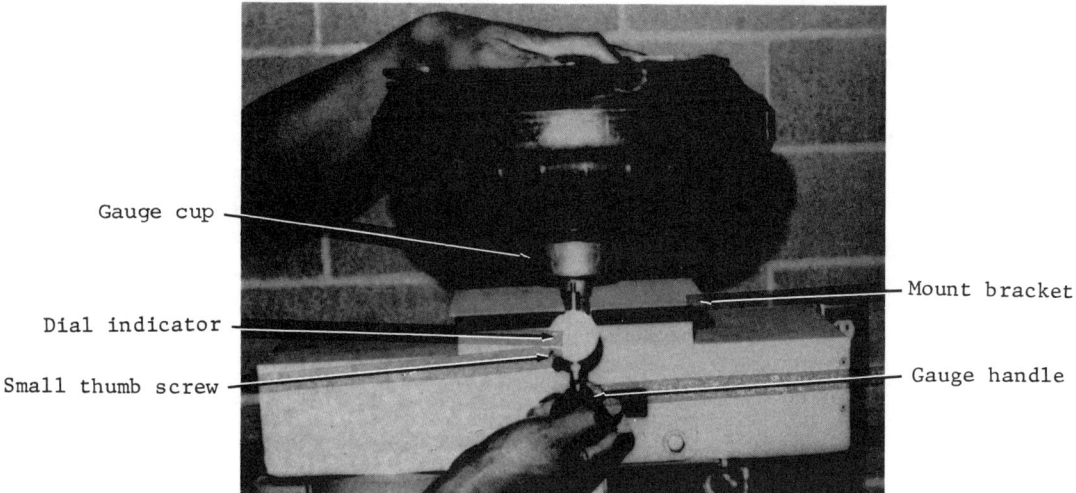

FIG. 7-10 Performing a turbine end-play test using the universal end-play gauge.

6. Push up on the gauge handle once more and read the dial indicator. If the reading is not to specifications, replace the torque converter with a new or rebuilt unit.

<u>converter leakage tests</u>

The purpose of this check is to determine if the welds around the torque converter housing are leaking. A technician will usually make this test only if (1) there has been an undetectable leak in the converter housing area; or (2) the serviceman has cut open the converter, rebuilt its components, and rewelded the housing back together.

Regardless of the reason for the test, it does require certain special tools in order to perform it. If the shop does not already have a special tool, build a substitute with the commonly available parts as shown in Fig. 7-11. When properly constructed, this device can test almost any converter with a drain plug. In order to use this tool on a converter without a drain plug, you will have to drill and tap the housing for a plug, or use another type of tool that has an air valve built into the expanding plug assembly.

To perform this test using the tool shown in Fig. 7-12, proceed as follows:
1. Remove one of the converter drain plugs and drain all the old fluid from the converter.
2. Clean the outside of the converter with solvent and a clean rag.
3. Install the air valve in the open drain plug hole and tighten securely.

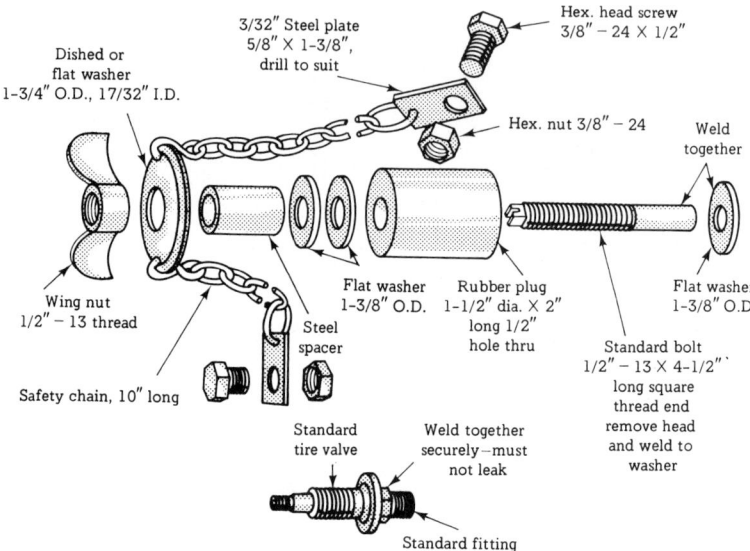

FIG. 7-11 A universal torque converter leakage tester can be made using these parts.

4. Install the expandable rubber plug assembly into the converter hub opening, and expand it by tightening the wing nut securely.

5. Introduce air pressure into the converter housing via the air valve. Check the air pressure with a tire gauge, and adjust the pressure to 20 psi.

6. Place the torque converter into a tank of water for about 10 minutes and then observe all the welded areas for signs of bubbles. If you observe no bubbles, assume that the welds are not leaking.

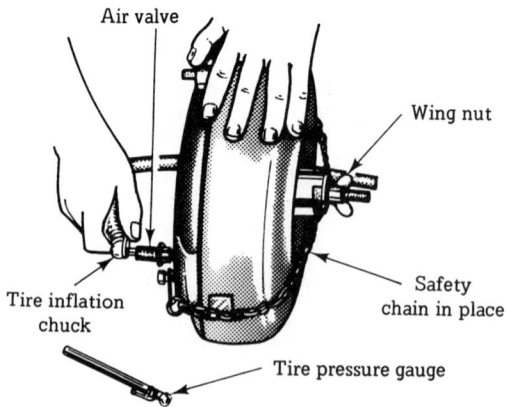

FIG. 7-12 Performing an air leak test on a torque converter, using a universal converter leakage tester.

Torque Converter and Hydraulic Pump Inspection, Testing, and Service

Converter Flushing

If the torque converter has passed all the above-mentioned tests, assume that it is still serviceable and is ready for an internal cleaning or flushing. In order to properly clean a sealed converter, a flushing machine like the one shown in Fig. 7-13 is necessary. As previously mentioned in Section 1, this machine, when in operation, cleans a converter by pumping solvent through the unit while at the same time rotating the turbine.

To set this machine up to clean a typical converter, perform the following procedures:

1. Check the converter for an available drain plug. Remove the plug and completely drain the old fluid out of the converter into a waste oil container. <u>Note</u>: If the converter has no drain plug, drill and tap it for one as explained later on in this section. Having a drain plug in the converter makes it easier to drain not only the dirty fluid out of the unit but, later on, the cleaning solvent as well. Any residual dirty fluid left in the converter will contaminate the machine's cleaning solvent, and any cleaning solvent left in the converter can circulate through the overhauled transmission where it can harm seals.

FIG. 7-13 A typical torque converter flusher.

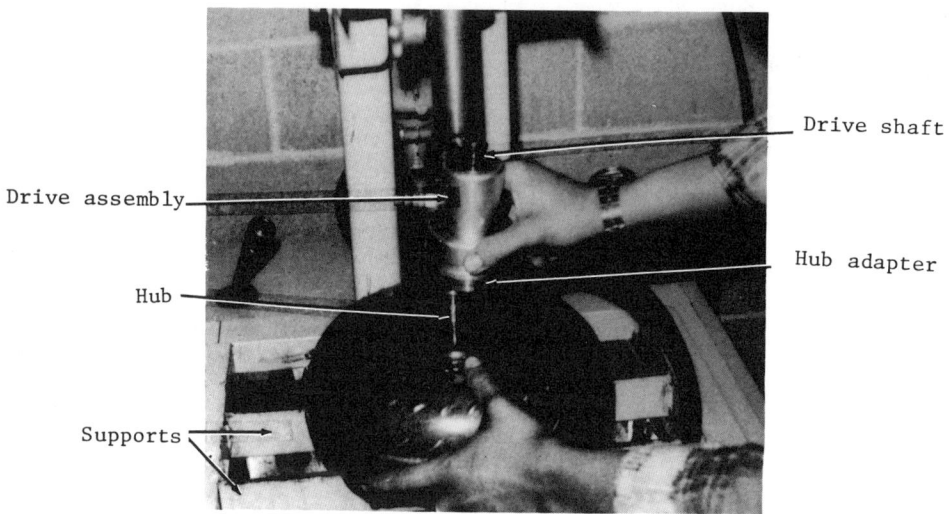

FIG. 7-14 Installation of the drive assembly into the converter hub.

2. After completely draining the converter, place it on the flushing machine supports over the sump tank with the hub side up.

3. Insert the drive assembly in the converter hub, and push the hub adapter down as far as it will go (Fig. 7-14).

4. Push down and turn the drive shaft in order to engage the drive tip with the turbine splines. <u>Note</u>: Always make certain that this shaft turns freely but with an inertia effect, caused by the

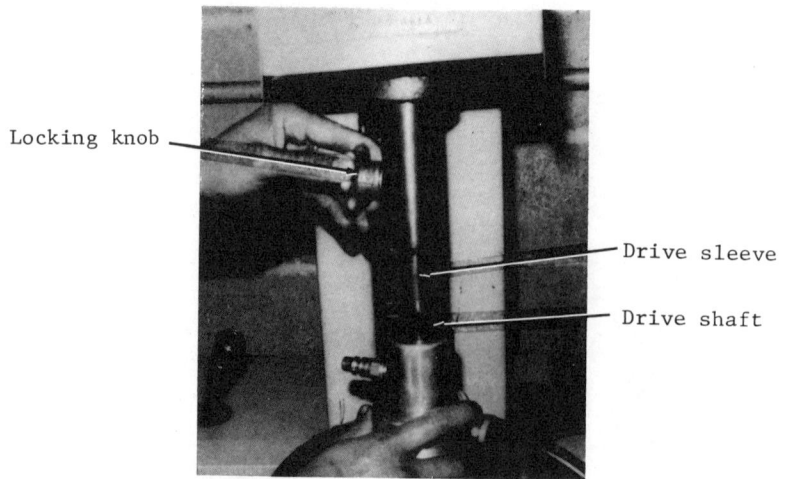

FIG. 7-15 Engaging the drive sleeve with the drive assembly.

Torque Converter and Hydraulic Pump Inspection, Testing, and Service

weight of the revolving turbine inside. If the turbine does not feel like it is revolving, or if the drive tip seems as if it's skipping, remove the drive assembly and visually check the turbine splines.

5. Place the converter assembly directly under the motor-driven shaft. Lower the drive sleeve and align it with the drive shaft of the drive assembly. Lock the drive sleeve in position with the locking knob (Fig. 7-15).

6. Attach the inlet pressure hose to the quick-disconnect fitting on the drive assembly (Fig. 7-16). Make sure that the drive assembly outlet hose is in position inside the sump tank.

7. Set the timer to the desired position, 10 minutes for a moderately dirty converter, and 30 minutes or more for an extremely contaminated converter (Fig. 7-17). Note: If the converter makes a racheting sound while the drive is turning, stop the machine and recheck the drive shaft to make certain it is securely held down in position. If the noise continues, check the splines of the turbine. If, on the other hand, the drive will not rotate the turbine at all, replace the converter with a new or rebuilt unit.

8. After flushing the converter, loosen the lock knob, raise the drive shaft, and disconnect the inlet hose from the drive assembly.

9. Tilt the converter to a vertical position with the drain plug hole at the bottom. This facilitates a complete draining of the cleaning solvent from the converter. Install the drain plug and torque it to specifications.

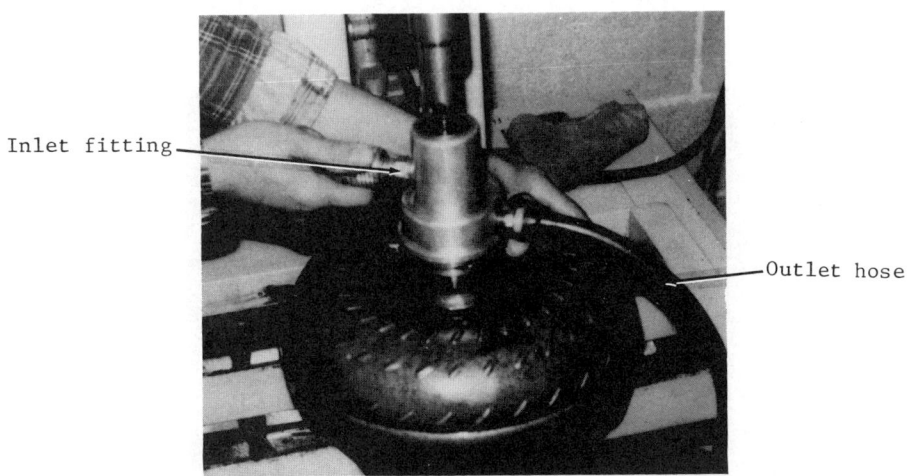

FIG. 7-16 Connecting the pressure, inlet hose to the drive assembly.

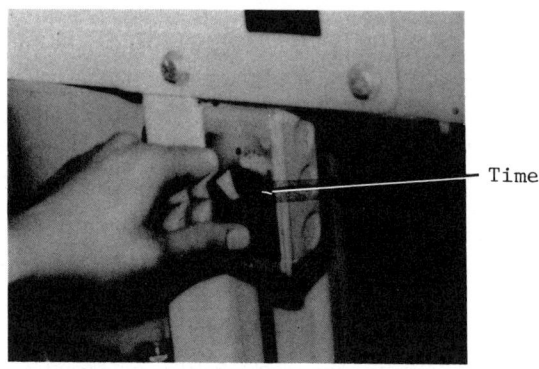

FIG. 7-17 Setting the timer to a desired position.

Other Converter Service
<u>drain plug installation</u>

As previously mentioned, some converters do not have factory installed drain plugs. Consequently, changing the fluid with the transmission and converter still in the vehicle, and flushing the converter is more difficult and time consuming. In the long run, then, it will save someone time and effort if whenever one of these converters is out of a vehicle, a mechanic drills, taps, and installs a drain plug in the unit.

Using the drill and tapping equipment referred to in Section 1, a technician can install a plug in a converter following these instructions:

1. Clean the outside of the converter with solvent and a clean rag; then, place the converter on a clean, wet-type work bench.

2. Install the mount bracket on the converter drive lugs or flanges with the hardware provided with the kit (Fig. 7-18). Tighten the mounting hardware securely.

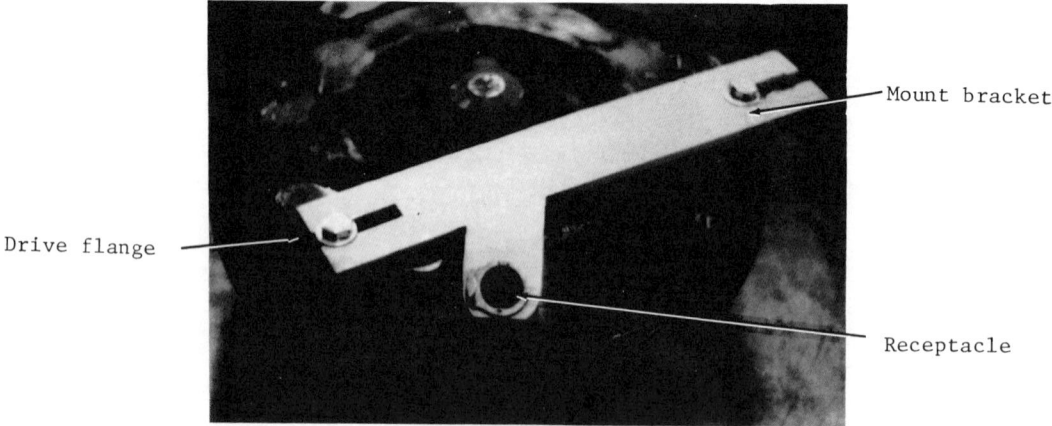

FIG. 7-18 Installation of the bracket onto the converter.

Torque Converter and Hydraulic Pump Inspection, Testing, and Service 209

3. Install the drill guide insert into its receptacle on the mount bracket. Lubricate the guide with oil.

4. Install the special drill in a drill motor. Then drill a hole through the outer converter housing (Fig. 7-19). <u>Note</u>: As the drill begins to cut through the last few layers of metal, ease off your pressure on the drill motor; otherwise, the drill may strike and damage the turbine as it breaks through the remaining layers of housing material.

5. With low pressure compressed air, blow the chips away from the hole and drill <u>before</u> removing the drill from the hole and drill guide. This procedure prevents excess metal chips from entering the converter.

6. Remove the drill guide insert and install the tap guide insert into the bracket receptacle.

7. Using the 1/8-inch pipe tap supplied with the kit and a tap wrench, cut the plug threads into the drilled hole (Fig. 7-20). <u>Note</u>: To prevent tap or thread damage, it may become necessary to break the chips up by occasionally reversing the direction of the tap wrench. If this becomes necessary, reverse the direction of the tap only one full revolution of the wrench at any given time.

8. Before removing the tap completely from the hole and guide, again blow any chips away from the area using low pressure compressed air.

9. Remove the hardware and bracket from the converter. Stand the converter up with the drain opening down and drain any remaining fluid out of it.

10. Thoroughly flush the converter as outlined earlier in this section. Install a new drain plug and torque to specifications.

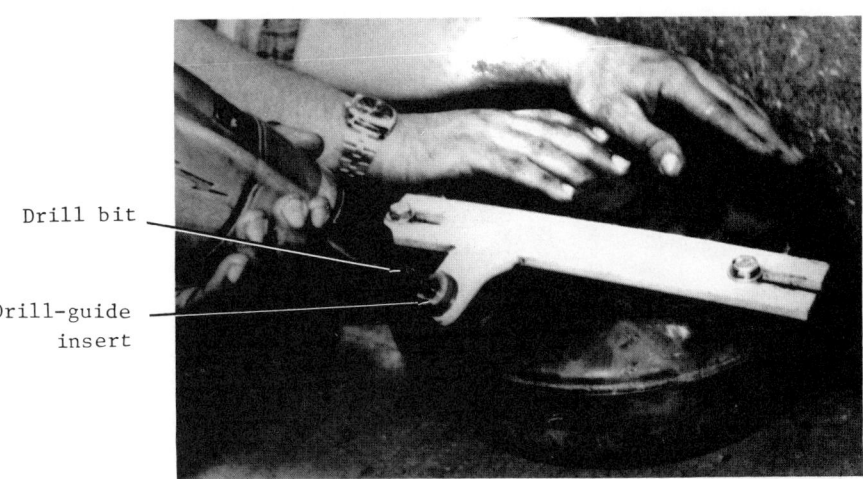

FIG. 7-19 Drilling the hole into the converter.

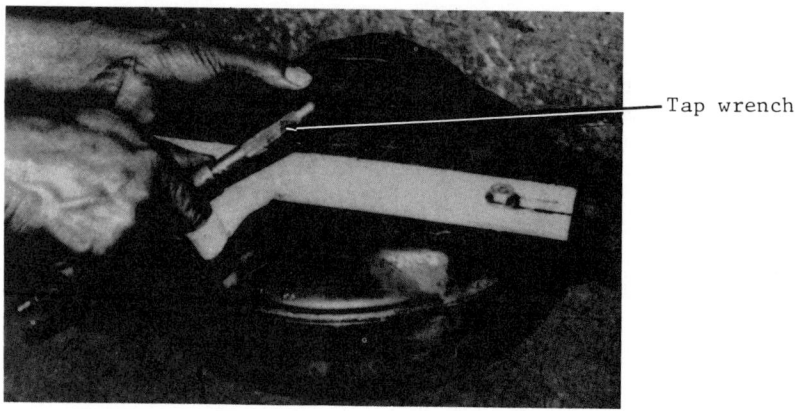

FIG. 7-20 Cutting the plug threads into the converter using a pipe tap wrench.

starter ring-gear replacement

Some modern torque converters have the starter ring gear directly mounted on the outer diameter of the converter front cover. When this gear is defective, a mechanic can remove and replace it. It is not necessary, then, to replace the converter unless it is also defective or leaking.

With the converter removed from the vehicle, remove the gear following these general procedures:

1. Cut through the weld material at the rear side of the ring gear, using a hack saw or grinding wheel (Fig. 7-21). Be very careful not to cut or grind into the front cover.

2. Scribe a heavy line on the front cover next to the front face of the ring gear as an aid in locating the new ring gear.

FIG. 7-21 Cutting through the ring gear weld material.

Torque Converter and Hydraulic Pump Inspection, Testing, and Service

3. Support the converter on the bench with the four drive lugs or studs resting on blocks of wood. Note: The converter must not rest on the front cover hub during this operation.

4. Using a blunt chisel or drift punch and hammer, tap downward on the ring gear near the welded areas in order to break away the remaining weld material (Fig. 7-22). Tap around the gear until it comes completely off the converter housing.

5. Smooth off the welded areas on the cover, using a suitable file.

Using any one of the following methods, heat and expand the starter ring gear before installation on the torque converter. Note: To prevent injury, wear heavy gloves when handling the hot ring gear.

1. Place the new ring gear in an electric or gas oven and set oven temperature to $200°F$. Keep the gear in the oven 15 to 20 minutes, or

2. Position the new ring gear in a shallow container, add water, and heat the water until it reaches its boiling point. Continue to heat the ring gear, using the boiling water, for about 8 minutes, or

3. Place the new ring gear on a flat surface, and then direct a steam flow around the gear for about 2 minutes, or

4. Position the new ring gear on a flat surface. Using a gas welder with a medium-sized tip, direct a slow flame evenly around the inner rim of the gear. Do not apply this flame to the gear teeth. During the heating process, at given intervals apply a few drops of water to the face of the gear. When the gear is hot enough to just boil the water, it is ready for installation on the converter.

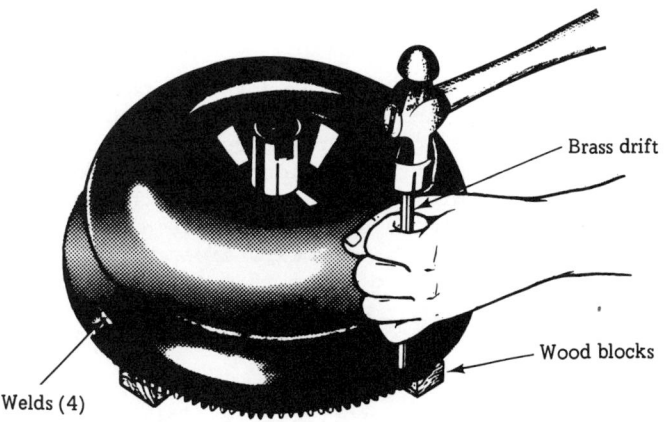

FIG. 7-22 Removing the ring gear with a chisel and hammer.

Regardless of what process you used to heat the ring gear, install it on the housing using this procedure:

1. Place the ring gear in position on the converter front cover. Then tap the gear onto the cover evenly with a plastic or rawhide mallet until the gear's face is even with the scribed line (made during gear removal) on the front cover. Make certain that the gear's face is even with the scribed line, running around the full circumference of the front cover.

2. Reweld the new ring gear to the torque converter front cover. Be careful to place, as nearly as possible, the same amount of weld material in exactly the same location as the original weld. This procedure is necessary in order to maintain torque converter balance. Also, place the welds alternately on opposite sides of the converter to minimize the possibility of distortion. The following suggestions may assist you in making the weld:

    a. Do not use a gas welder.

    b. Use a DC welder that is set at straight polarity, or an AC welder if the correct type of electrode is available.

    c. Use a 1/8-inch diameter welding rod, and set the welding current to 80 to 125 amps.

    d. Direct the arc at the intersection of the ring gear and front cover from an angle of 45 degrees from the rear face of the ring gear.

3. Inspect all the gear teeth and remove all nicks, raised metal, or weld-metal splatter. This will assure longer gear life and quieter starter operation.

HYDRAULIC PUMP INSPECTION AND SERVICE

Whenever an automatic transmission is overhauled, the hydraulic pump should be disassembled, cleaned, and carefully inspected. The reason for this is twofold: First, the pump gears or rotors must operate with given clearances in order to supply the transmission with the necessary fluid volume and pressure. If clearances become too large, the transmission can malfunction or even fail because of a decrease in fluid flow. Second, as previously stated, the pump is also a very expensive component; the technician, then, should do everything possible to reuse it.

Teardown and Inspection

To determine if the pump is still serviceable, follow these steps:

1. Remove the stator support-to-pump-housing attaching bolts and separate the pump assembly into three sections, the pump housing,

Torque Converter and Hydraulic Pump Inspection, Testing, and Service

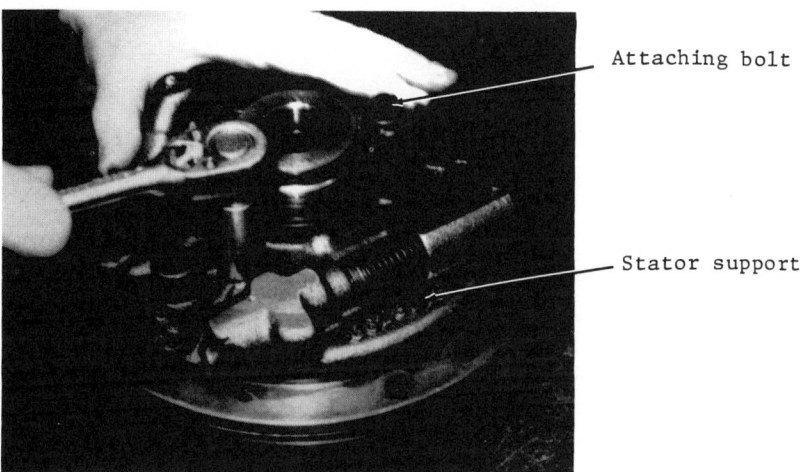

FIG. 7-23 Removing the stator support-to-housing attaching bolts in order to separate the main components.

pump gears or rotors, and the stator support (Fig. 7-23). <u>Note</u>: Before removing the pump gears or rotors from the pump housing, check them both for identification marks. If there are none, make an alignment mark across both gears or rotors and the housing using Prussian blue or an indelible marking pen (Fig. 7-24). <u>Never use a center punch or metal scribe</u>. These marks will assist

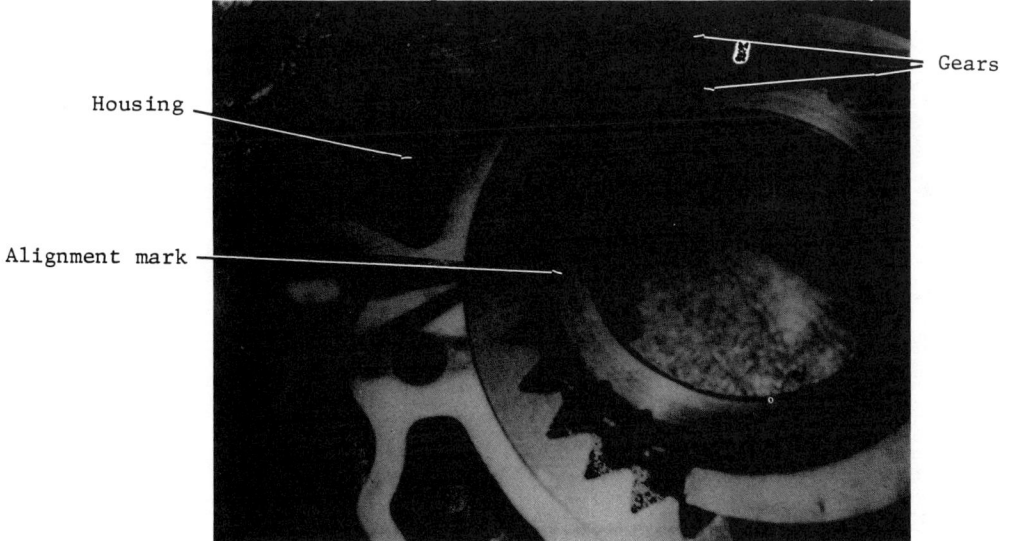

FIG. 7-24 Marking the gears and housing using Prussian blue.

in reinstalling the gears or rotors in the same relative position when you reassemble the pump.

2. With the tool shown in Fig. 7-25, remove and discard the front pump seal. Remove and discard the pump housing-to-case square-cut seal.

3. Clean the three components in a suitable solvent and blow dry with low pressure compressed air. <u>Make sure you thoroughly blow out fluid passages; any clogged or restricted passage can cause a malfunction.</u>

4. Check all the parts for burrs and light scoring; if you find any, use crocus cloth to clean and polish the surfaces. Then wash these parts again to remove any grit.

5. Inspect the gears or rotors carefully for broken teeth or lobes, which require replacing the pump.

6. Inspect the inner and outer diameter of the gears or rotors and also the teeth or lobes for excessive wear. <u>Note</u>: Moderate wear does not necessitate using a new pump. In fact, if the hydraulic pressure was normal during the entire diagnosis procedure (refer to Section 3), the pump very likely has enough output for reuse.

7. Inspect the mating surfaces of the gears or rotors, pump housing, and stator support for abnormal wear. Don't be concerned if the black lubrite coating has worn off during operation; this is a normal tendency. Pay special attention to the inside of the pump body (the counterbore) where the gears or rotors operate (Fig. 7-26). Check the depth of any scoring with your fingernail; if there is excessive wear in this area and pump pressure was low during diagnosis, replace the pump.

FIG. 7-25 Removing the front pump seal.

Torque Converter and Hydraulic Pump Inspection, Testing, and Service 215

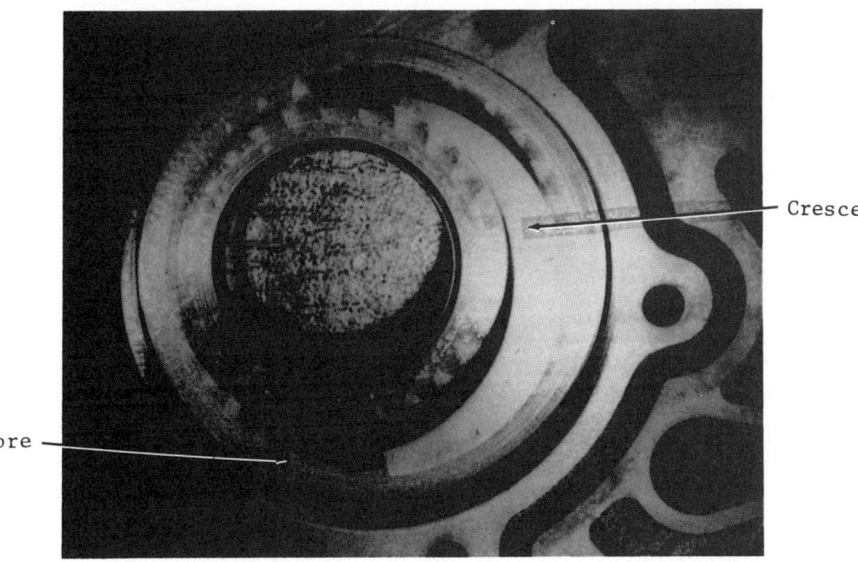

FIG. 7-26 Checking the pump counterbore for wear.

8. If everything has checked out so far, reinstall the gears or rotors in the pump body, aligning the marks made during disassembly.

9. Measure the pump body-to-gear or rotor end-play clearance using a straight edge and a feeler gauge (Fig. 7-27). This measurement must be within specifications.

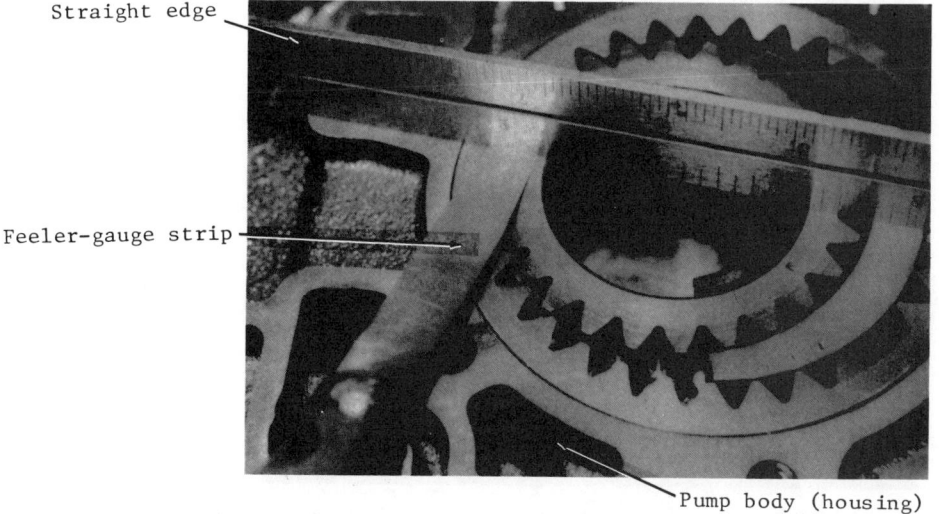

FIG. 7-27 Measuring the pump body-to-gear end clearance.

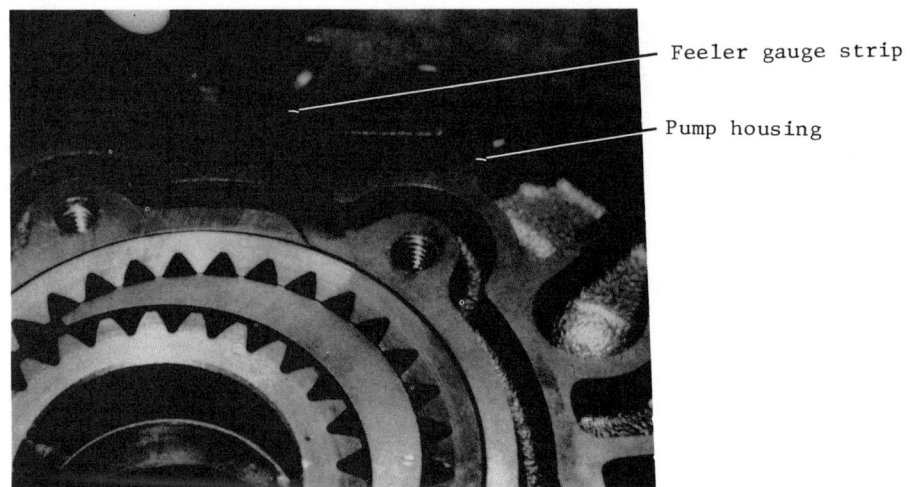

FIG. 7-28 Measuring the clearance between the outer gear and the pump housing.

10. Using a feeler gauge of the specified thickness, measure the clearance between the outer gear or rotor and the pump housing bore (Fig. 7-28).

11. On the gear-type pump only, measure, with the specified size feeler gauge, the clearance between the teeth of each gear and the pump crescent (Fig. 7-29).

12. On the rotor-type pump only, measure, with the specified size feeler gauge, the clearance between the tips of the inner and outer rotors

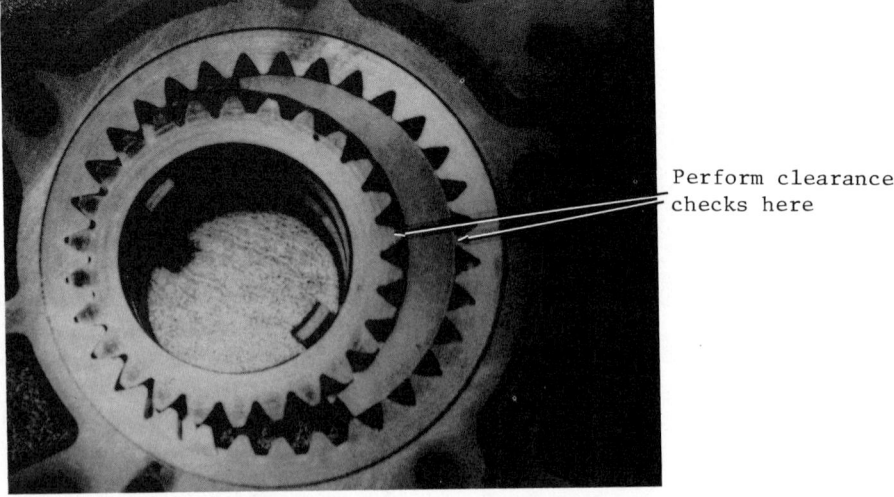

FIG. 7-29 Locations where you measure the clearance between both gears and the crescent.

Torque Converter and Hydraulic Pump Inspection, Testing, and Service   217

(Fig. 7-30). <u>Note</u>: <u>If any of these feeler gauge clearance checks are not to specifications, replace the pump assembly.</u>

13. Remove all the sealing rings from the stator support. Inspect the stator support for worn or damaged ring grooves, clutch drum bearing surfaces, thrust washer surfaces, or splines. If wear or damage exists in any of these areas, replace the stator support or pump assembly.

Service and Reassembly

1. Again remove the gears or rotors from the pump housing.

2. With the tools shown in Fig. 7-31, replace the front pump bushing. <u>Note</u>: For installation of the bushing, always refer to the transmission service manual for staking procedures, when used, and the bushing's proper location in the bore. Some bushings have grooves, slots, or oil passages which have to be located in a given area of the bore for proper bushing lubrication.

3. Install a new front pump seal using the tool shown in Fig. 7-32. <u>Note</u>: If the new seal does not have a rubber or resin coated backing, coat the backing with a nonhardening sealer before installation. Make sure the sealing lip faces inward toward the gears or rotors.

4. Lubricate the gears or rotors with clean hydraulic fluid and reinstall them into the pump housing, aligning their locating marks.

5. Position the stator support over the pump housing and install the stator support-to-housing attaching bolts. Tighten these bolts finger tight.

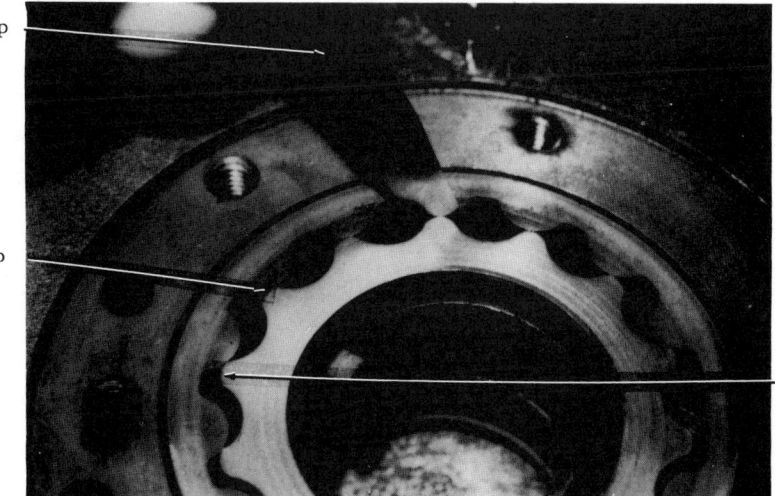

FIG. 7-30 Measuring rotor tip clearance.

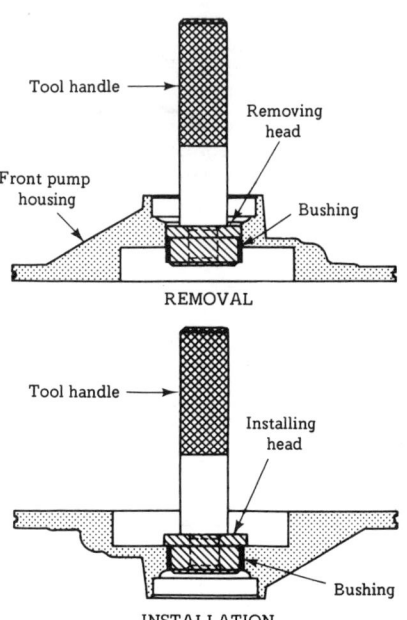

FIG. 7-31 Removing and replacing a front-pump bushing.

6. When specified by the manufacturer, install the housing-to-stator support aligning tool (Fig. 7-33) and torque the stator support-to-housing bolts to specifications.

7. Coat the selective thrust washers with Vaseline and install on the stator support.

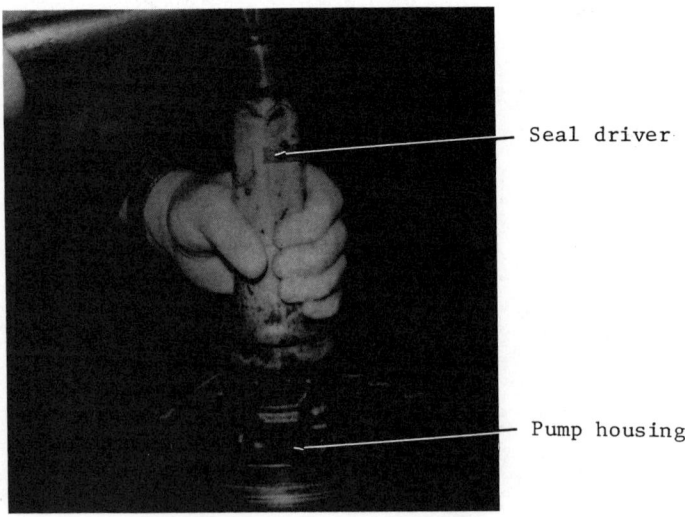

FIG. 7-32 Installing the front-pump seal.

Torque Converter and Hydraulic Pump Inspection, Testing, and Service

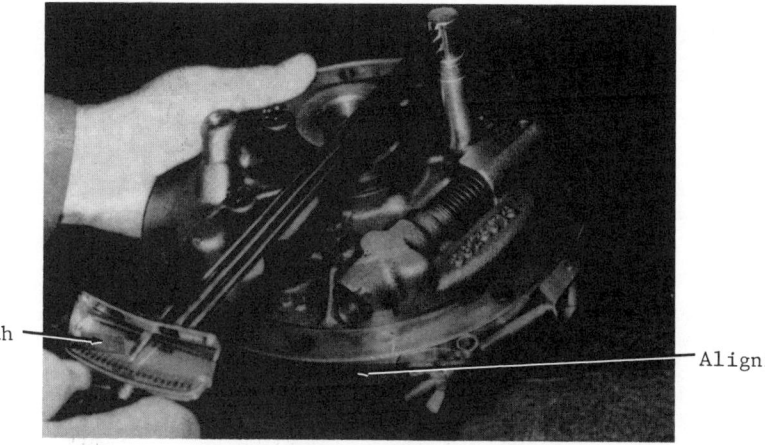

FIG. 7-33 Typical pump aligning band installed around the pump assembly, during the torquing procedure.

8. Install the new sealing rings in their proper locations on the stator support (Fig. 7-34). <u>Be careful, these rings may be of different diameters. If in doubt as to their proper locations, always refer to the appropriate transmission service manual.</u> After installation, check these rings for freedom of movement in their respective grooves.

9. Place a torque converter on the bench with its pump drive hub facing upward. Then position the pump assembly over the converter hub, engaging the hub with the pump drive gear or rotor. Rotate the pump housing in both directions to check the pump gears or rotors for

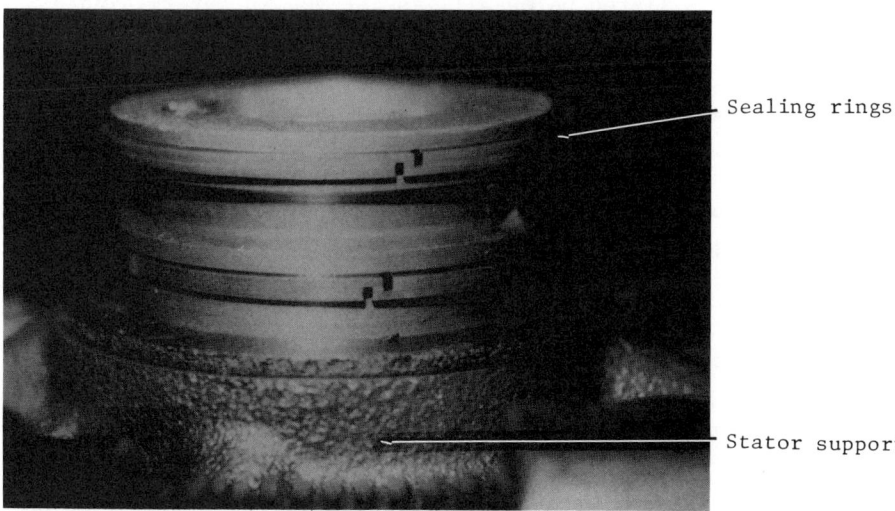

FIG. 7-34 Replacing the stator support sealing rings.

FIG. 7-35 Checking the pump gears or rotors after pump assembly.

freedom of movement within the pump housing and stator support (Fig. 7-35).

CHECK-UP QUESTIONS

The questions listed below will assist you in determining how well you remember the material contained in this section. Read each question carefully before adding the word or words necessary to complete the sentence. If you can't complete the sentence, review that portion of the section that covers the question.

1. It makes good sense to do everything practical to save the converter and hydraulic pump for reuse because they are very _____ components.
2. The most important service operation performed on a converter is to _____ it.
3. The mechanic should perform the serviceability checks _____ using the flusher on the converter.
4. Pilot shoulders permit the converter to run _____ with the engine and transmission.
5. There can be converter hub wear at _____ points.
6. Use your _____ to check the depth of the score marks on the converter hub.

Torque Converter and Hydraulic Pump Inspection, Testing, and Service   221

7. If the score marks are not too deep on the hub, _____ them away using _____ cloth.
8. If there is any scraping or binding felt during the stator-to-impeller interference test, _____ the converter.
9. The stator one-way clutch test indicates whether or not the clutch will hold the stator against _____ rotation.
10. The turbine end-play test measures the amount of wear on _____ _____ or _____.
11. The purpose of the converter leakage test is to see if the _____ around the housing are leaking.
12. To properly clean the inside of a sealed converter a _____ _____ is necessary.
13. If the torque converter does not have one, install a _____ _____ before flushing the unit.
14. It is not necessary to replace the converter if the starter _____ _____ is defective.
15. If the hydraulic pump is worn excessively, it may not be able to supply the _____ and _____ necessary to properly operate the transmission.
16. Before removing the gears or rotors from the pump housing, mark them with _____ _____ or a _____ _____.
17. When cleaning the pump housing and stator support, make sure you blow out all fluid _____.
18. Moderate teeth or lobe wear does not always necessitate using a _____ _____.
19. A feeler gauge and a _____ _____ are necessary to perform the body-to-gear or rotor end-play test.
20. When specified by the manufacturer, use an _____ _____ to hold the pump housing and stator support together while you torque their attaching bolts.

SECTION

8

Subassembly Cleaning, Inspection, and Service

REFERENCES: <u>Automatic Transmission Fundamentals</u>, Chapters 2, 3, 5, 6, 9, 10, 11, and 12.
<u>Automatic Transmission Service</u>, Sections 1 and 2.

In addition to the torque converter and hydraulic pump, the automatic transmission contains many other hard and soft parts which the mechanic must clean, inspect, service, or replace during an overhaul to make the job successful. The <u>hard parts</u> include such items as the transmission case, extension housing, drums, one-way (overrunning) clutches, servos and accumulators, governors, planetary gear trains, shafts, thrust washers, bearings, bushings, and valve bodies. The <u>soft (commonly-replaced) parts</u> include the clutch plates; bands; metal, Teflon, and rubber sealing rings; metal clad seals; and gaskets.

Since it would be impossible in the space available to cover all the varied service techniques used by all the transmission manufacturers, this section will present only an overview of the commonly used practices for cleaning, inspecting, and servicing the transmission's hard and soft parts and the cooler lines and cooler. If you are ever in doubt as to how to perform similar procedures on any particular make of transmission, always refer to the appropriate service manual.

Subassembly Cleaning, Inspection, and Service

TRANSMISSION HARD PARTS

Transmission Case

Because the transmission case itself forms the main framework or housing for the transmission components, and its many drilled passages make up a major portion of the transmission's hydraulic circuits, it is very important that the mechanic pay special attention to this assembly during the overhaul procedure (Fig. 8-1).

cleaning the case

To clean a transmission case properly, the cleaning process will have to remove many types of contamination, oil, grease, road dirt, varnish, or shellac. Of the five, the mechanic can easily remove the first three kinds by steam cleaning or washing the case thoroughly in a safety-type, solvent-filled parts washer. Varnish or shellac deposits are much harder to dissolve.

There are two general methods used to remove varnish or shellac from the case. The first method is to place the case in a jet-type cleaner (Fig. 8-2). This device sprays a solution of hot water mixed with a chemical that dissolves the contaminants without damaging the metal, while the case revolves inside the machine. The second method is to submerge the case in a cold or hot tank filled with a chemical that also cuts through the varnish and shellac without harming the metal itself. Cold tanks normally contain a chemical similar to carburetor cleaner while a hot tank contains a solution much like that used in a jet cleaner.

FIG. 8-1   Typical automatic transmission case.

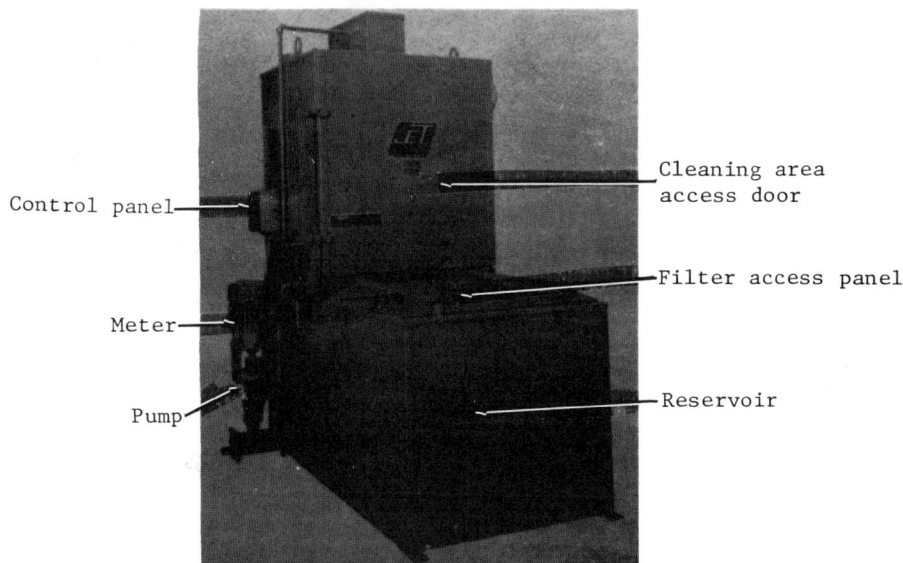

FIG. 8-2   Jet cleaner used to clean transmission parts.

No matter what chemical treatment the technician uses to clean up varnish or shellac built up inside a case, he should follow it up by rinsing and air drying the case. The serviceman can rinse a case by either steam cleaning it again or by washing it thoroughly with large quantities of water, preferably hot. After the rinsing process, he should then air dry the case with low pressure compressed air, making sure to blow out all passageways completely.

case inspection

With the case clean and dry, inspect it for the following problems:

1. Damaged or stripped threads.
2. Plugged or restricted fluid or ventilation passages.
3. Worn or damaged bushings.
4. Worn or damaged manual valve and parking pawl linkages and seal.
5. Worn or damaged clutch plate slots or lugs and snap-ring grooves.
6. Worn or damaged governor, band anchor pin, modulator valve, and speedometer gear bores.
7. Damage, cracks, or case porosity. If the case has damage, cracks, or porosity, it may be better to replace the case at this point because repairs for these defects are not always successful. But this section, later on, will cover a repair for case porosity.

Subassembly Cleaning, Inspection, and Service

thread repair

To repair damaged threads with the thread repair kit, referred to in Section 1, follow this simple procedure:

1. With the specified drill bit and drill motor, machine out all of the damaged threads (Fig. 8-3).
2. With a tap wrench and the special tap included in the kit, cut the threads for the heli-coil into the drilled hole.
3. Thread the installation tool into the heli-coil. Make certain that the coil's driving tang fully engages with the slot in the end of the installation tool.
4. Blow out all metal chips from the tapped hole, and thread the heli-coil into the tapped hole, using the installation tool. Stop coil installation when the insert's top is 1/4 to 1/2 turn below the metal surface of the hole.
5. Remove the installation tool from the coil, and break the tang off the coil at its notched point with a square-end punch or pliers.

bushing replacement

A transmission case bushing supports and guides the output shaft. This bushing must be in good condition or the output shaft will wobble in the bushing, causing noise and possible vibration. If upon inspection the bushing shows damage or excessive wear, replace it using the equipment mentioned in Section 1 and this procedure:

1. If using a bushing chisel, cut along the bushing's seam until the chisel breaks through the bushing wall (Fig. 8-4). Pry up the loose ends of the old bushing with an awl or needle-nose pliers and remove the bushing, or
2. With the handle and correct-size driving head, press or drive the old bushing from the case (Fig. 8-5).

Install the new bushing using the following procedure:

1. Inspect the bushing bore for cracks, wear, or damage.
2. Position the new bushing into its bore. Be sure to align, where used, any bushing lubrication holes or grooves with the corresponding passages or holes in the case. Always start a new bushing into the

Drill

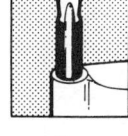
Tap

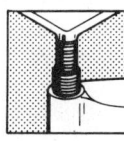
Install

FIG. 8-3   Repairing damaged threads using a Heli-coil repair kit.

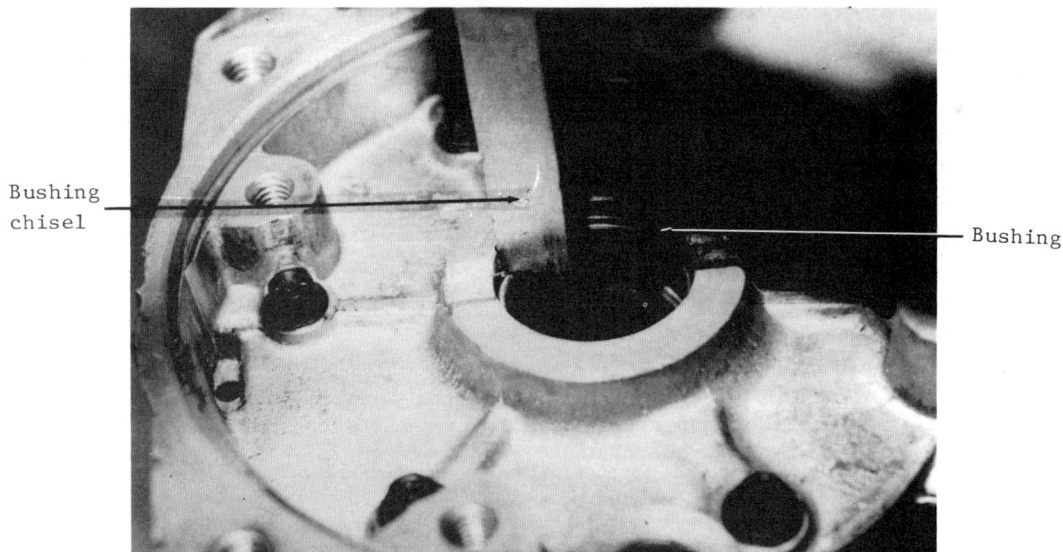

FIG. 8-4 Removing a transmission case bushing using a bushing chisel.

FIG. 8-5 Pressing or driving a transmission case bushing from its bore.

Subassembly Cleaning, Inspection, and Service

chamfered side of the bore. This chamfer will help prevent the bushing from cocking in the bore as you press it into place.

3. Using a press or hammer, the handle, and the correct-size installing head, insert the bushing into its case bore. Be sure to press in the bushing so that its ends are the same distance from each end of the case bore, or the distance specified by the manufacturer (Fig. 8-6).

4. Carefully insert the shaft that rides in the bushing, and check it for free rotation within the bushing. If the shaft will not go into the bushing or is hard to turn, the pressing operation damaged the bushing, and you will most likely have to replace it.

seal replacement

The shifter shaft seal prevents external fluid leakage around where both the throttle and manual valve shafts protrude through the case. The mechanic should always change this seal during overhaul, or a leak may develop later on.

In order to replace a shifter shaft seal in the side of a typical transmission case, the mechanic should follow this general procedure:

1. From inside the case, loosen the throttle lever attaching bolt and remove this lever from its shaft (Fig. 8-7).

2. Also from inside the case, loosen the manual valve lever attaching bolt and remove this lever from its shaft.

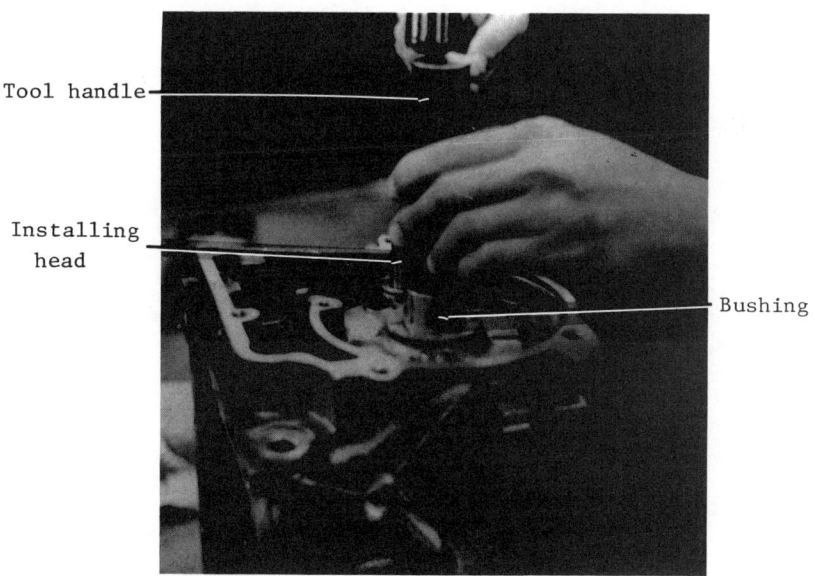

FIG. 8-6   Installing a typical transmission case bushing.

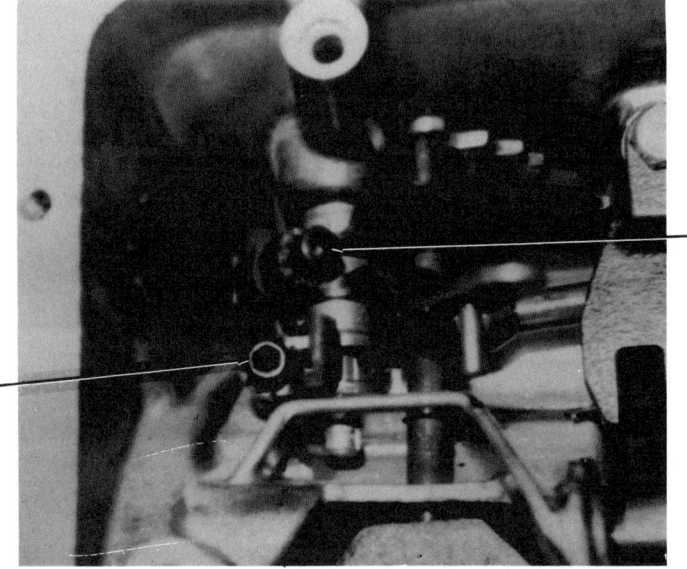

FIG. 8-7 Removing the throttle-and manual-valve levers from inside the transmission case.

3. From the outside of the case, remove both the manual and throttle lever shafts from the case (Fig. 8-8).

4. With a suitable puller or screw driver, pull or pry the seal from its case bore.

FIG. 8-8 After removing the throttle-and manual-valve levers, just pull the two shafts from the case.

Subassembly Cleaning, Inspection, and Service

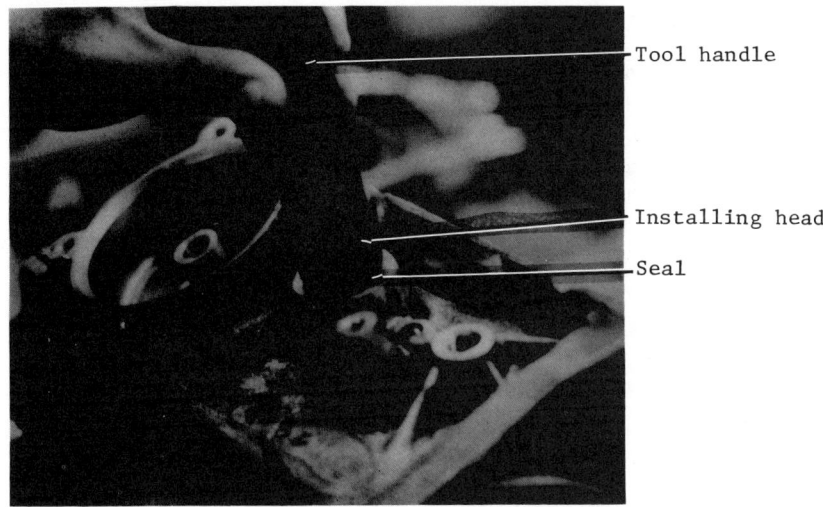

FIG. 8-9 Installing the shifter shaft seal.

Use the following procedure to install the new seal:

1. Using a hammer, handle, and correct-size installing head, drive the new seal into the case (Fig. 8-9). Be certain that the seal firmly seats into the case counterbore, and its lip faces inward or toward the inside of the case.

2. Lubricate both the manual and the throttle valve shafts, and carefully reinstall them through the seal and into their case bore.

3. From inside the case, reinstall the manual valve lever over the manual lever shaft and torque its attaching bolt to specifications.

4. Also from the inside the case, reinstall the throttle valve lever over the throttle lever shaft, and torque its attaching bolt to specifications.

repair of case porosity

Case porosity is a condition in which a pressurized or static fluid leak has developed through the outside of the metal case in certain places. It is possible, in some situations, to repair a leak of this nature without removing and or replacing the case itself through the application of a special epoxy cement to the porous area. While this method does not always work, it is sometimes worth attempting before the transmission is removed from the vehicle.

To repair a case with a porous-type leak, follow this recommended procedure:

1. Road test the vehicle in order to bring the transmission's temperature up to operating range, approximately 180°F.

2. With another mechanic in the vehicle, raise it with an overhead hoist or floor jack. If you use a floor jack, position the jack stands under the frame at suitable locations, and lower the vehicle onto the stands.

3. Locate the leak by having the other mechanic start the engine and operate the transmission in all driving ranges. The use of a mirror and flashlight is helpful in finding a leak of this nature.

4. Shut the engine off and thoroughly clean the porous area with a cleaning solvent and brush. Then blow the area dry with compressed air.

5. Using the manufacturer's instructions, mix a sufficient amount of epoxy to make the repair. Observe all the precautions of the manufacturer in handling this material.

6. While the transmission case is still hot, apply the epoxy to the area under repair. A clean, dry, soldering acid brush works well to not only clean but apply the epoxy cement to the area. <u>Make sure to completely cover the porous area</u>.

7. Allow the cement to cure for three hours before starting the engine to perform a leak test.

8. Lower the vehicle and road test. Then recheck the repaired area for leakage. If the case still leaks, remove the transmission and replace the case.

Extension Housing

The extension housing itself attaches to the back end of the transmission case. Its function is to enclose and protect the components that extend from the rear of the case such as the output shaft, governor, and speedometer drive assembly (Fig. 8-10). If the mechanic removes this housing from the case, he should also clean, inspect, and service it before reinstallation.

cleaning

Before cleaning the extension housing with the same cleaning procedures used on the case, the technician should first remove the extension housing oil seal. The reason for this is that most solvents or cleaning agents, used to dissolve varnish or shellac, will deteriorate this seal. But if the seal is in good condition and the mechanic is not going to change it, <u>he can wash the housing in a cleaning detergent that will not harm</u> the seal, and then blow it dry with compressed air.

inspection

With the extension housing removed and cleaned, inspect it for the following defects:

1. Cracks or signs of other visible damage.
2. Worn or hard rear oil seal.

Subassembly Cleaning, Inspection, and Service

FIG. 8-10 A typical extension housing.

3. Worn or damaged bushing.
4. Stripped or damaged threads.
5. Warpage at its mating surface.

bushing and oil seal replacement

The extension housing bushing supports the output shaft and the slip yoke of the drive shaft. This bushing is a precision type bearing that requires no reaming or finishing after installation.

To replace a worn or damaged bushing along with the rear oil seal, follow this typical procedure:

1. With a chisel or suitable puller, remove the oil seal from the housing (Fig. 8-11).
2. With a bushing chisel, cut along the bushing's seam until the chisel breaks through the bushing wall. Pry the loose ends of the old bushing up with an awl or needle-nose pliers and remove the bushing. Or,
3. With the handle and correct-size head, press or drive the old bushing from the housing (Fig. 8-12).

To install the new bushing:

1. Inspect both the bushing and seal bores for cracks, wear, or damage.

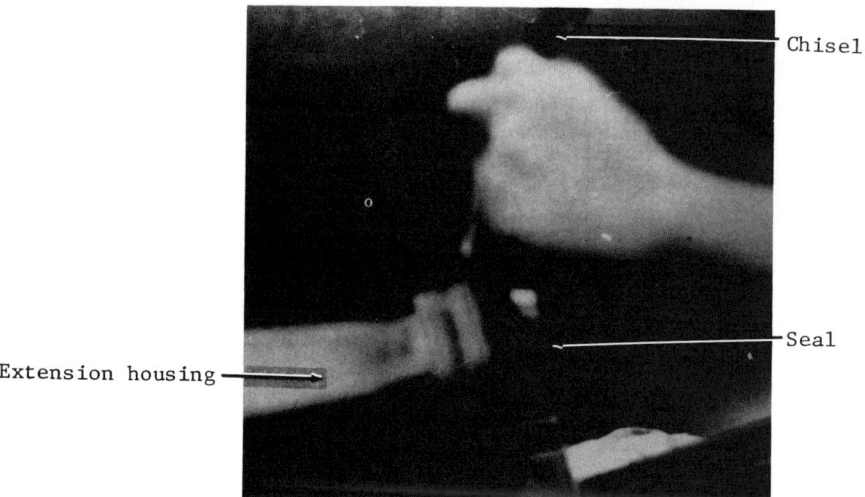

FIG. 8-11 Using a hammer and chisel to remove the extension housing seal.

2. Position the new bushing into its bore. Be sure to align, where used, any bushing lubrication holes or grooves with the corresponding passages or holes within the housing.

3. Using a press or hammer, handle, and correct-size installing head install the bushing into its housing bore (Fig. 8-13). Be certain

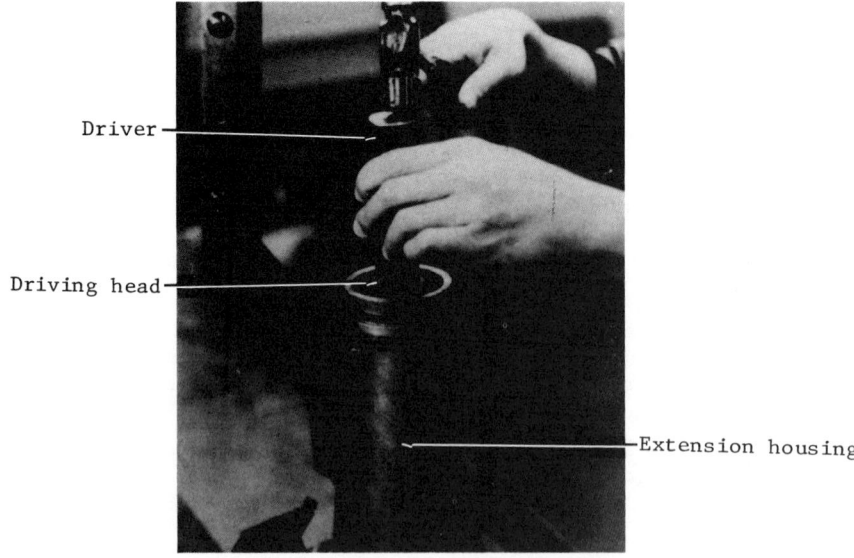

FIG. 8-12 Removing the extension housing bushing.

Subassembly Cleaning, Inspection, and Service

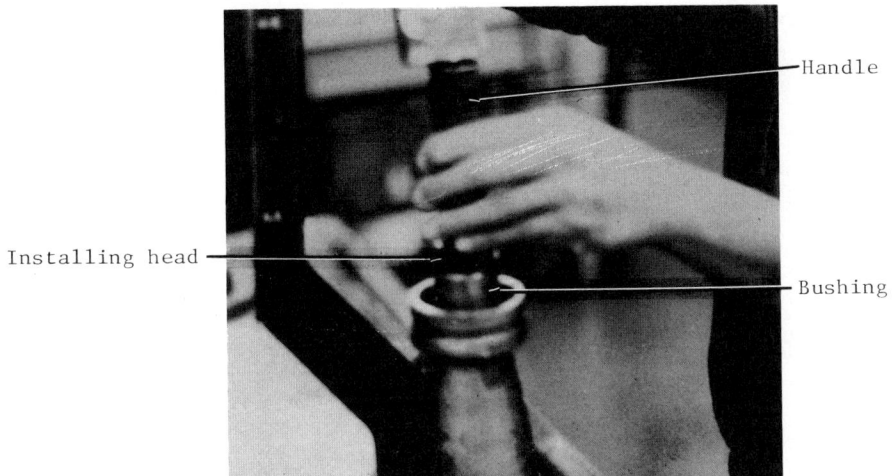

FIG. 8-13 Installing the extension housing bushing.

to press the bushing in far enough so that its end is flush with the top of the bore, or the distance specified by the manufacturer.

4. Carefully insert the slip yoke that rides in the bushing, and check it for free rotation within the bushing. If the yoke will not go into the bushing or is hard to turn, the pressing operation damaged the bushing, and you will most likely have to replace it.

5. Remove the slip yoke. Using a hammer, and the driver, install a new oil seal into the housing. Be certain that the seal firmly seats into the housing counterbore, and its lip faces inward or toward the bushing (Fig. 8-14).

Drums

A drum may serve several functions within the automatic transmission. For example, it can act as the housing for a multidisc clutch assembly used to connect a planetary gear train member to the input shaft, or along with a brake band, it can hold a planetary gear train member stationary.

Since the drum can serve several different functions, each independent of the other, its cleaning, inspection, and service requirements will also be slightly different. To differentiate the different drum usages and their service requirements, this section will refer to them as either a clutch drum or brake drum.

clutch drum disassembly

Before cleaning and inspecting a typical clutch drum, the mechanic must first remove its internal components. To accomplish this, the mechanic should follow this procedure:

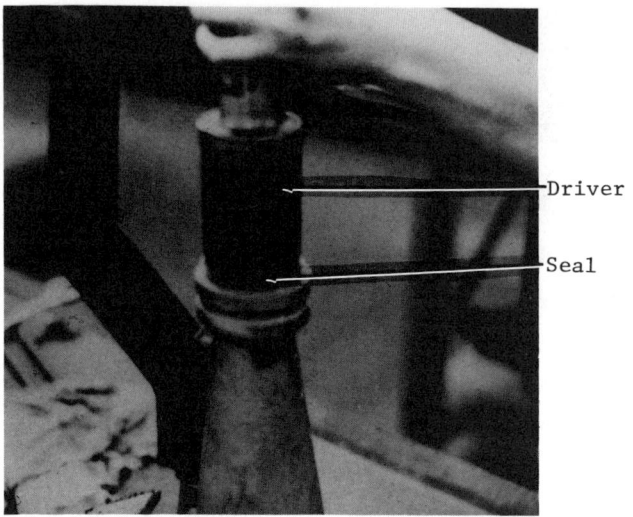

FIG. 8-14 Installing the extension housing seal.

1. Remove the retainer ring with a screwdriver and lift the cover or flange assembly from the clutch drum (Fig. 8-15).
2. Remove the clutch hub's rear thrust washer.
3. Lift out the clutch hub. Remove the clutch pack and the hub's front washer. Tie these components together with a piece of fine wire as an aid to reassembly.

FIG. 8-15 Removing the retainer snap ring from a typical clutch drum.

Subassembly Cleaning, Inspection, and Service

4. Install a spring compressor (refer to Section 1) over the clutch spring retainer, and slowly compress the springs until you can easily remove the retainer snap ring with a scribe, screwdriver, or snap-ring pliers (Fig. 8-16). Carefully release the tool's pressure on the springs and remove the retainer, and the springs.

5. Lift up on the piston with a twisting motion to remove it from the drum. Note: On some clutch assemblies, it is necessary to direct air pressure to the clutch apply port in order to force the piston from its bore. Remove the inner piston seal from the drum guide and the outer seal from the piston.

clutch drum and component cleaning

To prepare the clutch drum and its various components for inspection, wash or clean all the metal parts as outlined under the section on case cleaning. Do not wash or submerge any rubber seals or friction plates in solvent or a varnish dissolving cleaning agent. These materials will deteriorate the rubber seals and friction plates.

clutch drum and component inspection

With all the components clean and dry, follow these steps:

1. Check the drum bushing for excessive wear or scoring.
2. Check the steel residual check ball in the clutch drum. Be sure that this ball is free to move in its bore and that the orifice leading to the front of the drum is open. Note: On some clutch assemblies this check ball's location is in the piston itself. In either case, if

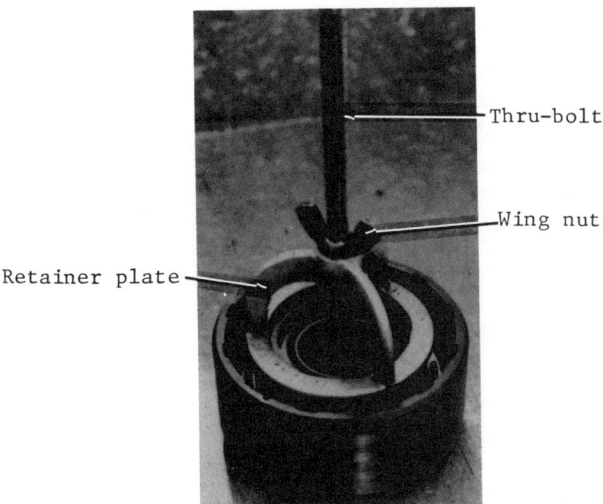

FIG. 8-16 Using a clutch-spring compressing tool to compress the clutch-piston return springs.

this check ball is loose enough to come out or not loose enough to rattle, replace the clutch drum or the piston assembly. <u>Do not attempt to replace or restake the check ball.</u>

3. Check the fit of the clutch flange (cover) in the drum slots. There should be no appreciable radial play between the two components.
4. Check the drum for burred, scored, or damaged thrust surfaces.
5. Check the inside of the drum for rough or badly worn clutch plate slots or grooves.
6. Inspect the drum for nicked or deeply worn sealing ring surfaces and for obstructed or restricted fluid passages.
7. Check the piston bore for wear and scoring.
8. Check the piston for cracks, wear, or damage.
9. Inspect the piston return springs for wear, damage, or signs of overheating.
10. Check the friction plates for excessive wear; cracked lining material; charred, burned, or glazed lining; pitted or scored lining; or distortion.
11. Check the steel plates for distortion, surface scuffing or scoring, and damaged drive lugs.
12. If the clutch drum or any of its various parts are defective, replace them as necessary.

<u>clutch drum assembly</u>

1. Prepare the steel and friction plates as outlined in this section under soft parts--inspection and service.
2. Install a new piston inner seal over the hub of the clutch drum.
3. Install a new outer seal in the clutch piston. Refer to the portion of the section entitled "Transmission Soft Parts" for further details on the installation of these inner and outer seals.
4. Lubricate these seals and the piston bore in the drum generously with transmission fluid. Reinstall the piston in the clutch drum with a twisting motion. <u>Note</u>: If the inner and outer seal are the lip type, it may also be necessary to use a feeler gauge strip as an aid in installing the piston. If this is necessary, slowly work the gauge strip around the circumference of the piston to avoid damage to the outer seal (Fig. 8-17).
5. Position all the springs on the installed piston. Then place the retainer in position on the springs.
6. Using the compressor tool shown in Fig. 8-16, slowly depress the retainer plate and the springs far enough to permit the installation of the retainer snap ring in its groove on the clutch hub guide.
7. Reinstall the clutch hub thrust washer with its lip facing down or toward the clutch drum. Install the clutch hub over this washer.

Subassembly Cleaning, Inspection, and Service   237

FIG. 8-17 Using a feeler gauge as an aid in installing a clutch piston.

8. Install the steel clutch plates and faced friction plates alternately, beginning with a steel plate.

9. Install the rear, clutch hub thrust washer with its flange upward or toward the cover flange. Then, reinstall the cover flange assembly and secure with its snap ring (Fig. 8-18). <u>Note</u>: When installed, the openings in the snap ring should be adjacent to one of the lands of the clutch drum.

FIG. 8-18 Installing a clutch-drum retainer snap ring.

10. Check the clutch assembly by turning the clutch hub to be certain that it is free to rotate. <u>Note</u>: On some types of clutch assemblies, it is also necessary to check clutch plate operating clearance. When in doubt as to whether this check is necessary, refer to the appropriate transmission service manual.

Brake Drum Cleaning and Inspection

Before inspecting a brake drum for defects, clean it as outlined in the section on case cleaning.

With this drum clean and blown dry, inspect it for the following defects:

1. Worn, scored, or damaged thrust washers.
2. Worn or damaged splines.
3. Worn or damaged drive lugs.
4. Worn, scored, or damaged band application surface.
5. Worn or scored bushing.
6. Cracks or damaged drum surfaces.

<u>brake drum service</u>

If the drum is still usable, perform the following service operations on it:

1. Remove and replace the drum bushing, if so equipped, using the tools and procedure as outlined under the section on transmission case or extension housing service.
2. Deglaze the portion of the drum where the band rides to restore its friction characteristic. For instance, for drums that use a paper-lined band, you deglaze with a 120-180 grit sandpaper or emery cloth by sanding around the drum but never in a front to back direction (Fig. 8-19). For drums that have an asbestos-lined band, you deglaze with 40-60 grit emery cloth or sandpaper by sanding the drum front to back (Fig. 8-20).

One-Way (Overrunning) Clutches

Automatic transmissions have one or more one-way clutches for the main purpose of holding a component stationary in one direction; but when necessary, this device allows the same component to rotate freely in the opposite direction. The stator one-way clutch, within a torque converter, is a good example of one practical use for this device. In this situation the one-way clutch locks the stator against any counterclockwise rotation, but it permits the stator assembly to turn freely in a clockwise direction.

Subassembly Cleaning, Inspection, and Service

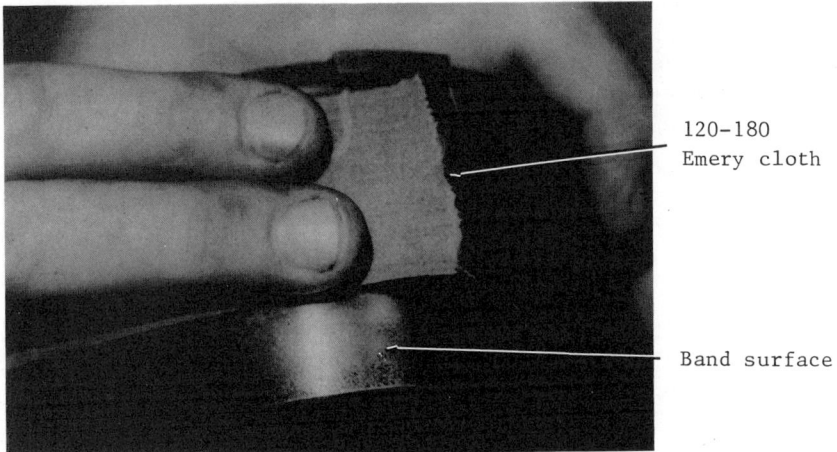

FIG. 8-19 Deglazing a drum that uses a paper-lined band.

Within the transmission itself, this clutch acts as a holding and free wheeling mechanism for a given planetary gear train member. In other words, the one-way clutch also holds a planetary member against any counterclockwise rotation, but it permits the member to spin freely in a clockwise direction.

There are two basic types of one-way clutches used to hold a planetary member, the sprag and roller. Although these clutches each have a different design, both of them do the same job but in a slightly different manner. Because of their somewhat different design, each one will have special inspection and service requirements.

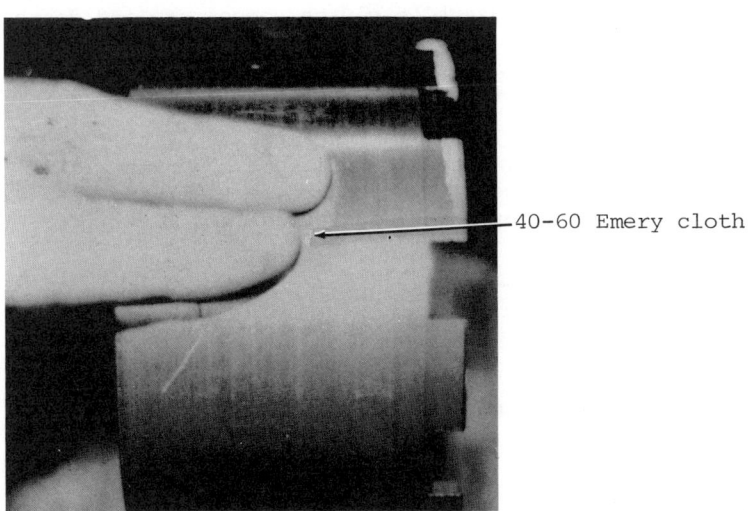

FIG. 8-20 Deglazing a drum that uses an asbestos-lined band.

FIG. 8-21 Inspecting the components of a typical sprag clutch.

cleaning

Because of the size and, in the case of the roller-type clutch, the number of individual parts, you must follow certain precautions when cleaning a one-way clutch assembly to prevent the loss of parts. With this in mind, take the disassembled clutch components and do the following:

1. Place them in a small parts basket and submerge them in an agitated solvent tank or cold tank solution.
2. After soaking the parts for a period of time in the solvent bath, blow them dry, using low pressure compressed air. <u>Note</u>: The parts cleaned in a cold tank require neutralizing with water before blowing them dry with compressed air.

inspection--sprag type

To inspect a sprag type clutch (Fig. 8-21) check for these conditions:

1. Scored or damaged surfaces on both the inner and outer races.
2. Excessively worn or damaged sprag segments.
3. Bent or damaged sprag spring retainer.

inspection--roller type

To inspect a roller type clutch (Fig. 8-22) check for these conditions:

Subassembly Cleaning, Inspection, and Service

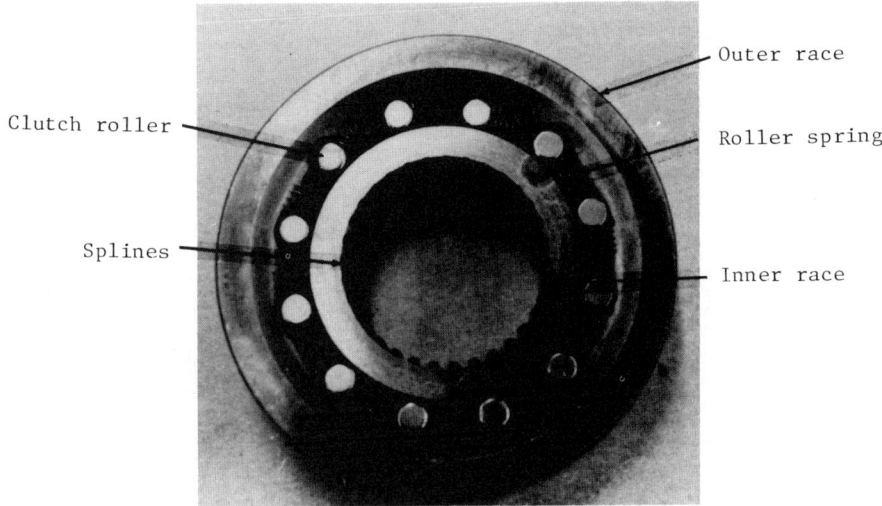

FIG. 8-22 Inspecting the components of a common, roller-type overrunning clutch assembly.

1. Scored or damaged surfaces and signs of brinelling on the inner race.
2. Scored or damaged roller ramp surfaces on the outer race.
3. Flat spots, chipped edges, or signs of excessive wear on each of the clutch rollers.
4. Bent or damaged spring retainer.
5. Worn, bent, or damaged roller springs.
6. Wear or damage to splined areas.
7. Stripped or damaged threads in any tapped holes in the races.

Servos and Accumulators

The servo and accumulator assemblies in many transmissions look very much alike, but they each have different functions (Fig. 8-23). The servo is responsible for applying a band around a drum. The accumulator cushions the application of a band or clutch by absorbing servo or clutch-apply pressure.

Not only do the servo and accumulator components resemble one another, their service requirements are also about the same. The service procedures that apply to a servo assembly for the most part apply to the accumulator as well.

FIG. 8-23 Typical servo and accumulator assemblies.

cleaning

Because of the size and number of smaller components, you must follow certain precautions when cleaning servo and accumulator components to prevent the loss of parts. With this in mind, take the disassembled servo or accumulator components and perform the following steps:

1. Remove all rubber seals from them because most cleaning agents will cause seal deterioration.
2. Place all pieces in a small parts basket and submerge them in an agitated solvent tank or a cold tank solution.
3. After soaking the parts for a period of time in the solvent bath, blow them dry, using low pressure compressed air. Note: Any parts cleaned in a cold tank require neutralizing with water before blowing them dry with compressed air.

inspection

After cleaning and drying all the components along with their bores, inspect for the following defects:

1. Cracked, scored, or worn servo or accumulator bores.
2. Worn or damaged pistons and piston rods.
3. Worn, damaged, or hard seals. Discard all the seals after this inspection, and install new seals on all the components as outlined later in this section.
4. Cracked, broken, distorted, or worn return springs.

Subassembly Cleaning, Inspection, and Service

5. Restricted piston movement in its bore.
6. Damaged or distorted servo- or accumulator-cover mating surfaces.
7. Closed or obstructed case or housing fluid passages.
8. Frozen or stiff movement of any servo apply linkage or lever. If necessary, remove the linkage or lever and inspect the assembly for worn or frozen support bearings, worn support shaft, or worn or damaged lever. Replace worn or defective parts and reassemble the linkage.
9. Worn or damaged band adjusting screw, adjusting screw threads, and adjusting screw lever threads. Before reassembly into their respective bores, replace any servo or accumulator parts found to be defective.

Governors

The governor assembly is a device that produces a hydraulic pressure signal which is in proportion to vehicle speed. This signal is responsible for automatically upshifting the transmission at a given road speed. At a given road speed, the pressure signal from the governor valve will be great enough to open a shift valve that directs pressure to a clutch or band to upshift the transmission. It should be very obvious then that whenever the mechanic removes this device during a transmission overhaul, he must give it very special attention, or the transmission may not upshift properly.

The location of the governor itself is within the extension housing area. On some transmission styles, the assembly mounts to and rotates with the output shaft. While on still others, the transmission case supports the governor assembly, and a gear on the output shaft spins it.

Because of the different methods of mounting and driving governor assemblies, their construction and inspection and service requirements will vary among transmission manufacturers. When servicing a given governor assembly, therefore, it may be necessary to refer to the transmission service manual for specific instructions.

cleaning the typical governor assembly

With the governor assembly removed from the transmission and disassembled, do the following:

1. Place all of its components into a small parts basket and submerge them in an agitated solvent tank or a cold tank solution.
2. After soaking the parts for a period of time in the solvent bath, blow them dry, using low pressure compressed air. Note: Any parts cleaned in a cold tank require neutralizing with water before blowing them dry with compressed air.

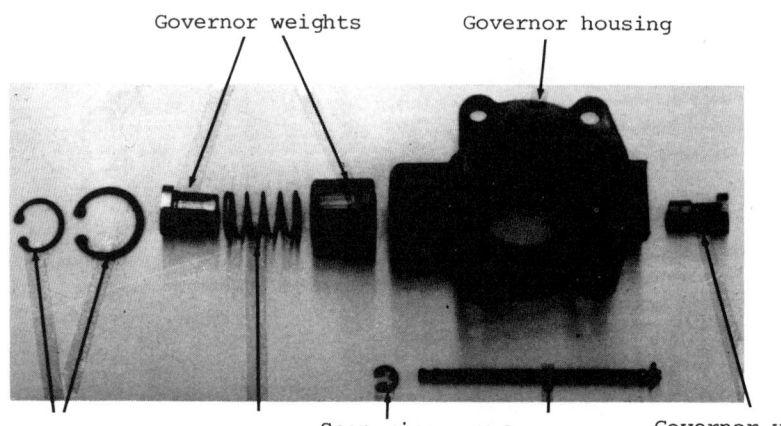

FIG. 8-24 Inspecting the components of a typical governor assembly.

component inspection

With all governor parts clean and dry, inspect them for the following defects (Fig. 8-24):

1. Scored, worn, or rusted valve housing bore.
2. Scored, nicked, or rusted governor valve spool.
3. Binding or sticking of the governor valve in its bore. With the valve clean and dry, it must slide freely back and forth within its bore.
4. Scored, nicked, or rusted governor weights.
5. Binding or sticking of the governor weights in their bores. With the weights clean and dry, they must slide freely back and forth in their bores.
6. Broken or distorted governor spring.
7. Broken snap rings and damaged snap ring grooves.
8. Damaged or distorted governor housing mating surfaces.
9. Damaged or restricted governor housing screen. Note: The location of this screen may also be in either the valve body or main transmission case.
10. Worn or broken sealing rings.
11. Plugged or restricted governor fluid passages in the governor housing, output shaft, and transmission case.

component service

1. Replace all defective components before reassembly.

Subassembly Cleaning, Inspection, and Service

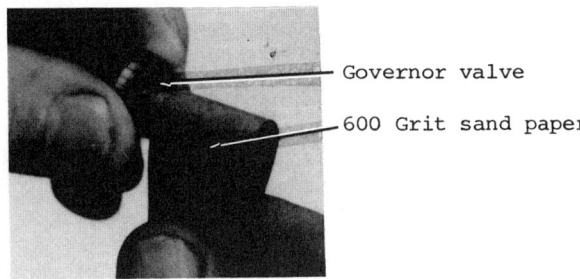

FIG. 8-25 Polishing a governor-valve spool with fine sandpaper.

2. Polish a sticky governor valve or weight, using 600 grit sandpaper or emery cloth (Fig. 8-25). Polish evenly around the circumference of the valve spool or weight. <u>Do not round off any square edges or corners</u>. These edges are necessary to cut through any varnish or shellac that may build up within the valve's or weight's bore.

3. Clean thoroughly any polished valve or weight and blow dry. Wet them with clean transmission fluid prior to their installation in their respective bores.

4. Install all snap rings or C-clips making sure to secure them tightly into their respective grooves.

Planetary Gear Trains

All domestic automatic transmissions have some form of planetary gear train. This device is responsible for actually providing the vehicle with its various forward gear ratios and a reverse. To provide these ratios, the gear train will always consist of one or more ring gears, carriers, and sun gears.

Because of its design, function, and number of components the planetary gear train is a very expensive transmission hard part. The technician should do everything possible, therefore, to reuse it when overhauling the transmission. But like so many other transmission components, there are various types of planetaries with each type having its own inspection and service requirements. Therefore, this section will attempt to explain to the reader the inspection and service techniques only for one design (Fig. 8-26). If in doubt as to how to service another type of planetary gear train, always refer to the appropriate transmission service manual.

<u>cleaning</u>

To clean the planetary gear train components, do the following:

1. Wash or soak the parts as outlined under the section on case cleaning and rinse them off with water as necessary.

FIG. 8-26 Inspecting a typical planetary gear train assembly.

2. Blow all components dry using low pressure compressed air. <u>Do not allow the compressed air to spin the pinion gears; this will damage their support bearings.</u>

<u>inspection</u>

With the gear train components clean and dry, follow these steps:

1. Inspect all the planet pinions for wear, nicks, tooth damage, and binding or looseness in their support bearings.
2. Check the end-play clearance of each planet pinion gear (Fig. 8-27). This clearance should be .006 to .030 inch.
3. Check the input sun gear for wear, nicks, or tooth damage. Inspect the input sun gear's thrust washer for damage.
4. Inspect the low sun gear for wear, nicks, or tooth damage.
5. Inspect the ring gear for wear, nicks, or tooth damage. Check the ring-gear clutch-hub splines for excessive wear or damage.
6. Inspect the carrier for cracks or damage. Check its attached output shaft for worn or damaged splines, obstructed lubrication passages, and worn or nicked bearing surfaces. <u>Note</u>: Individual parts on most planet carriers are now no longer serviceable. Therefore, if the carrier, pinions, thrust washers, or bearings do not pass inspection, replace the carrier assembly.

<u>service</u>

Before reinstalling the planetary gear train into the case, follow these procedures:

Subassembly Cleaning, Inspection, and Service

FIG. 8-27 To check pinion-gear end play, insert a feeler gauge between the end of each gear and its thrust washers.

1. Replace any worn or damaged ring or sun gears, thrust washers, bearings, or carriers.
2. Coat all thrust washers or bearings with Vaseline to hold them in place.
3. Lubricate all other components with clean transmission fluid.

Transmission Shafts

While the average automatic transmission has two rotating shafts, others will have as many as four. But no matter how many shafts the transmission has, they all do the same thing--they carry engine torque through the various gear train members and then to the drive shaft. An input shaft brings the torque into the transmission and to the planetary gear train. The output shaft delivers this torque from the planetary to the drive shaft (Fig. 8-28).

These shafts, which bushings support within the transmission, operate independently of one another. And under normal conditions, they never touch each other. Furthermore, splines mate each independent shaft with the appropriate gear train member. These splines provide not only a positive method of locking the two components together but also provide a way to quickly separate the two during service operations.

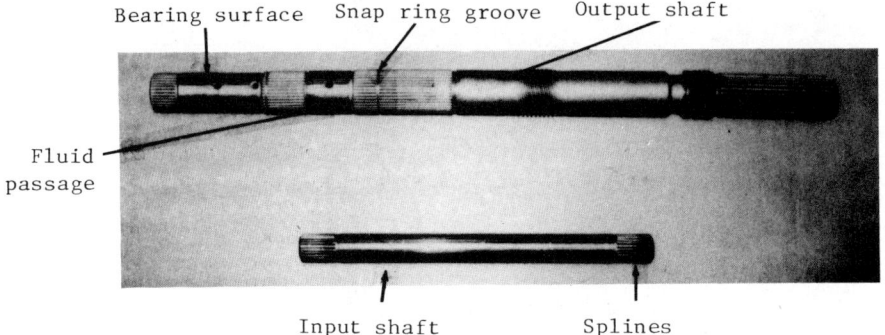

FIG. 8-28 Inspecting a typical input and output shaft.

cleaning and inspection

With the shafts removed from the transmission, clean and dry them following the same procedure as outlined previously for other large hard parts, like the case and planetary gear train. Then, inspect the shafts for these conditions:

1. Worn or damaged shaft splines.
2. Shaft distortion or cracks.
3. Burrs, scores, or other damage to the bearing or thrust washer surfaces.
4. Worn, broken, or damaged sealing rings.
5. Restricted or plugged fluid passages.
6. Worn or damaged snap-ring grooves.
7. Worn or damaged speedometer- or governor-drive gear teeth.
8. Scored, worn, or damaged bushings.

service

1. Replace any shaft that does not pass inspection.
2. Remove and replace any worn or damaged bushings.
3. Replace all sealing rings.
4. Lubricate the shaft generously with clean transmission fluid before reinstalling it into the transmission.

Thrust Washers

The automatic transmission also usually contains a number of thrust washers (Fig. 8-29), which separate the many moving parts within the transmission. The moving components rotate against the thrust washers with the thrust washers themselves absorbing component endwise movement.

Subassembly Cleaning, Inspection, and Service

FIG. 8-29 Typical transmission thrust washer.

cleaning and inspection

With the transmission disassembled, clean and dry all thrust washers as you would any small transmission component like the overrunning clutch or governor. With the thrust washers clean and dry, inspect them for the following defects:

1. Nicked, burred, or scratched thrust washer mating faces.
2. Nicked, burred, worn, or scratched thrust washer bearing surfaces on gears, shafts, or clutch drums.
3. Worn thrust washers. Use a 0- to 1-inch micrometer to measure the thickness of the washer (Fig. 8-30). Compare this reading to the manufacturer's specifications for each thrust washer.

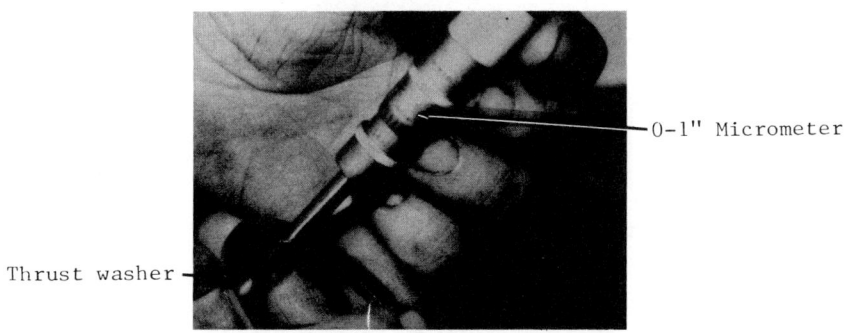

FIG. 8-30 Measuring the thickness of a thrust washer with a micrometer.

service

1. Replace any worn or damaged thrust washers and, if necessary, the component it fits against.
2. To hold the thrust washer in place during transmission assembly, coat its backing with Vaseline.
3. Coat all other thrust washer bearing surfaces generously with clean transmission fluid.

Bearings

The automatic transmission may also contain one or more needle- or ball-type bearings (Fig. 8-31). These devices serve several functions. First, a bearing assembly can act as a thrust washer to absorb endwise movement of components. Second, a bearing can support a rotating shaft, like the output shaft.

cleaning and inspection

No matter which function the bearing serves, clean it as you would a thrust washer and blow it dry with low pressure compressed air. <u>Be careful not to allow the air blast to spin any of the bearing races at high speeds; this can damage the bearing's rollers or balls.</u> With the bearing clean and dry, check the bearings and both races for the following conditions:

1. Binding. Hold one of the bearing races and rotate the other; you should <u>not</u> feel any catching or binding between the two races. If you do, the bearings are still dirty or are defective.
2. Missing or damaged balls, needles, or rollers.
3. Badly worn thrust surfaces.
4. Worn, damaged, or distorted races and bearing cage.

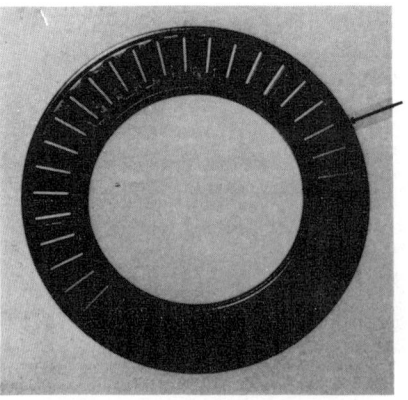

Roller-type bearing thrust washer

FIG. 8-31 Typical roller-type thrust washer.

service

1. Replace any worn or defective bearings.
2. Lubricate a shaft support bearing before installation with clean transmission fluid.
3. To hold needle-type thrust bearing in position during transmission assembly, coat its various components with Vaseline.

Bushings

The bushing serves the same function as a shaft supporting needle or roller bearing (Fig. 8-32). In other words, it also supports rotating shafts and other transmission components. However, the bushing has a much different design that necessitates different inspection and service procedures.

The bushing itself is a precision nonadjustable bearing with a steel backing lined with either copper or a babbit-type material. Once the technician presses the bushing into its bore, there is a given clearance established between this lining and the shaft or component the bushing supports. Over a period of operating time, the lining can wear or become damaged, necessitating a bushing replacement.

cleaning and inspection

Since the bushing presses into another component, the mechanic cleans the bushing at the same time he does the component itself. Once the bushing is clean and dry, inspect it for scoring, pits, and wear. If the bushing requires replacement, inspect the shaft or component that rides in the bushing. It may also show signs of scoring or wear.

In some situations, shaft mounted Teflon or metal sealing rings ride inside the bushing. Wear on these sealing rings can also damage the

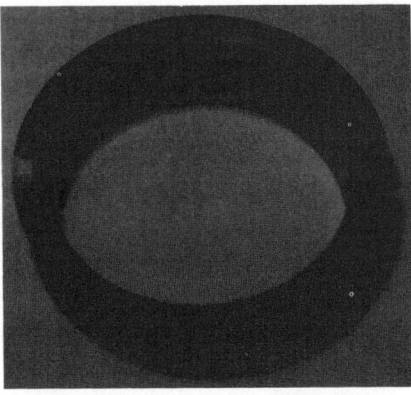

FIG. 8-32 Typical transmission bushing.

FIG. 8-33 Using a bushing chisel to cut through a bushing's wall.

bushing. Therefore, check the bushing closely for ring grooving; if this condition exists, check the shaft seal-ring lands for wear and damage. Replace the bushing, seals, or shaft as necessary.

<u>service</u>

If bushings are worn or damaged, the mechanic should replace them using the tools and equipment mentioned in Section 1. The choice of tools to remove and replace the bushing depends, of course, on its location. To remove a bushing in a closed or blind bore, follow this procedure:

1. Use a cape or bushing chisel and cut along the bushing's seam until the chisel breaks through the bushing wall (Fig. 8-33). Pry the loose ends of the old bushing up with an awl or needle-nose pliers and remove the bushing.

2. To pull a bushing using the threaded remover and cup assembly:

    a. Thread the remover into the bushing as far as possible by hand (Fig. 8-34).

    b. Using a wrench, screw the remover into the bushing 3 to 4 additional turns to firmly engage its threads into the bushing.

    c. Install the cup assembly over the remover and down onto the component that houses the bushing.

    d. Thread the hex nut down on the remover until it contacts the cup.

    e. Tighten the hex nut with a wrench to pull the bushing from its bore.

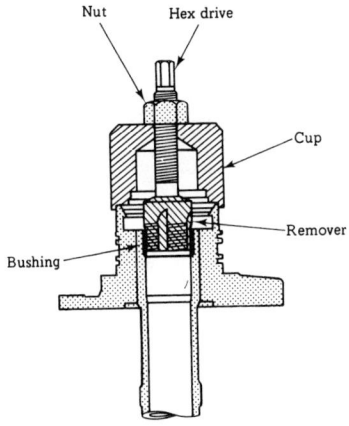

FIG. 8-34 Pulling a bushing using a threaded remover.

3. To press or drive a bushing from its bore using a tool handle and removing head,

   a. Place the correct-size removing head into the bushing, and install the tool handle into the opening in the head (Fig. 8-35).

   b. Drive or press the bushing straight down and out of its bore. Be careful not to cock the tool while it is in the bore.

4. To install a bushing in either a closed or open bore, clean and check the bore for damage; then, follow these steps:

   a. Position the new bushing onto the proper-size installing head.

   b. Install the tool handle into the head.

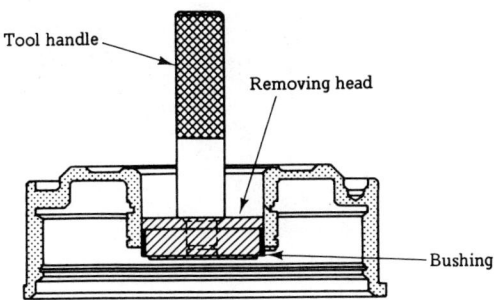

FIG. 8-35 Removing a bushing using a tool handle and removing head.

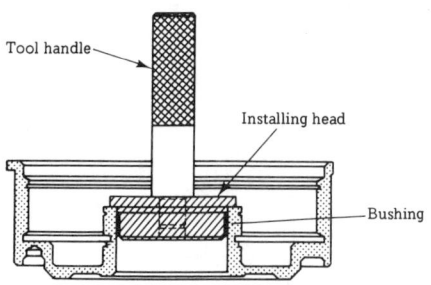

FIG. 8-36 Installing a typical bushing using the tool handle and installing head.

    c. Start the new bushing into its bore, making certain to align all oil holes or grooves in the bushing with the corresponding oil passages in the bushing bore.

    d. Press or drive the bushing into the bore (Fig. 8-36). <u>Be sure the bushing bottoms in its bore, or drive it into the position, specified by the transmission manufacturer.</u>

5. Lubricate the bushing thoroughly with clean transmission fluid before installing the shaft or component it supports. Note: If the component will not go into the bushing or it is hard to turn, the installing process damaged the bushing, and it will most likely have to be replaced.

Valve Bodies

    The inside of the average valve body (Fig. 8-37) contains a great number of springs and valves. Each of these many parts performs a small but important

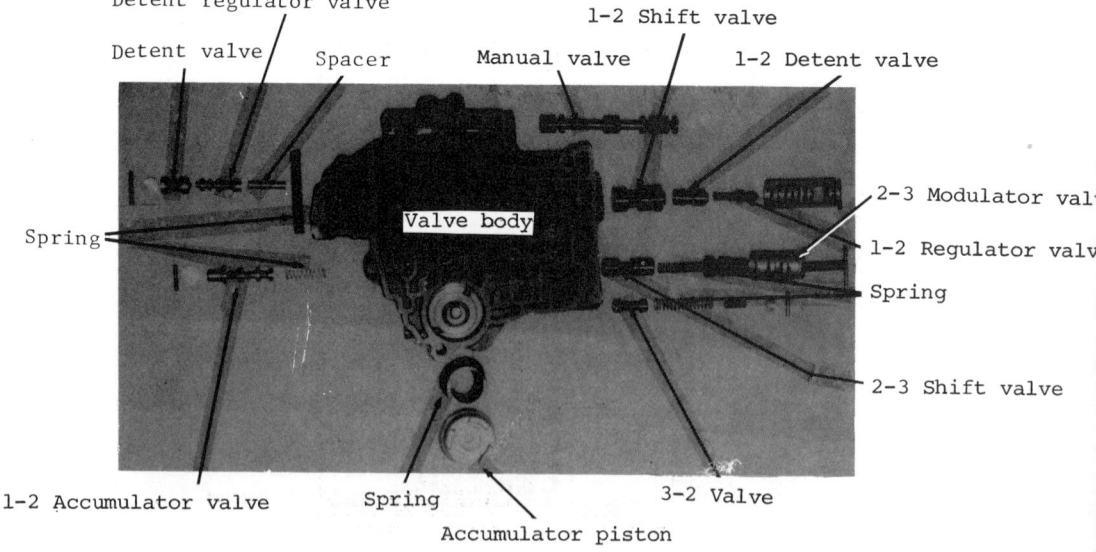

FIG. 8-37 The illustrated parts-breakdown of a typical valve body.

Subassembly Cleaning, Inspection, and Service

function during the operation of the transmission. When working together, they convert the valve body into a hydraulic computer that is responsible for upshifting or downshifting the transmission according to the will of the driver, vehicle speed, and engine load.

Because of its many precision components and the very important job it performs, the mechanic must <u>carefully</u> and <u>thoroughly</u> clean and inspect the valve body as part of every transmission overhaul, or a malfunction may develop. This procedure is usually the most difficult of all overhaul tasks because of the wide variety of valve body designs and variations in internal component structure. When servicing the valve body for any particular transmission model, therefore, the mechanic must follow these important general rules:

<u>disassembly</u>

1. Make certain your hands, tools, and workbench are clean before working on a valve body.
2. Use low pressure compressed air to clean or dry a component. <u>Never use rags for this purpose. A single particle of lint can cause a valve to stick in its bore.</u>
3. Be careful never to drop or nick a valve spool during the service work. A nick can **also** cause a **valve** to stick in its bore.
4. When disassembling the valve body, lay the parts out on a clean surface in the order in which you remove them.
5. Whenever possible, obtain an illustrated parts breakdown for the valve body you are servicing to assist you in reassembling the unit.
6. Using the parts breakdown, identify all disassembled parts; number each spring on the diagram; and install each individual spring on a numbered peg of a spring holder (Fig. 8-38). This procedure will

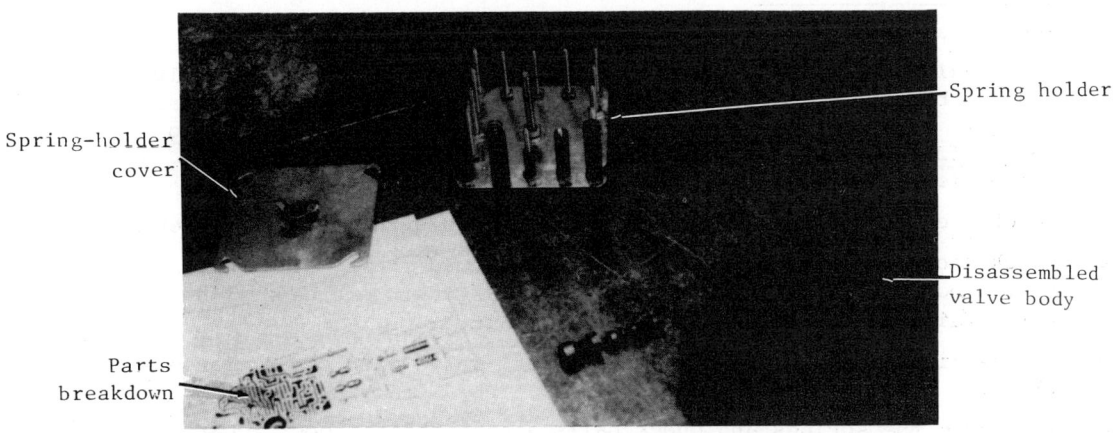

FIG. 8-38 Before cleaning the springs, install them on a numbered-peg, valve-body spring holder.

assist you in identifying all the components and springs during valve body reassembly.

7. Save the old gaskets to match up with the new ones.
8. To clean a typical valve body, follow these steps:
   a. Soak all valve body parts in a cold tank filled with a solution such as carburetor cleaner that dissolves varnish or shellac along with dirt and grease from all the components.
   b. Neutralize or rinse all the parts in quantities of hot water.
   c. After rinsing, immerse the parts in clean mineral spirits or solvent to separate the drops of water from the cleaned parts.
   d. Blow all the parts dry with low pressure compressed air.
9. Inspect a typical valve body and its internal components for the following defects:
   a. Bent manual valve.
   b. Bent or damaged separator plate.
   c. Cracked valve body castings and distorted or damaged mating surfaces.
   d. Plugged or restricted fluid passages in the valve body castings.
   e. Scored or rusted valve bores.
   f. Broken, bent, or worn valve springs.
   g. Scored, cracked, or burred valves, valve sleeves or plugs, and valve lands for shiny areas.
   h. Stuck or worn check valves or balls.

<u>service and reassembly</u>

1. Replace all damaged or worn parts.
2. Shiny valve land areas indicate friction between the affected valve spool and its bore. To correct this problem polish these areas away using 600-grit sandpaper or emery cloth (Fig. 8-39). <u>Be careful not to round off any edges of the valve land when polishing.</u> Rewash and dry the polished valve.
3. Match the old gaskets to the new ones to insure they are exact replacements.
4. Check each clean valve in its bore. It should freely slide back and forth in its bore due to its own weight without sticking or hesitating. If a valve fails this test, polish or replace it.
5. Before final installation, lubricate each valve with clean transmission fluid.
6. Make sure the springs and check balls are in their proper locations.
7. Always follow the manufacturer's torquing sequence and specified torque values when assembling the valve body.

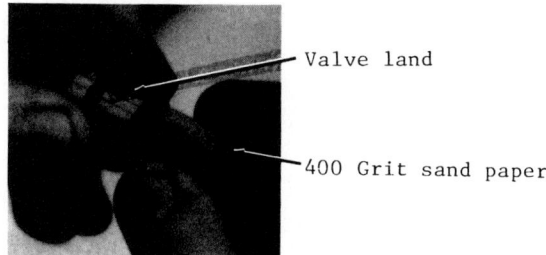

FIG. 8-39 Polishing a valve land with fine sandpaper.

TRANSMISSION SOFT PARTS

The soft parts of the automatic transmission are those less expensive components that the technician usually replaces during a complete transmission overhaul. Soft parts include such items as the clutch plates; bands; metal, Teflon, and rubber sealing rings; metal-clad seals and gaskets. Soft parts are those parts that usually wear out or deteriorate and cause transmission malfunctions.

The only parts mentioned above that are sometimes reused are the clutch steel plates and the bands. But as a rule, it requires more time and effort to prepare steel plates for reuse in the rebuilt transmission than it is worth. And as far as the bands are concerned, the general rule is to replace all paper lined bands, but reuse an asbestos lined band if it is still in good condition. Therefore, with the exception of the steel plates and asbestos lined bands, the mechanic will usually not clean and inspect the soft parts; he will just replace them. The technician will not waste time examining these items unless he is looking closely for the cause of a certain malfunction which he has, up to this point, had some difficulty in finding.

Clutch Plates

steel clutch plate service

As previously mentioned, it is far less expensive to use new steel clutch discs because of the time it takes to prepare old ones for reuse. But if the technician is going to reuse the old plates, he should bead blast or refinish them on both sides with 120 to 320-grit emery cloth or sandpaper (Fig. 8-40). When sanding the plates, rub until sanding marks are visible over the entire area on both sides of each plate. This bead blasting or sanding process removes polished surfaces and restores new roughness to the disc, which is necessary for proper clutch application and break-in. After servicing the plates, the mechanic should clean and then lubricate the steel plates with transmission fluid before installation.

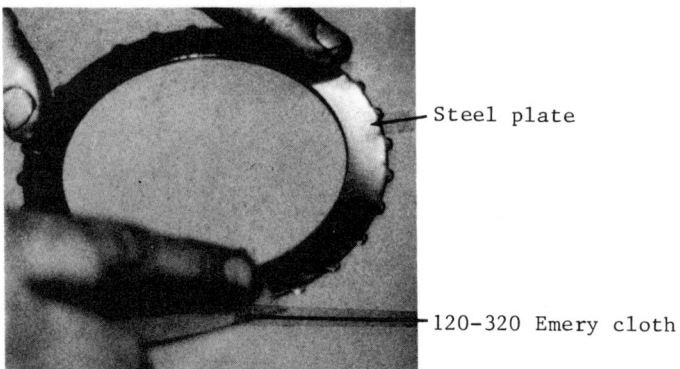

FIG. 8-40 Deglazing a typical steel clutch plate.

friction clutch plate service

The technician should presoak all new friction-type clutch plates in clean automatic transmission fluid for no less than 30 minutes before installation for two reasons: (1) The composition lining on the clutch plates acts as an insulator that holds friction heat on the plate surface, causing premature clutch failure due to surface glazing or burning; and (2) a thoroughly presoaked plate quickly dissipates surface heat through the fluid retained in the composition lining. Note: If for any reason, friction-type plates are reused, they should also be soaked in fluid for a few minutes before installation to prevent plate damage caused by dry friction.

Bands

cleaning and inspection

If a band appears to be serviceable at first glance, wash it in a cleaning detergent and blow it dry with compressed air. Do not soak the band in solvent or cold tank solution, because these chemicals can deteriorate the lining material. Then inspect the band for these conditions:

1. Distortion.
2. Cracked anchor ends.
3. Excessive or uneven lining wear.
4. Burned, charred, or glazed lining.
5. Loose lining.
6. Flaking or pitted lining.

Replace a band that shows any sign of these defects.

Subassembly Cleaning, Inspection, and Service

service

Before installing a new or used band, follow these instructions:

1. For quick surface heat dissipation, soak all new paper and asbestos lined bands in clean transmission fluid for at least 30 minutes.
2. Before reusing an asbestos lined band, scrape the lining with a knife or bearing scraper to remove glaze or carbon (Fig. 8-41).

Metal and Teflon Sealing Rings

Metal and Teflon seals, when installed on rotating shafts and pistons, prevent excessive losses in fluid pressure and flow (Fig. 8-42). Manufacturers install this type of seal in high temperature areas of the transmission, where a positive seal is not necessary, or where a seal is necessary on rotating units.

metal seal installation

When installing a metal seal, follow these procedures:

1. Compare the old sealing ring with the new one to make sure you have the correct size.
2. Never expand a sealing ring any further than necessary to install it in its groove.
3. Make sure the installed sealing ring is free to rotate in its groove.
4. When installing the locking end sealing ring, make sure that the ring ends properly lock together.
5. Always lubricate the sealing rings with clean transmission fluid before installing the rings over or into the component they seal.

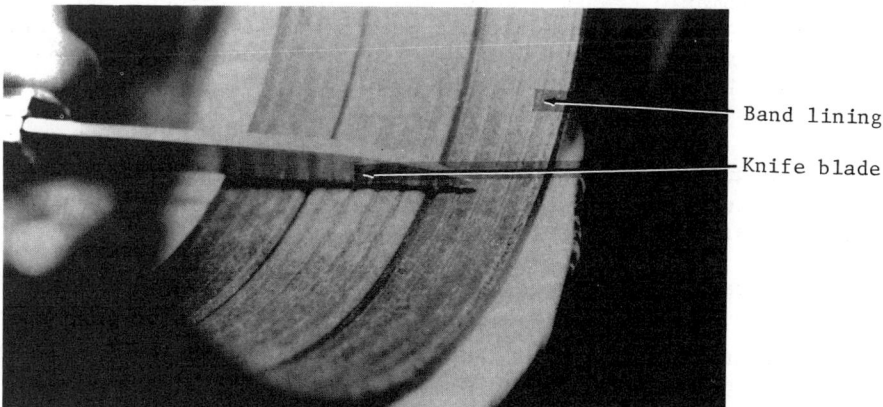

FIG. 8-41 Scraping the carbon and glaze off an asbestos-lined band.

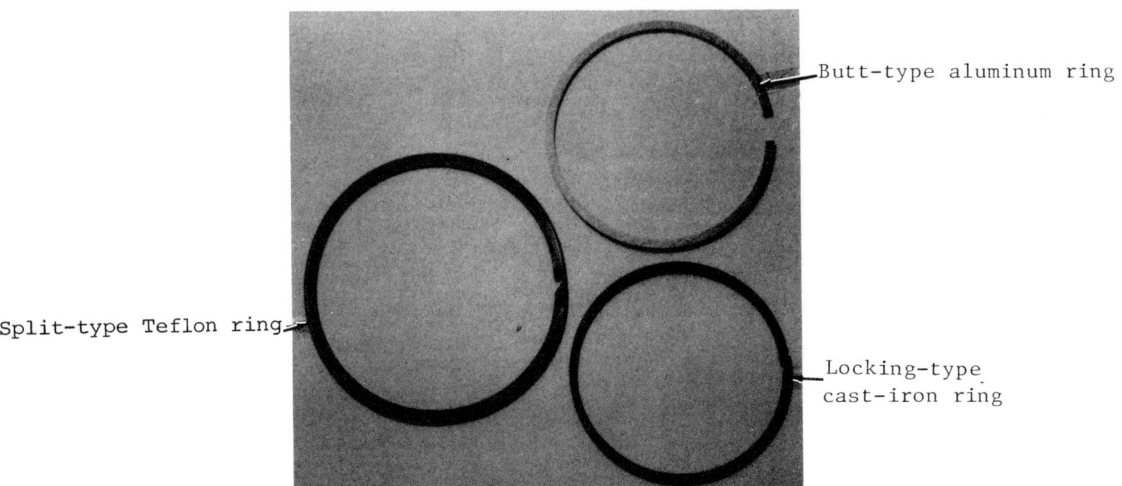

FIG. 8-42 Typical metal and Teflon sealing rings.

Teflon seal installation

1. Always replace Teflon seals with the metal type when overhauling the transmission.
2. While replacing the Teflon seals with the metal type, use the same procedures as outlined under the section on metal rings.

Rubber Sealing Rings

The mechanic will encounter three types of neoprene rubber sealing rings in automatic transmissions, the O-ring, lathe-cut, and lip seal (Fig. 8-43). All of these devices prevent fluid pressure and flow losses around such components as clutch, servo, and accumulator pistons. Although the service procedures for all the seals are nearly the same, the lip seal requires special attention during its installation.

O-ring and lathe-cut seal installation

When installing an O-ring or a lathe-cut seal, follow these general instructions:

1. Compare the old seal with the new one to make certain you have the same size and type.
2. Never expand the seal any farther than necessary to install it in its groove.
3. Make certain the seal fully seats in its groove.
4. Always generously lubricate the seal and the bore it fits into with transmission fluid before attempting to install the sealed component.

Subassembly Cleaning, Inspection, and Service

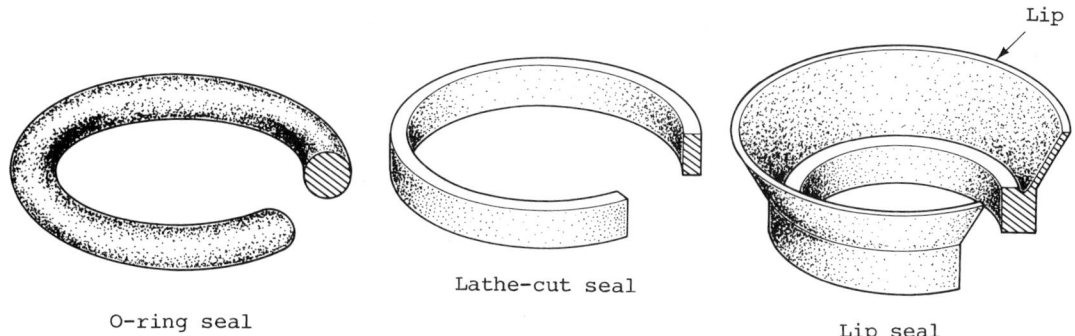

FIG. 8-43 Typical O-ring, lathe-cut, and lip seals.

into the bore. Note: Some mechanics prefer to use petroleum jelly, mineral oil, or Door-Ease on the seal instead of fluid as an aid to piston installation.

lip-seal installation

When installing lip seals be sure to follow these instructions:

1. Always adhere to the installation procedures set forth earlier for O-ring and lathe-cut seals.

2. Always install the seal so that the lip faces toward the fluid source. Note: Some clutch assemblies will use two of these seals, one inside the drum and the other around the outer circumference of the piston. The lips of these two seals must face down into the drum or toward the fluid source that will apply the clutch piston. On another type of clutch assembly that has three lip seals, the fluid source enters the inside of the drum between the lower and two upper seals. Therefore, the lowest seal's lip faces upward toward the fluid source, and the lips of the inner and outer piston seals face downward.

Metal Clad Seals

The metal clad seal is nothing more than a neoprene lip seal encased in a metal housing (Fig. 8-44). This seal design also controls fluid leakage around rotating components such as the converter hub and output shaft.

installation

To install a metal clad seal properly, follow these general instructions:

1. Always install its lip facing toward the fluid source.
2. If the seal's backing does not have a rubber or resin coating, coat the steel backing with nonhardening sealer just prior to installation.

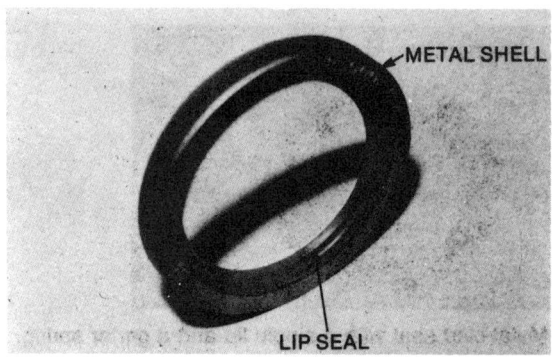

FIG. 8-44 Common-type metal-clad seal.

3. Always use a seal driver to install the seal into its bore.
4. After installation, always lubricate the seal's lip with clean transmission fluid.

Gaskets

Gaskets are a necessary sealing device, found in many locations, in the automatic transmissions. These devices provide a positive seal between two components, thus preventing losses of fluid pressure and flow.

When working with transmission gaskets, follow these simple rules:

1. Never reuse an old gasket.
2. Always compare the old gasket to the new one to make sure you have the correct one.
3. Never use any form of sealer on any transmission gasket.
4. Always clean off the gasket mating surfaces thoroughly.

TRANSMISSION COOLING SYSTEM INSPECTION AND SERVICE

Every automatic transmission has some form of fluid cooling system. This system, as its name implies, reduces the temperature of the transmission fluid, thereby increasing its overall life. Therefore, if this system should fail to function properly, the fluid's life is cut short, which can result in severe damage to the transmission.

Subassembly Cleaning, Inspection, and Service   263

There are three basic types of fluid cooling systems: a water cooled type, an air cooled type, or an after market air cooled type. The water or coolant type system (Fig. 8-45) includes two cooler lines and a fluid cooler located in the lower or side tank of the radiator. The air cooled design consists of a finned converter and a ducted or drilled converter housing (Fig. 8-46). The after market, air cooled system uses a special finned, auxiliary cooler (Fig. 8-47) connected to the transmission by two cooler lines. Manufacturers also produce special finned fluid pans to increase the fluid capacity of the transmission and cool the fluid normally contained in the pan.

Before a mechanic reinstalls the rebuilt transmission into the vehicle, he should inspect and service the transmission's cooling system. The reason for this is twofold. First, the system may now contain foreign material, metal and friction plate particles, carried into the system by the fluid. These particles, if allowed to remain inside the cooling system, will damage the newly rebuilt transmission as the new fluid carries them back to the transmission. Second, the system may be partially or completely restricted, which would cause the new fluid to overheat and eventually cause damage to the transmission because of improper lubrication. Furthermore, if the water cooled system has internal or external leakage, this can damage the transmission by allowing coolant to enter the transmission or by permitting transmission fluid to leak out.

Inspection

water-cooled type

Referring to Fig. 8-45, inspect this system for the following defects:

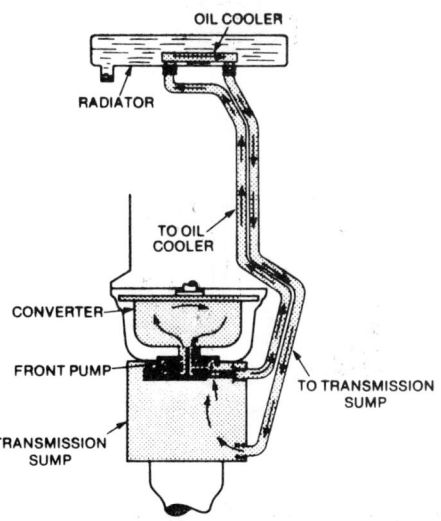

FIG. 8-45 Typical water-type fluid-cooling system.

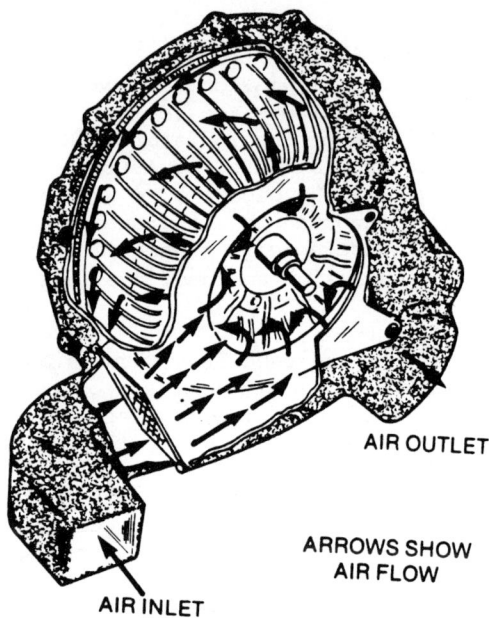

FIG. 8-46 Typical air-cooled fluid-cooling system.

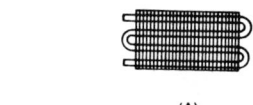

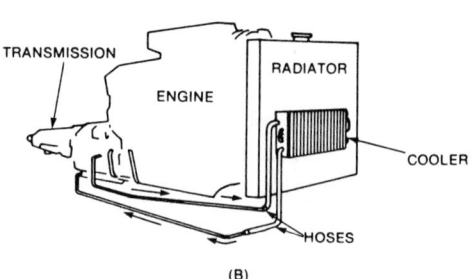

FIG. 8-47 An auxilary cooler is shown at A. B shows the cooler connected directly to the transmission's cooler lines.

Subassembly Cleaning, Inspection, and Service

1. Kinked cooler lines.
2. Cracked, broken, or leaking lines.
3. Deteriorated flexible cooler hoses and broken clamps.
4. External leakage in the radiator.
5. Internal leakage in the transmission cooler. <u>Note</u>: If the cooler lines or transmission, upon teardown for overhaul or at any time, contains <u>any</u> radiator coolant, perform this cooler leakage test:

    a. Remove the radiator cap to release cooling system pressure on the cooler itself.

    b. Disconnect both cooler lines at the radiator, and cap one of them.

    c. Attach an air coupling or valve to the other cooler fitting.

    d. Apply 25 to 50 psi air pressure to the cooler and check coolant for air bubbles.

    e. If the cooler will not hold air pressure or the coolant shows the presence of bubbles, replace or repair the cooler.

6. Restricted or plugged cooler lines or cooler. <u>Note</u>: Make this check while you are flushing the system, using the procedure outlined under cooler flushing.

<u>air-cooled type</u>

Referring to Fig. 8-46, inspect this system for the following defects:

1. Cracked, loose, or damaged air fins on the torque converter.
2. Plugged or restricted air ducts or ports in the converter housing.

<u>after-market type</u>

Referring to Fig. 8-47, inspect this system for the following problems:

1. Kinked cooler lines.
2. Cracked, broken, or leaking lines.
3. Deteriorated flexible cooler hoses and broken clamps.
4. External cooler for leaks.
5. External cooler fins for obstructions.
6. Restricted or plugged cooler lines or external cooler. <u>Note</u>: Check this type of system for restrictions also while flushing the system.

Service

<u>line repair</u>

If a cooler line becomes kinked or badly damaged, it is better to replace it whenever possible. But a small pin hole leak can be repaired by using a

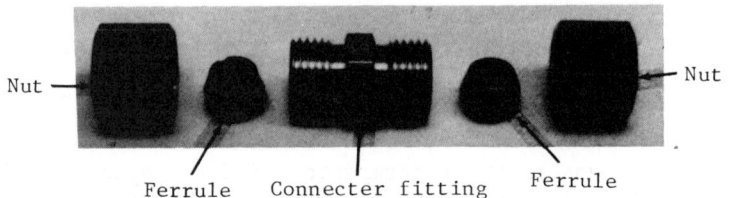

FIG. 8-48 A connecter assembly used to repair a leak in a transmission cooler line.

connecter fitting or a piece of high-pressure hose. To repair a leaky line using a connecter (Fig. 8-48), follow these steps:

1. Using a tubing cutter, cut away the damaged section of the tube (Fig. 8-49). If the damaged area of the tube is longer than the distance inside the connecter where the cut-off end of each tube will bottom in the connecter, use a pressure hose to repair the damage. Otherwise, the repaired cooler line will no longer be the correct length.

2. Install a nut and ferrule over each cut-off end of the tubing.

3. Install one of the tubing ends into the connecter, making sure to push it in as far as it will go. Slide the ferrule against the connecter, and run the nut down on the connecter's threads finger tight.

4. Using two wrenches, tighten the nut securely. Position one of the wrenches on the connecter and the other on the nut while tightening the nut to prevent damage to the line or fitting (Fig. 8-50).

5. Repeat the same process on the other portion of the line.

6. Check the line and connecter for leaks by operating the transmission.

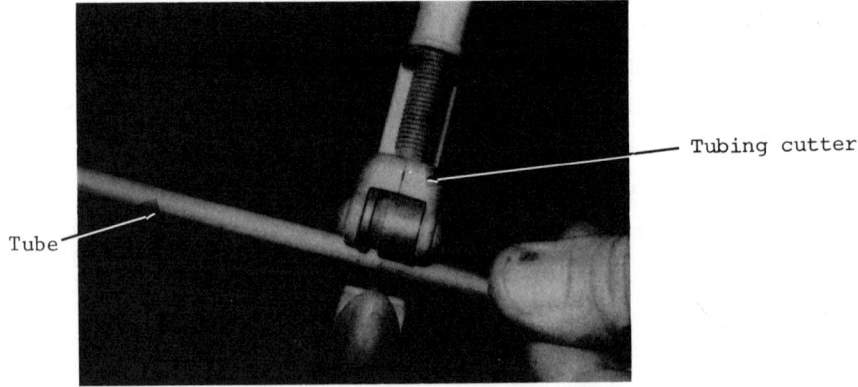

FIG. 8-49 Cutting out the damaged section of a cooler line using a tubing cutter.

Subassembly Cleaning, Inspection, and Service   267

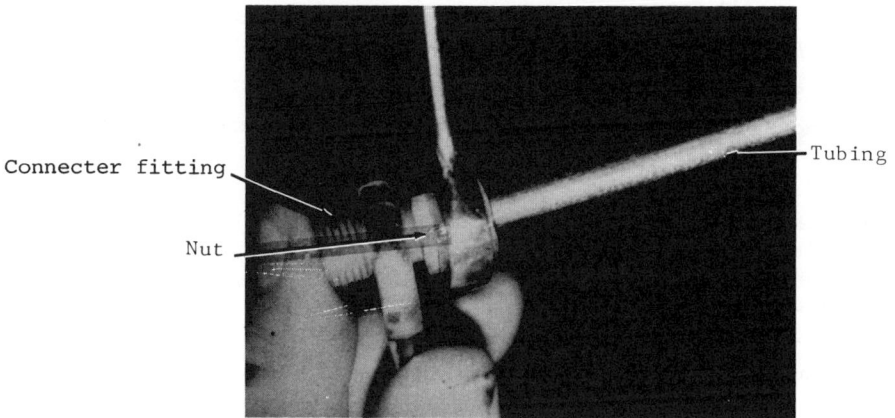

FIG. 8-50 Using two wrenches, tighten the connecter nut securely.

To repair a damaged cooler line using high pressure neoprene hose:

1. Using a tubing cutter, remove the damaged section of line.
2. With a flaring tool (Fig. 8-51), double-flare each cut off end of tubing. The flare will prevent the hose from slipping off the line after the clamp is tight.
3. Install a clamp loosely over each end of tubing.
4. Cut a piece of high pressure neoprene hose to a length approximately 2 inches longer than the piece of damaged pipe.

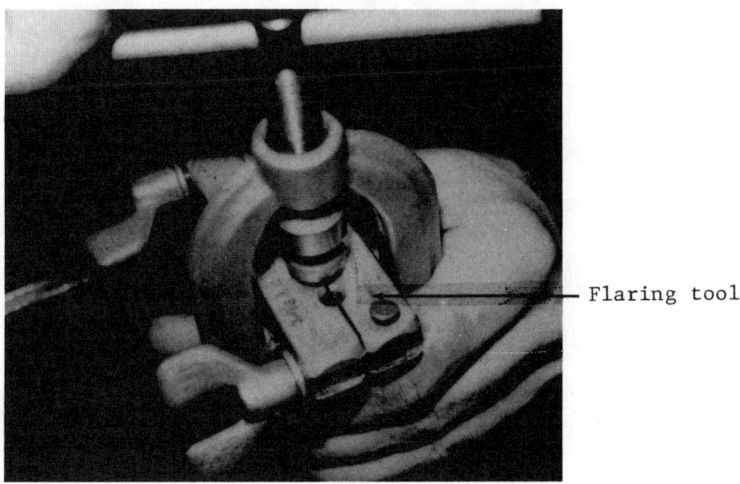

FIG. 8-51 Flaring the end of a cooler line, using a flaring tool.

5. Insert each end of the hose over the double-flared ends of the line. Make certain to push about 1 inch of hose onto each tubing end.

6. Slip the clamps over the ends of the hose and tighten it securely.

7. Check the hose and line for leaks by operating the transmission.

cooler and line flushing

With the flushing machine mentioned in Section 1, flush out the cooler and lines using this procedure:

1. Roll the machine to the vehicle. Be careful not to tip the machine over or slosh the solvent from the machine.

2. With the two auxiliary hoses and adapters supplied with the machine, connect the machine's pressure line to either of the two cooler lines (Fig. 8-52). Then connect the long auxiliary return hose to the remaining cooler line.

3. Place this return line in a waste container.

4. Turn the timer on for 2 minutes to pump a few quarts of solvent through the system and into the waste container (Fig. 8-53). This procedure prevents excessive amounts of dirty fluid, particles, or sludge from entering the machine's sump.

5. Remove the drain hose from the waste container and place it into the machine's sump.

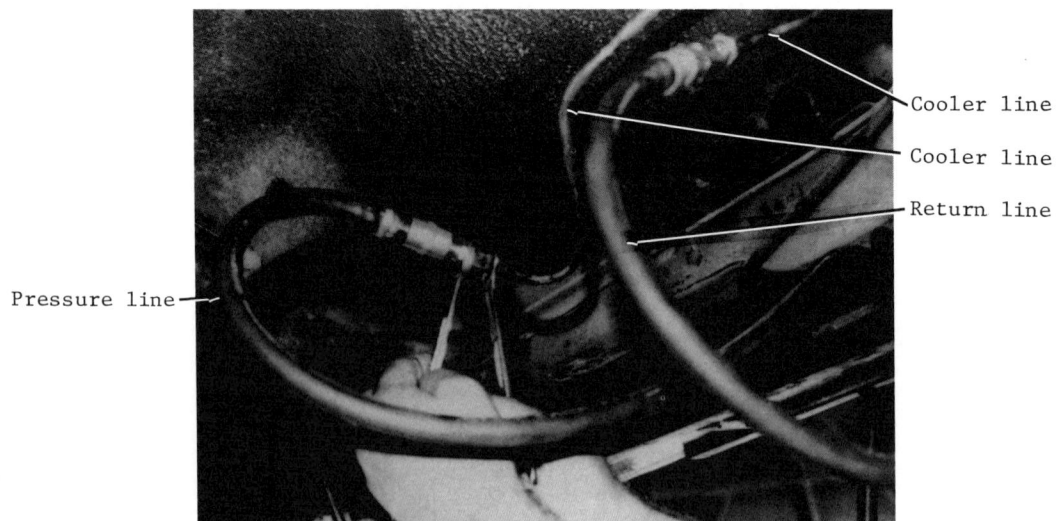

FIG. 8-52 Connecting the flushing machine's pressure and return hoses to the cooler lines.

Subassembly Cleaning, Inspection, and Service

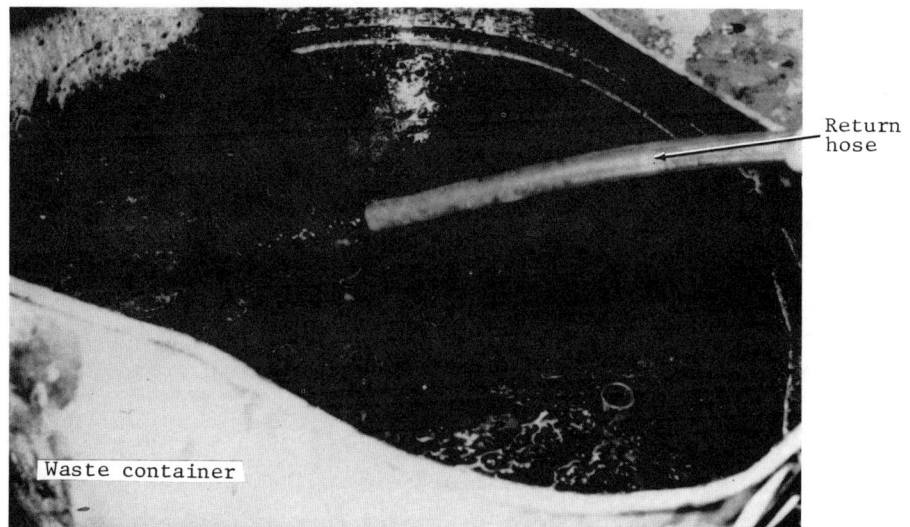

FIG. 8-53 Pumping a few quarts of solvent through the system and into a waste container.

6. Set the timer for a 5 to 10 minute flushing and occasionally check the sump return hose to see if solvent is returning freely from the system. If the solvent is not flowing freely, the system, cooler or lines, have a restriction.

7. Disconnect the pressure line from the machine and remove the pressure auxiliary hose from the line. Place the return line once more into a waste container.

8. Using low pressure compressed air, less than 50 psi, blow the remaining solvent from the system. Apply the air pressure to the cooler line where the machine's auxiliary pressure hose had previously been connected.

9. Disconnect the auxiliary return hose from the cooler line and return the machine to its storage area.

CHECK-UP QUESTIONS

The questions listed below will assist you in determining how well you remember the material contained in this section. Read each question carefully before adding the word or words necessary to complete the sentence. If you can't complete the sentence, review that portion of the section that covers the question.

1. The transmission case is the main _____ of the transmission.

2. _____ or _____ deposits are much harder to dissolve than any other type of contaminant from the transmission case or its components.
3. To dry the transmission case after cleaning it, use _____ _____ _____.
4. If the transmission case has cracks or porosity, it is better to _____ it.
5. Damaged threads are repairable using a _____ _____.
6. A _____ in the transmission case usually supports the output shaft.
7. A mechanic can press a bushing out or _____ it in two with a chisel.
8. If leakage occurs around the shift linkage shafts, replace the _____.
9. Case porosity is a condition where a _____ _____ develops through the metal of the transmission case.
10. A(n) _____ _____ is necessary to repair case porosity.
11. The component that encloses the components behind the transmission case is the _____ _____.
12. A(n) _____ _____ _____ usually supports both the output shaft and the slip yoke.
13. When installing the extension housing seal, make sure its lip faces _____ or toward the bushing.
14. A drum along with a multidisc clutch connects a planetary gear train member to the _____ _____.
15. After removing the clutch pack, tie its components together with _____.
16. It is sometimes necessary to use air pressure to force the _____ _____ from the drum.
17. The location of the residual check ball will be either in the clutch _____ or _____.
18. If the lip seal makes it difficult to reinstall a piston in a clutch drum, work a _____ _____ around the circumference of the piston.
19. When deglazing a drum that has a fabric lined band, sand it with _____ grit emery cloth or sandpaper.
20. An overrunning clutch holds a component stationary in _____ _____.
21. The _____ cushions the application of a band or clutch.

Subassembly Cleaning, Inspection, and Service   271

22. The _____ is a device that produces a pressure signal in proportion to vehicle speed.
23. When polishing a valve spool do not round off its _____ _____ or corners.
24. Transmission _____ carry engine torque through the various gear train members and then to the drive shaft.
25. _____ _____ separate the many moving parts within the transmission.
26. A _____ can also support a shaft or act as a thrust washer.
27. The device that has a steel backing lined with copper or babbit-type material is a _____.
28. To remove a bushing from a blind bore, use either a bushing _____ or a threaded _____.
29. The hydraulic computer of the transmission is the _____ _____.
30. When cleaning any transmission component, never use a _____ to dry parts.
31. Use _____ grit emery cloth or sandpaper to sand a steel clutch plate.
32. Soak all new friction clutch plates in clean automatic transmission fluid for _____ minutes.
33. Before reusing an asbestos lined band, _____ the lining with a knife or bearing scraper.
34. Always replace _____ seals with the metal type.
35. Always install the lip seal so that the lip faces _____ the fluid source.

SECTION 9

Transmission Overhaul

REFERENCES: Automatic Transmission Fundamentals, Chapters 2, 3, 5, 6, 8, 11, and 12.
Automatic Transmission Service, Sections 1, 2, 5, 7, and 8.
Vehicle or C-4 transmission service manual.

This section provides a step-by-step procedure for the overhaul of the Ford C-4 transmission. Although this section covers only one style of transmission, the repair sequences presented here are typical of those used to repair other units. As you proceed through this section and perform the various repair procedures, it will be helpful to refer back to the various other sections of this manual for further explanation of the various pieces of equipment, tools, and procedures presented here.

GENERAL INSTRUCTIONS

During the teardown, repair, and assembly of the various subassemblies of this transmission, the mechanic must follow certain general instructions. By following these instructions carefully, the mechanic will avoid damaging transmission components and the unnecessary repetition of certain repair operations.

Transmission Overhaul

1. After removing the transmission from the vehicle, plug all of its openings and steam clean the outside of the case thoroughly. This action prevents dirt from entering the mechanical components during disassembly.
2. Handle all transmission parts carefully to avoid nicking or burring the bearing or mating surfaces.
3. Lubricate all internal transmission parts with clean Type F Fluid. <u>Do not use any other lubricants except on thrust washers and gaskets. The mechanic can coat these with Vaseline to facilitate assembly.</u>
4. Always install new gaskets, sealing rings, and seals when rebuilding the transmission.
5. Always tighten all bolts and screws to the manufacturer's specifications.
6. When rebuilding this transmission, the mechanic must remove and reinstall 10 thrust washers. It is very important that each of these washers be in its correct position during the assembly process. To properly locate and identify these washers and their locations in the upcoming illustrations, the washers are numbered from 1 to 10. For example, No. 1 is the first thrust washer located on the front pump; the last washer--location No. 10--is at the parking pawl ring gear.
7. With the exception of bands and friction clutch plates, wash all parts with a suitable, clean solvent.
8. Do not dry parts with a rag; instead, use low pressure compressed air.

DISASSEMBLY

1. Remove the torque converter from the transmission and the converter housing. <u>Note</u>: If the converter housing attaches to the case via the front pump bolts, leave it installed until you remove the pump.
2. With a modulator wrench, remove the vacuum unit and control rod.
3. With a pencil magnet, remove the primary throttle valve from the opening at the rear of the case (Fig. 9-1).
4. Remove the transmission pan attaching bolts, pan, and gasket. Discard the gasket.
5. Remove the valve body attaching bolts, and remove the valve body from the case.
6. Loosen the intermediate band adjusting screw, and remove the intermediate band struts from the case (Fig. 9-2).
7. Loosen the low reverse band adjusting screw and remove the low reverse band struts.

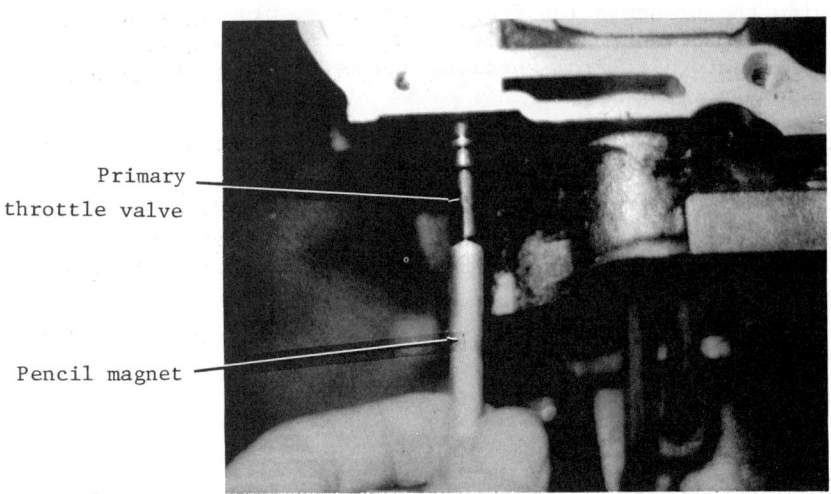

FIG. 9-1   Removing the primary throttle valve with a pencil magnet.

Transmission End-Play Checks

   1. To maintain the output shaft in alignment during the end-play check, install an extension housing oil-seal replacing tool, a front universal yoke, or a splined sleeve from the dynamometer (Section 1) into the extension housing and over the splined output shaft (Fig. 9-3

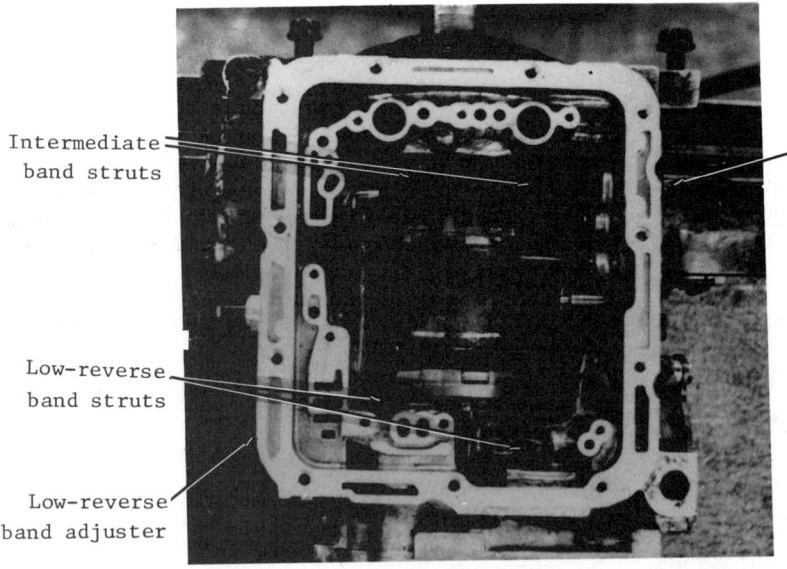

FIG. 9-2   Removing the band-adjusting screws and struts.

Transmission Overhaul

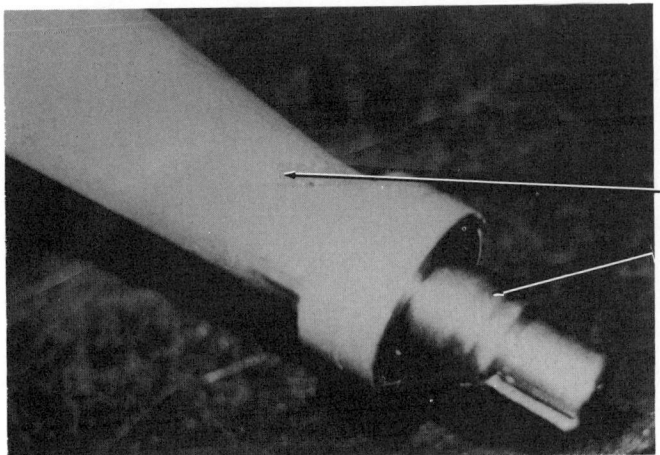

FIG. 9-3 The output shaft held in alignment with a splined sleeve from a dynamometer.

2. Move the input shaft and gear train toward the rear of the case.
   Note: The input shaft is a loose part that the mechanic must first engage in the splined front clutch hub before making the end-play check.
3. Remove one of the front pump-to-case attaching bolts, replace it with a long bolt or stud, and then mount a dial indicator as shown in Fig. 9-4.

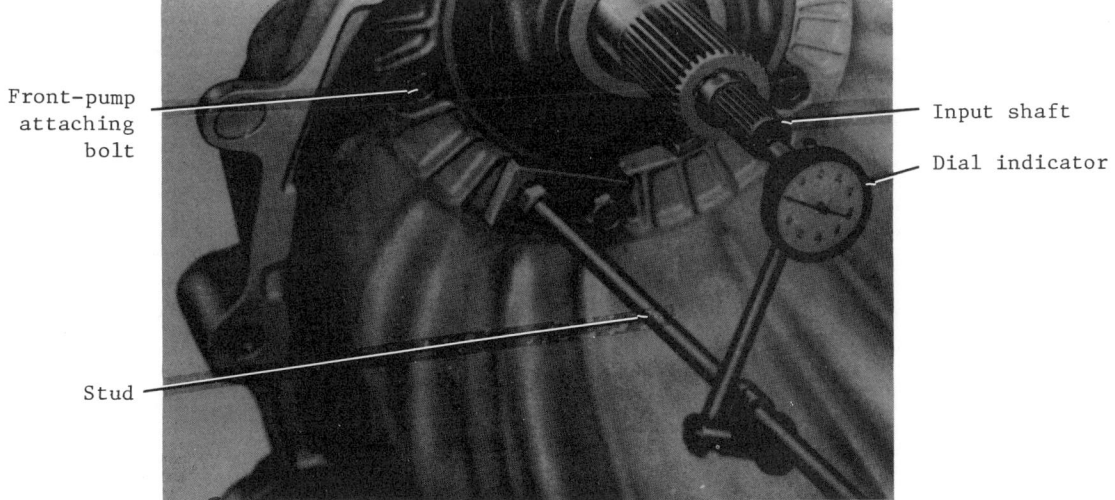

FIG. 9-4 Dial indicator in position to measure the end-play.

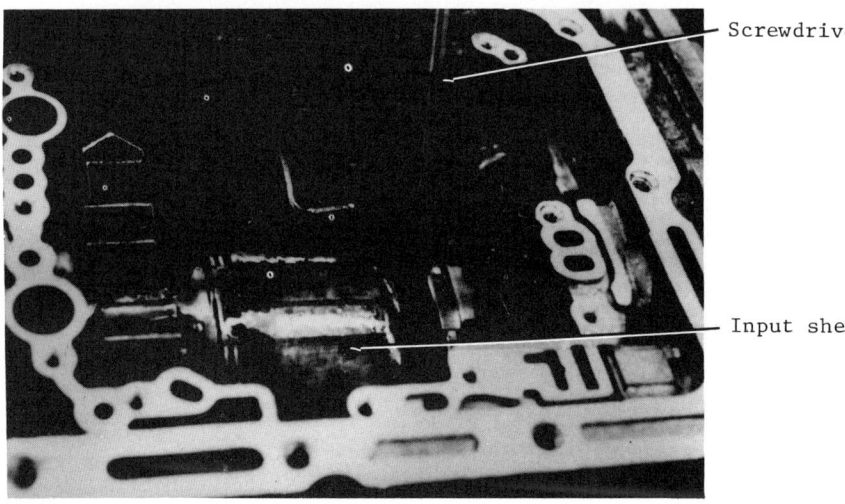

FIG. 9-5 Screwdriver in position behind the input shell in readiness to move the gear train forward.

4. With the dial indicator plunger contacting the end of the input shaft, set its indicator dial face to zero.

5. Place a screwdriver behind the input shell, and then move the front part of the gear train toward the front pump, or forward (Fig. 9-5).

6. Record the dial indicator reading. The reading should be between .008 to .042 inch. If the end-play is not within these specifications, the mechanic must replace the selective thrust washers (Fig. 9-6) as necessary during the reassembly process. The manufacturer makes both of these washers in various sizes, and the mechanic can replace them to obtain the specified end-play (Fig. 9-7).

7. Remove the dial indicator and its attachments and then pull the input shaft out of the front pump stator support (Fig. 9-8).

Removal of Case and Extension Housing Components

1. Remove the front pump attaching bolts. If these bolts secure the converter housing to the pump, also remove this housing from the case.

2. Insert a screwdriver behind the input shell (Fig. 9-5), and pry the input shell forward until the front pump seal is beyond the edge of the case.

3. Remove the front pump and gasket from the case; discard the gasket. If the No. 1 selective thrust washer did not come out with the front pump, remove it from the front of the reverse high clutch.

Transmission Overhaul

FIG. 9-6 The location of Nos. 1 and 2 thrust washers.

4. Remove the intermediate and low reverse band adjusting screws from the case. Discard the seal-type lock nuts.

5. Turn the intermediate band so that the band ends align with the clearance hole cut into the case (Fig. 9-9) and remove the band. Inspect the band for wear and loose or pitted lining material. If the band is still serviceable, clean it with a detergent solution and then blow it dry with low pressure compressed air.

6. Using a screwdriver again between the input shell and rear planet carrier, move the input shell forward and remove the front section of the gear train as an assembly from the case (Fig. 9-10).

7. Place the front part of the gear train in the holding fixture shown in Fig. 9-11.

8. With the gear train in the holding fixture, remove the reverse high clutch drum from the forward clutch (Fig. 9-12).

Thrust washer No. 1		Thrust washer no. 2	
0.053-0.0575	Red	No. stamped washer	Metal thrust washer
0.070-0.074	Green		
0.087-0.091	Natural	1	0.043-0.041
0.104-0.108	Black	2	0.058-0.056
0.121-0.125	Yellow	3	0.075-0.073

FIG. 9-7 Chart of selective thrust washer sizes.

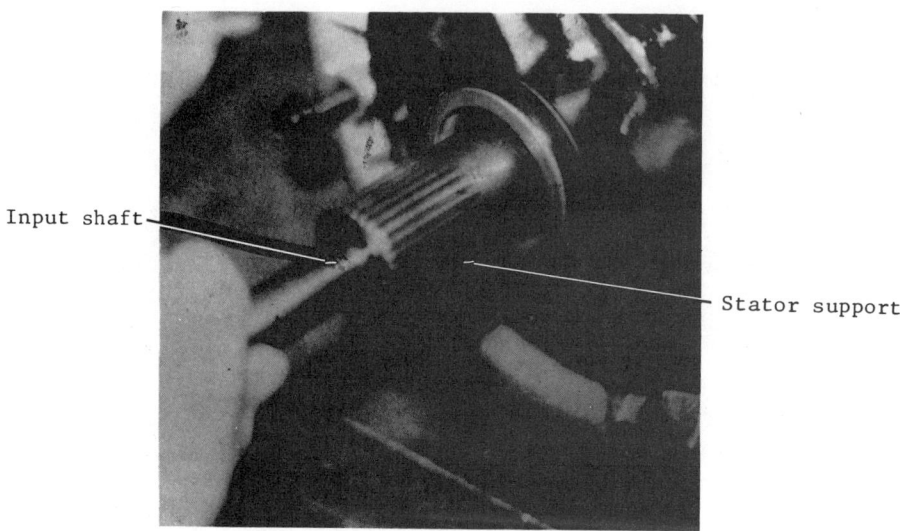

FIG. 9-8   Removing the input shaft.

9. If No. 2 thrust washer (Fig. 9-6) **did** not come out with the pump, remove it from the forward clutch cylinder. If a selective spacer was used, remove it also. Then remove the forward clutch from the forward clutch hub and ring gear (Fig. 9-13).

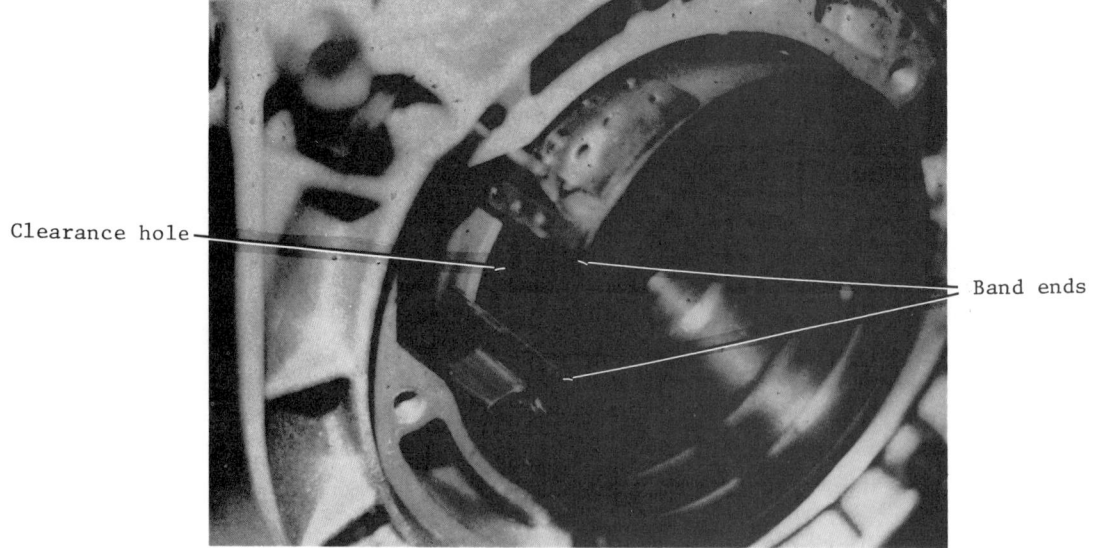

FIG. 9-9   Removing the intermediate band from the case.

Transmission Overhaul

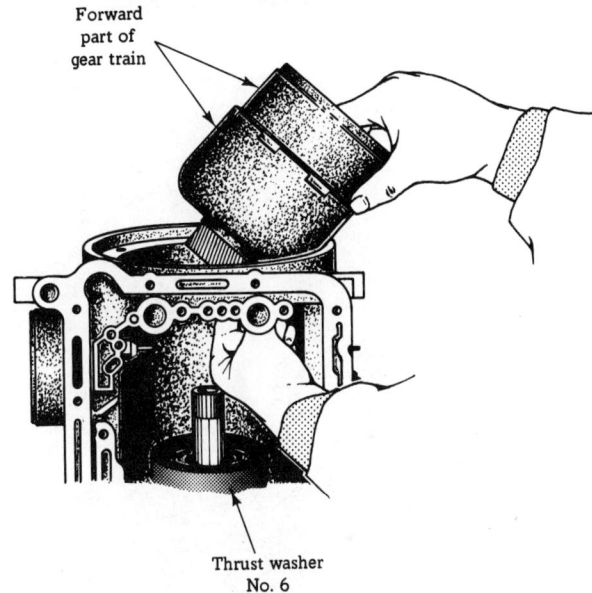

FIG. 9-10 Removing the input shell and gear train.

10. If No. 3 thrust washer did not come out with the forward clutch, remove it from the forward clutch hub.

11. Remove the forward clutch hub and ring gear from the forward planet carrier (Fig. 9-14).

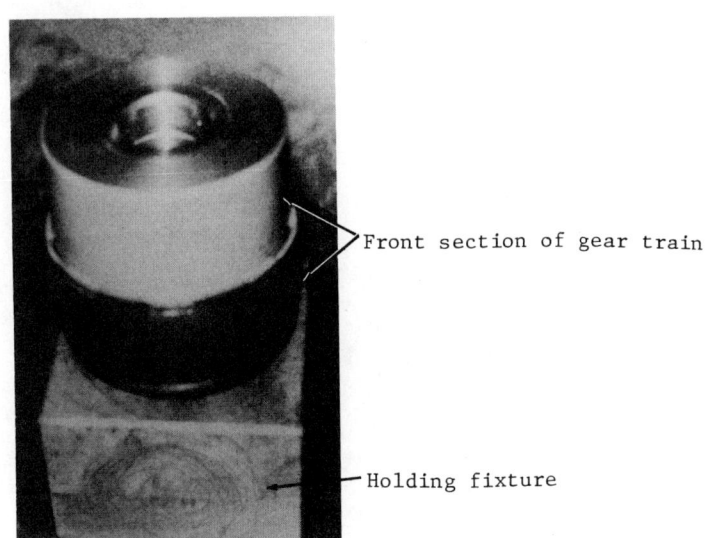

FIG. 9-11 Forward part of gear train positioned in holding fixture.

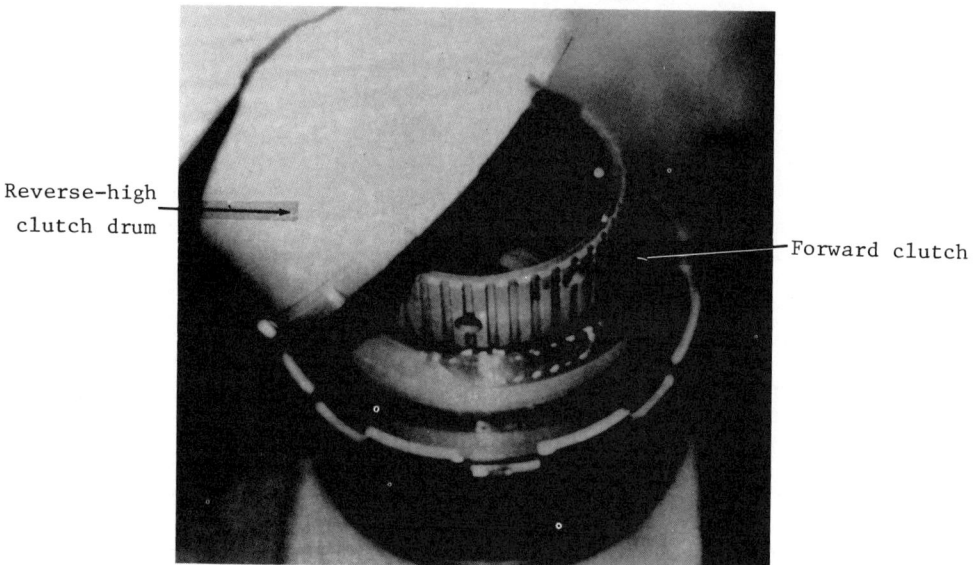

FIG. 9-12 The removal of reverse-high clutch drum from the forward clutch.

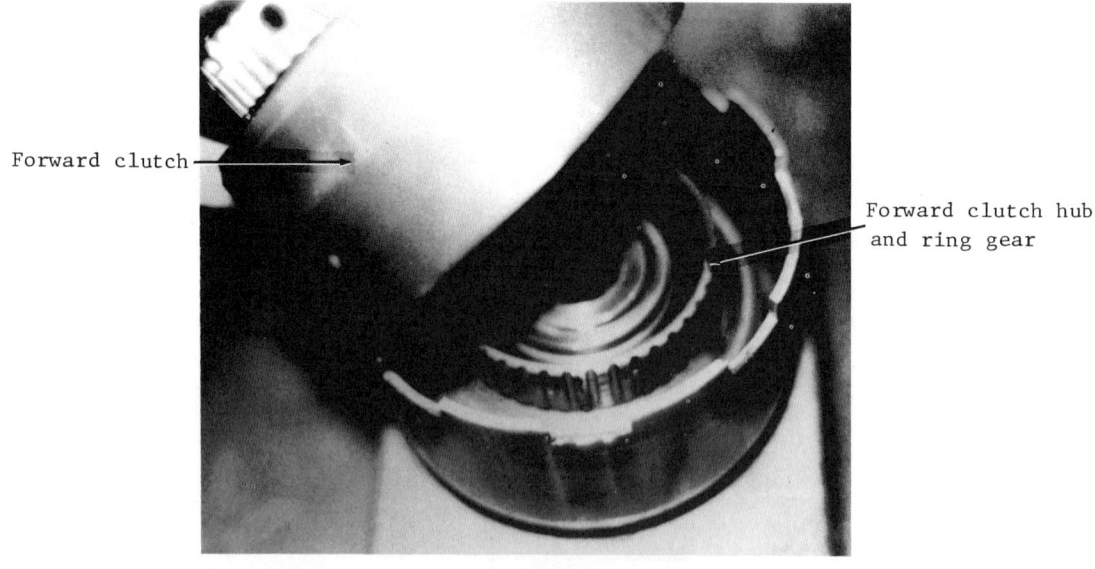

FIG. 9-13 Removal of forward clutch from forward clutch hub and ring gear.

Transmission Overhaul

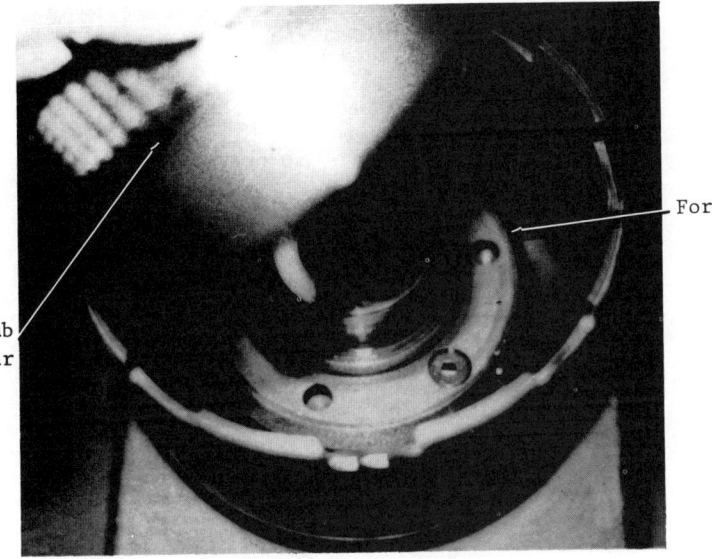

FIG. 9-14 Removal of forward ring gear from the forward carrier.

12. Remove No. 4 thrust washer and the forward planet carrier from the input shell (Fig. 9-15).
13. Remove the input shell, sun gear, and thrust washer No. 5 from the holding fixture.

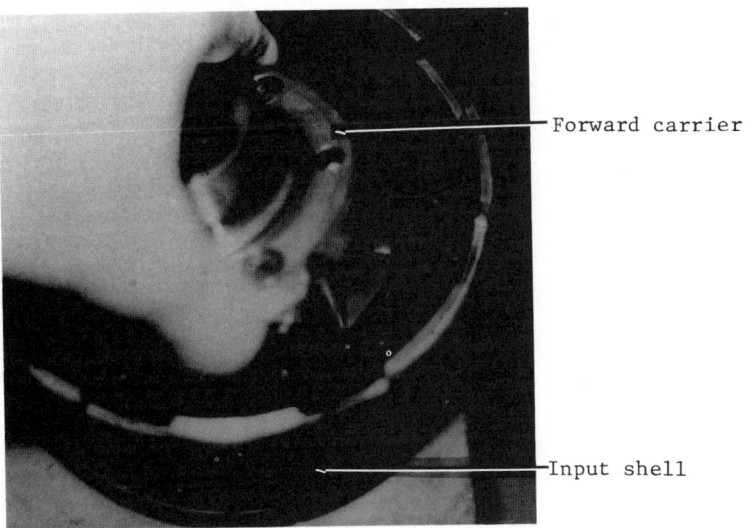

FIG. 9-15 Removal of forward carrier from the input shell.

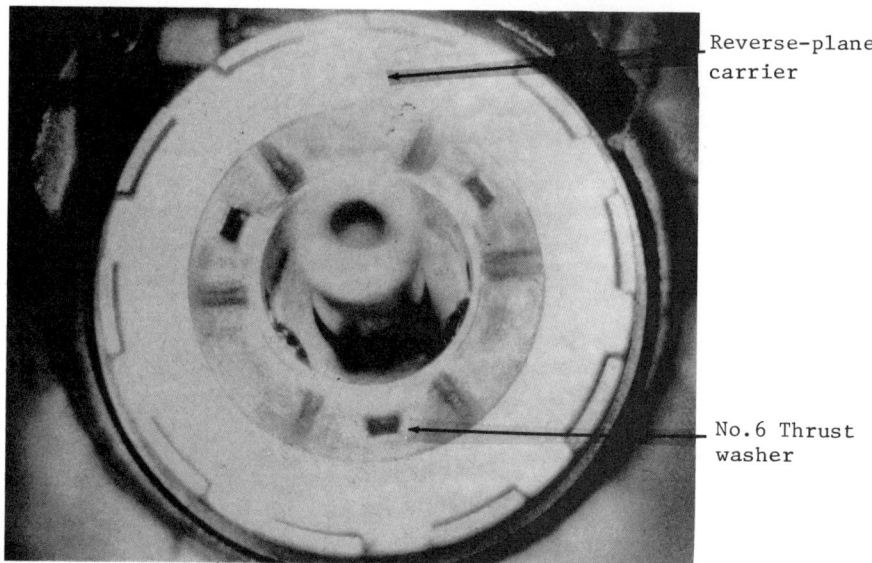

FIG. 9-16   Location of the No. 6 thrust washer.

14. From inside the transmission case, remove No. 6 thrust washer from the top of the reverse planet carrier (Fig. 9-16).

15. Remove the reverse planet carrier and No. 7 thrust washer from the reverse ring gear (Fig. 9-17).

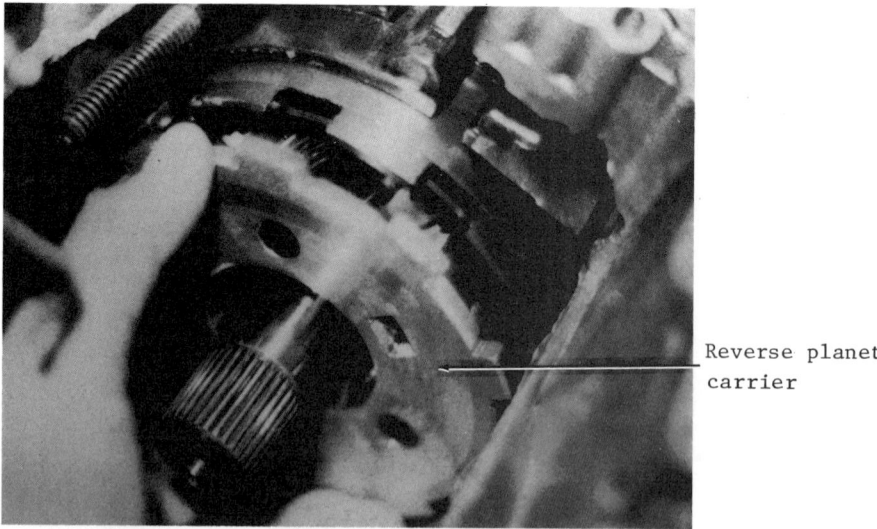

FIG. 9-17   Removing the reverse planet carrier and the No. 7 thrust washer.

Transmission Overhaul

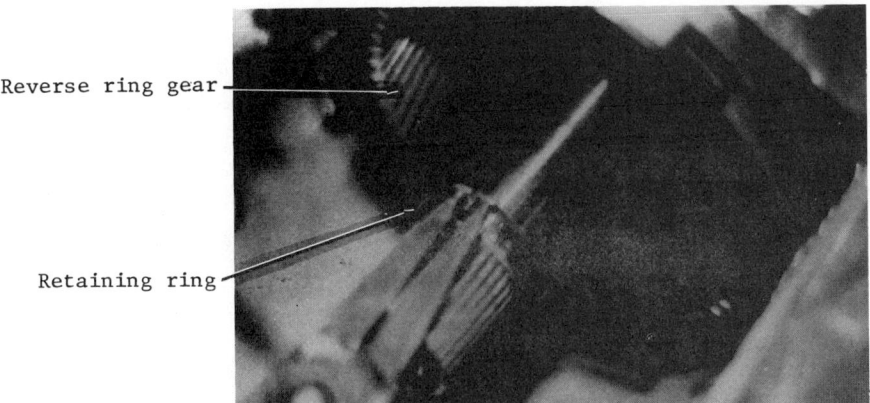

FIG. 9-18 Removing the reverse ring gear hub retaining ring.

16. Remove the reverse ring gear hub-to-output shaft retaining ring with snap-ring pliers (Fig. 9-18).
17. Remove the reverse ring gear and hub from the output shaft. Also remove the No. 8 thrust washer from the low reverse drum (Fig. 9-19).
18. Remove the low reverse band from the case (Fig. 9-20).
19. Remove the low reverse drum from the one-way clutch inner race (Fig. 9-21).

FIG. 9-19 No. 8 thrust washer location.

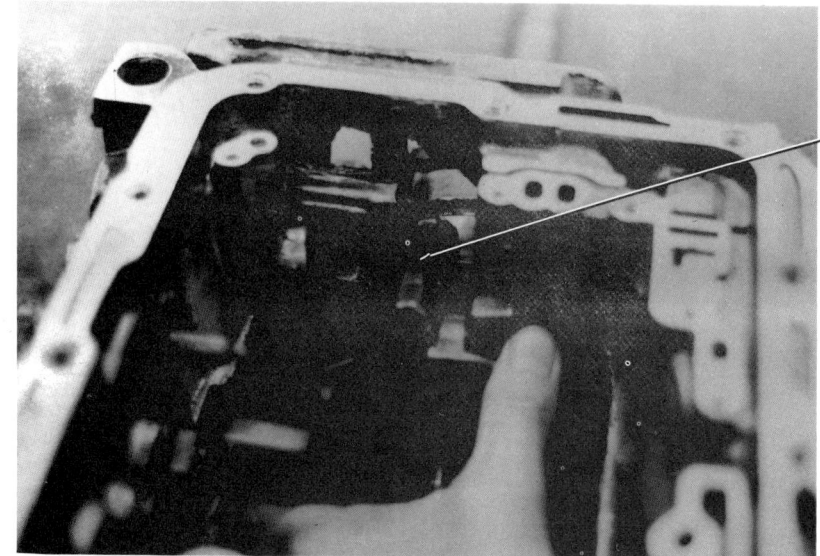

FIG. 9-20 Removal of low-reverse band from the case.

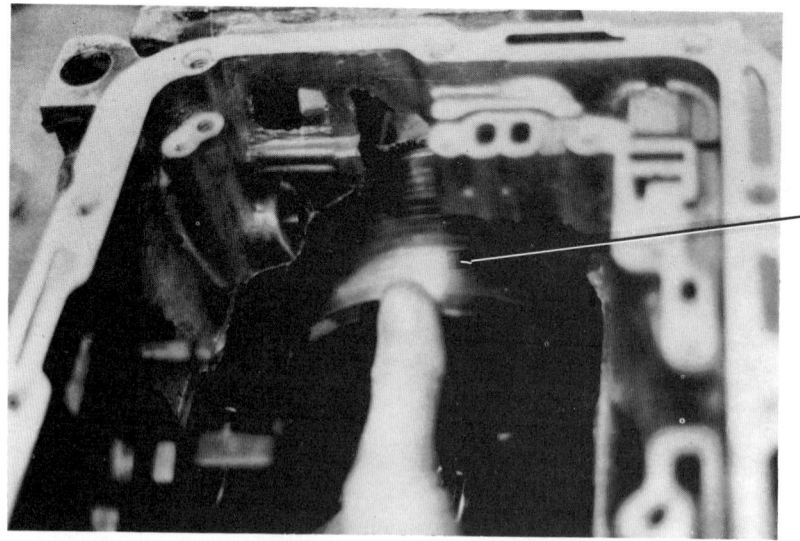

FIG. 9-21 Removal of low-reverse drum from the case.

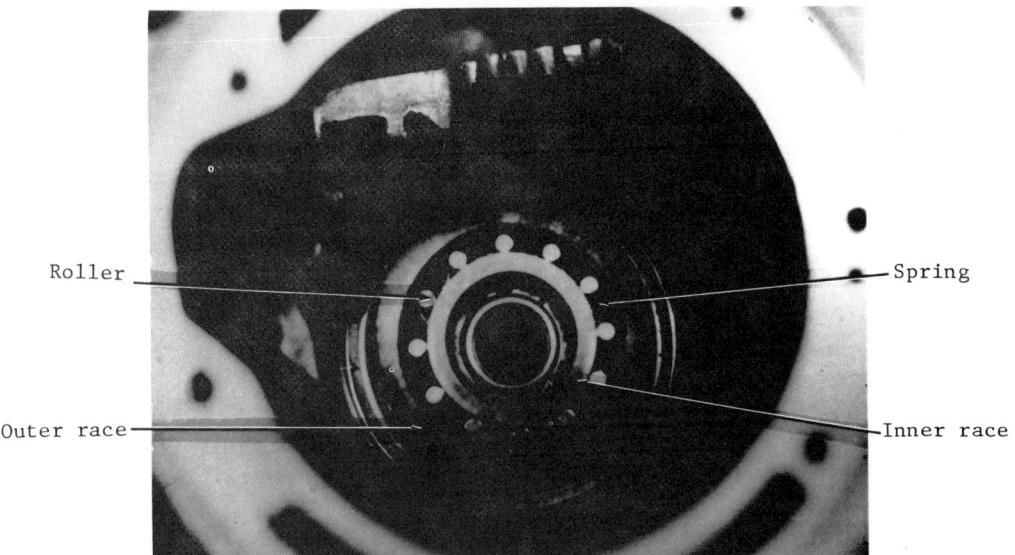

FIG. 9-22 Removing the components of the one-way clutch assembly.

20. Remove the one-way clutch inner race by turning it clockwise and pulling it outward at the same time.

21. Remove the 12 one-way clutch rollers, springs, and the spring retainer from the outer race (Fig. 9-22). <u>Do not lose or damage any of the 12 rollers or springs.</u> <u>The outer race of this one-way clutch is detachable from the case only after the extension housing, output shaft, and governor distributor sleeve are removed from the back of the case.</u>

22. Position the transmission on the bench in a vertical position with the extension housing up, and then remove the extension housing-to-case attaching bolts. Remove the extension housing and gasket; discard the gasket. Remove the governor-to-housing attaching bolts, and remove the governor from the fluid distributor.

23. Pull outward on the output shaft and remove the shaft and governor distributor assembly from the governor distributor sleeve, mounted in the back of the case (Fig. 9-23).

24. Remove the governor distributor attaching ring from the output shaft with snap-ring pliers (Fig. 9-24). Slide the governor distributor off the output shaft.

25. Remove the four distributor sleeve-to-case attaching bolts, and then <u>carefully</u> remove the distributor sleeve from the case. <u>Do not bend or distort the fluid tubes as they pull from the case with the distributor sleeve.</u>

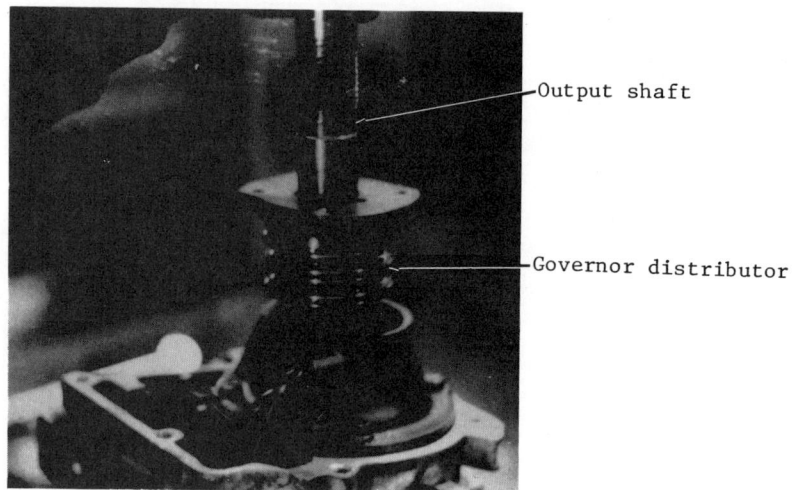

FIG. 9-23 Removing the output shaft and governor distributor.

26. Remove the parking pawl return spring, pawl, and retaining pin from the back of the case (Fig. 9-25).
27. Lift the parking gear and No. 10 thrust washer from the case.
28. Loosen and remove the six, one-way clutch outer race-to-case attaching bolts (Fig. 9-26). As you remove these bolts, hold the

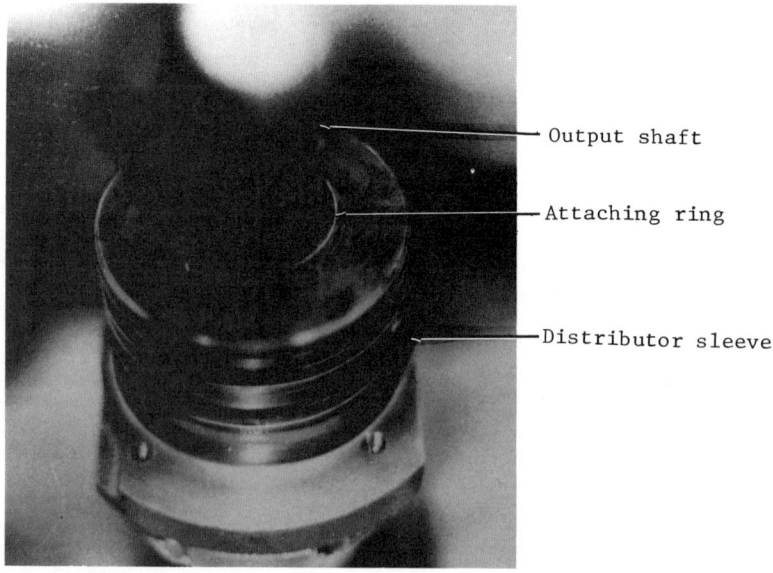

FIG. 9-24 Removing the governor distributor snap ring.

Transmission Overhaul

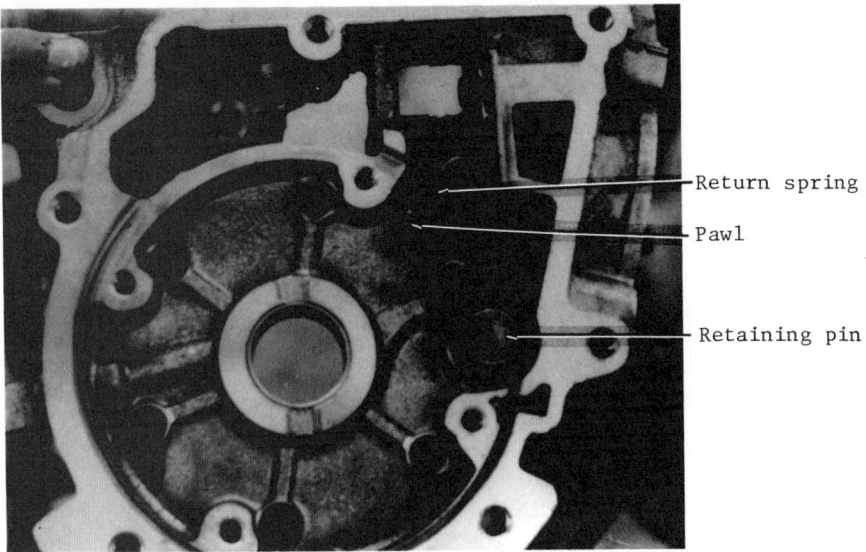

FIG. 9-25 Removing the parking pawl return spring, pawl, and retaining pin.

outer race, located inside the case, in position. Finally, remove the outer race and No. 9 thrust washer from the case (Fig. 9-27).

29. Remove the intermediate servo cover, gasket, piston, and return spring (Fig. 9-28). Discard the gasket and remove the piston from the cover.

FIG. 9-26 Removing the one-way clutch outer race-to-case attaching bolts.

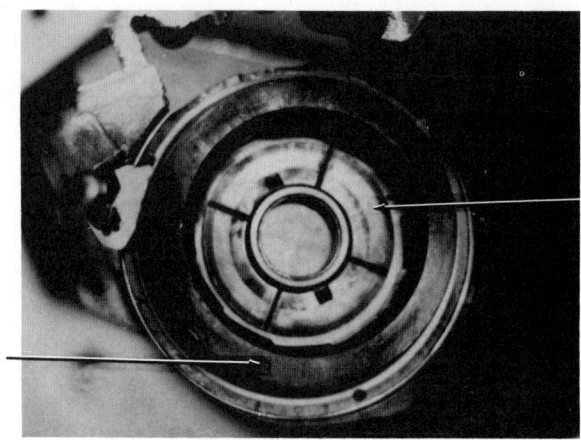

FIG. 9-27 No. 9 thrust washer location.

30. Remove the four low-reverse servo cover-to-care attaching bolts (Fig. 9-29). Remove the cover, seal, and piston from the case.

SUBASSEMBLY OVERHAUL

Case

1. Inspect the rear transmission case bushing for wear or damage; replace it as necessary (Fig. 9-30).
2. Inspect all tapped holes within the case for missing or damaged threads. Repair as necessary with Heli-coil inserts.

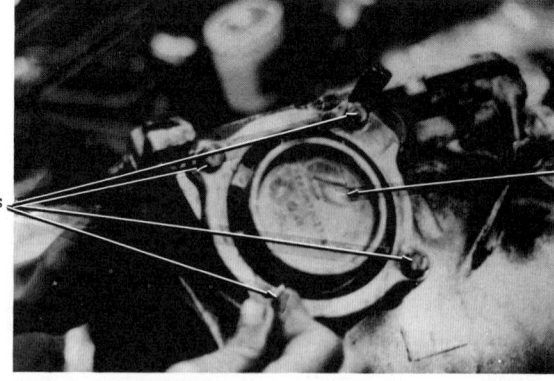

FIG. 9-28 Removing the intermediate servo cover.

Transmission Overhaul

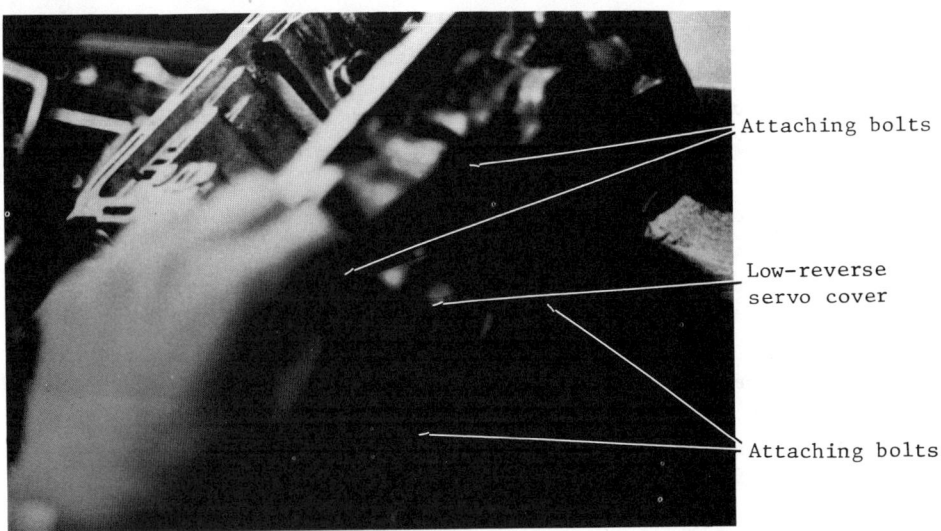

FIG. 9-29 Removing the low-reverse servo cover.

3. Thoroughly clean the inside of the case with a suitable cleaning solvent and blow out all passages with low pressure compressed air.

Case Manual Lever Seal

1. Apply penetrating oil to the outer lever attaching nut to prevent breaking the inner lever shaft.

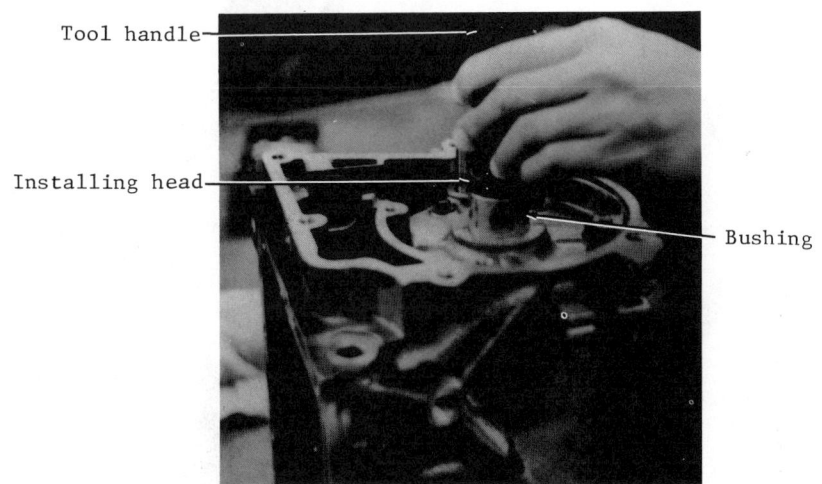

FIG. 9-30 Replacing the transmission case bushing.

2. Remove the downshift outer lever nut, and remove the outer lever arm (Fig. 9-31). On vehicles equipped with a case mounted neutral safety switch, remove its attaching screws and pry the switch carefully off the inner lever.

3. From inside the case, slide the downshift inner-lever from the outer manual lever shaft.

4. From inside the case, remove the inner manual lever attaching nut and lever (Fig. 9-32).

5. From the outside of the case, slide out the outer manual lever shaft and lever.

6. Using an appropriate puller, remove the seal from the case. Inspect all parking pawl linkage for wear and damage. Remove and repair these linkage components as necessary.

7. To install the new seal, use a driver and driving head as shown in Fig. 9-33.

8. Lubricate the outer manual lever shaft and insert it into the seal and case.

9. Install the inner manual lever and attaching nut with the chamfer on the nut facing the lever. Torque the nut to 30 to 40 pounds-foot. Reinstall the neutral safety switch, if so equipped, and torque its attaching bolts to 55 to 75 pounds-inch.

10. Lubricate and install the inner downshift-lever shaft into the outer manual lever shaft.

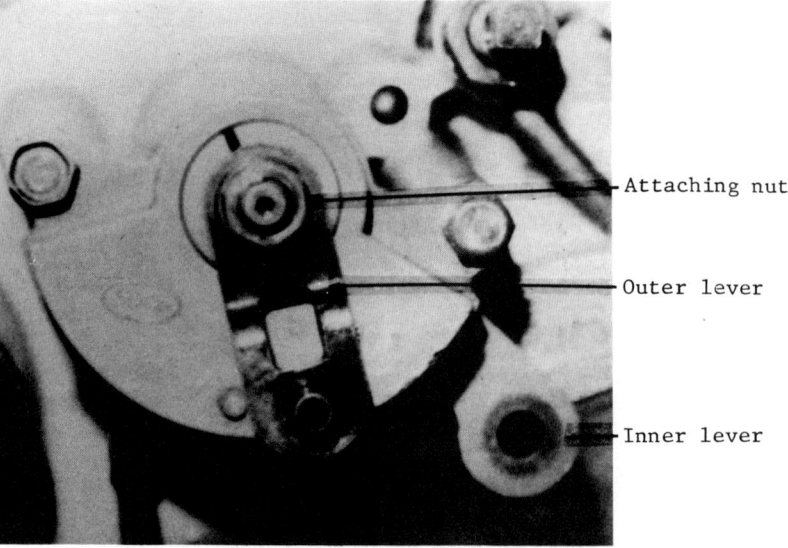

FIG. 9-31 Removing the outer-lever arm and attaching nut.

Transmission Overhaul

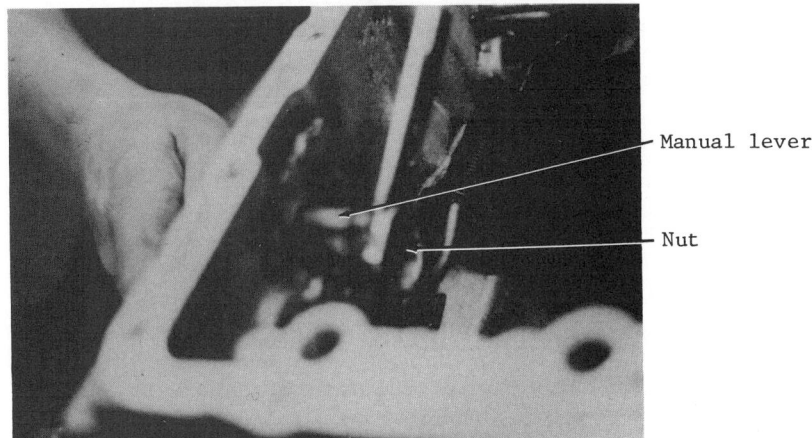

FIG. 9-32 Removing the inner manual lever and attaching nut.

11. Install the O-ring over the outer end of the downshift lever shaft. Install the outer downshift-lever and nut. Torque the nut to 12 to 16 pounds-foot.

Intermediate Servo

1. Remove the seal rings from the servo piston and cover.
2. Clean the piston and cover and install a new seal on the cover and servo piston.

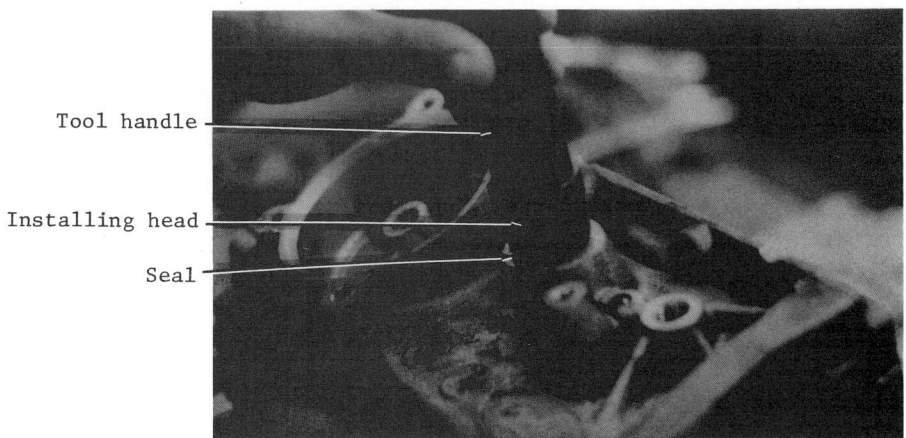

FIG. 9-33 Installation of case manual-shaft seal.

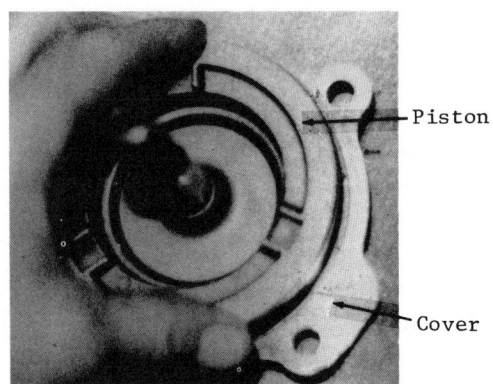

FIG. 9-34 Installation of the intermediate servo piston into its cover.

3. Lubricate the seals with clean transmission fluid. Install the piston into its bore in the cover; be careful not to damage the piston seal (Fig. 9-34).

4. Install the servo piston return spring in its bore within the transmission case.

5. Coat a new gasket with Vaseline and place it in position over the servo cover. Position the servo piston and cover assembly into the transmission case with the piston stem slot positioned horizontally to engage the strut.

6. Using two 5/16-18 bolts, 1-1/4 inches long, and 180 degrees apart to position the cover against the case, install two standard cover attaching bolts.

7. Remove the two 1-1/4-inch bolts and install the other two cover attaching bolts; torque these bolts to 16 to 22 pounds-foot (Fig. 9-35).

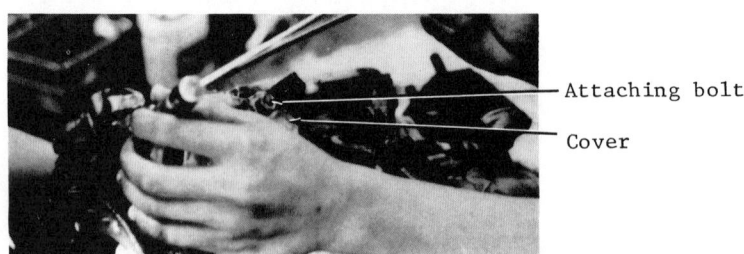

FIG. 9-35 Torquing the intermediate servo cover screws.

Transmission Overhaul

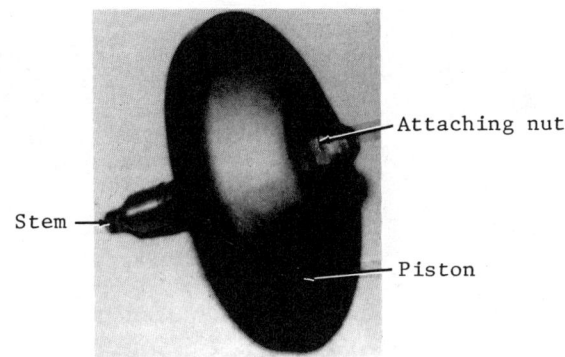

FIG. 9-36 Rear servo piston with detachable stem.

Low Reverse Servo

1. The manufacturer bonds the low reverse servo-piston seal to the piston. If the seal is worn, therefore, the mechanic must replace the piston assembly, which includes the seal and stem. <u>Note</u>: On some models, the piston stem is detachable from the servo piston by removing the piston attaching nut (Fig. 9-36).

2. If the stem is the detachable type, reinstall it on the new servo piston and torque the attaching nut to 12 to 20 pounds-foot.

3. Clean and reinstall the servo-piston return spring into its bore in the transmission case. Lubricate the piston seal with clean transmission fluid and install the piston into its bore within the case.

4. Place a new seal on the cover and install the servo cover. Reinstall the four servo cover-to-case attaching bolts; torque these bolts to 12 to 20 pounds-foot (Fig. 9-37).

Front Pump

1. Remove the four metal sealing rings from the stator support.

2. Remove the bolts which attach the stator support to the front pump housing (Fig. 9-38). Remove the stator support from the pump housing.

3. Remove the drive and driven gears from the front pump housing. Remove the front pump seal, using a suitable puller.

4. Wash all front pump components in a suitable cleaning solvent and blow dry with low pressure compressed air. Inspect the gears, housing, and support for wear and damage, replacing worn parts as necessary.

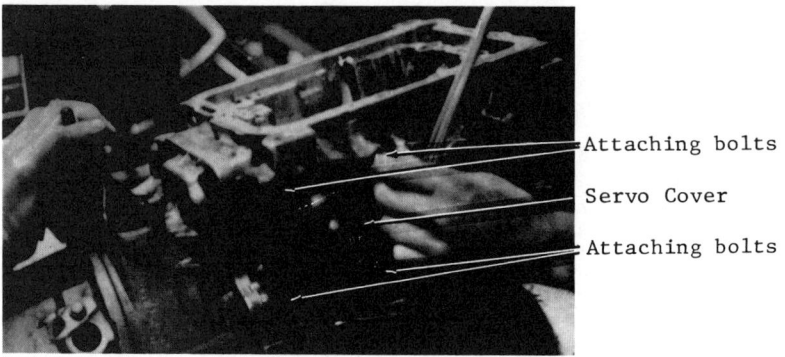

FIG. 9-37 Torquing the low-reverse servo cover.

5. Inspect the front and rear stator bushings for wear and damage. If replacement is necessary, use a cape or bushing chisel to cut along the bushing's seam until the chisel breaks through the bushing wall. Then pry the loose ends of the old bushing up with an awl or suitable tool and remove the bushing.

6. Using the correct driver and driving head, press the new bushings into place in the stator support (Fig. 9-39). When installing the rear bushing, make sure the hole in the bushing is in line with the lube hole in the stator support.

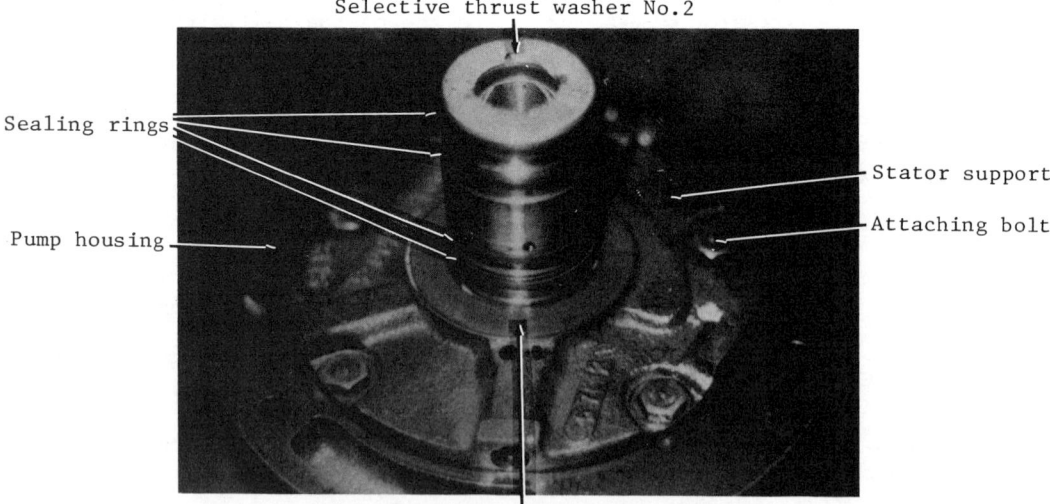

FIG. 9-38 Removing the front pump housing-to-stator support bolts.

FIG. 9-39 Installation of the stator support bushings.

7. Drive or press the old, converter-support bushing from the front pump housing (Fig. 9-40), using a driver and the correct-size driving head.

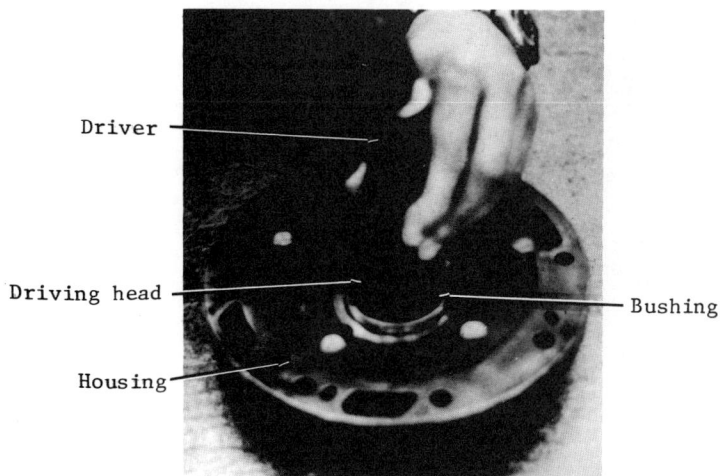

FIG. 9-40 Removal and replacement of the converter support bushing.

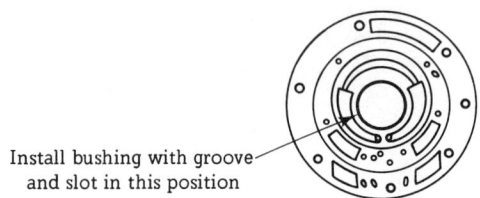

FIG. 9-41 The correct converter support bushing location in the pump body.

8. Press or drive a new bushing into the pump housing with the same tools as shown in Fig. 9-40. Make sure you install the bushing with the slot and groove positioned to the rear of the pump body and 60 degrees below the horizontal center line (Fig. 9-41).

9. Install a new front pump seal using the tool shown in Fig. 9-42.

10. Lubricate the drive and driven pump gears and reinstall them into the pump housing. Note: Each gear has an identification mark on the side of each gear where the gear teeth are chamfered. Position the chamfered side with the identification mark downward against the face of the pump housing.

11. Place the stator support in the pump housing and reinstall the attaching bolts. Torque these bolts to 12 to 20 pounds-foot (Fig. 9-43).

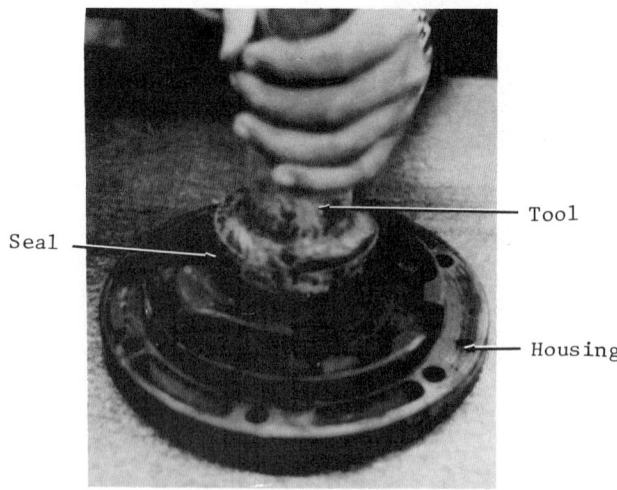

FIG. 9-42 Installation of the front-pump seal.

Transmission Overhaul

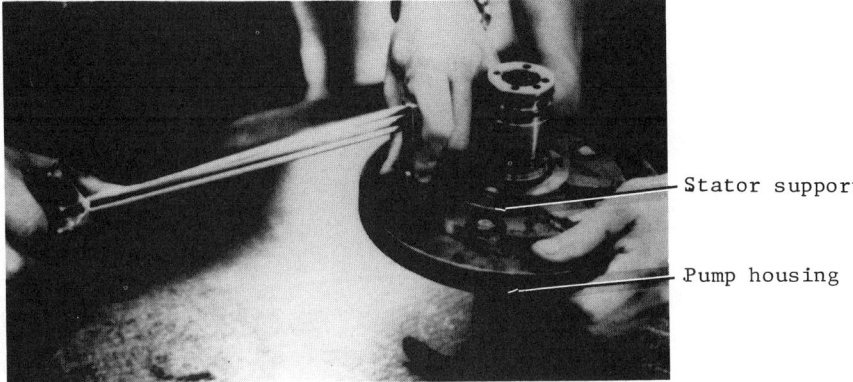

FIG. 9-43 Torquing the pump housing-to-stator support bolts.

12. Install four new metal sealing rings onto the stator support. Assemble the two larger rings first in the oil-ring grooves toward the front of the stator support (Fig. 9-44).

13. Check the pump for free rotation by placing the pump on the converter drive hub, in its normal operating position, and turning the pump housing.

Reverse High Clutch

1. Remove the pressure plate retaining ring with a screwdriver (Fig. 9-45).

FIG. 9-44 Installing the metal sealing rings on the stator support.

FIG. 9-45 Removing the reverse-high clutch retaining ring.

2. Remove the pressure plate, drive, and driven clutch plates.

3. To remove the piston-spring retainer snap ring, use the spring compressor tool shown in Fig. 9-46. With this tool in position, slowly compress the return spring(s) and remove the snap ring. Now, slowly release the compressor tool while guiding the spring retainer so that it clears the snap-ring groove cut into the piston guide.

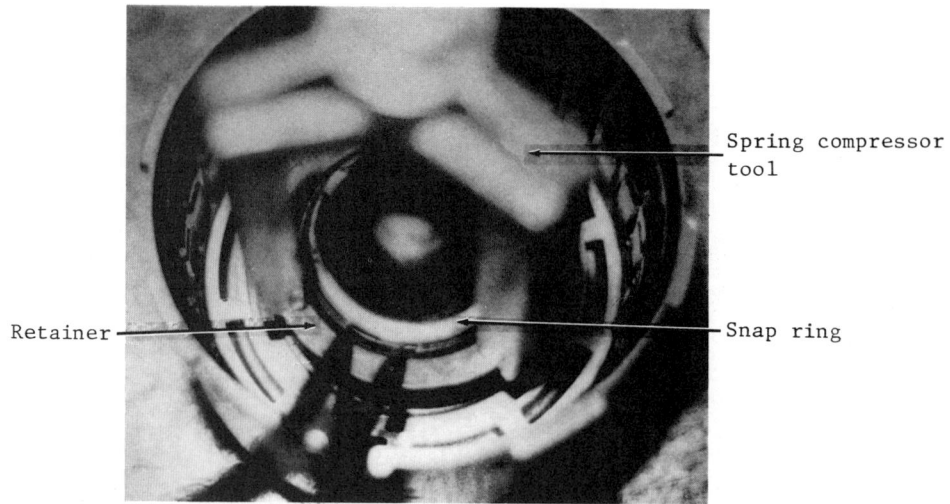

FIG. 9-46 Removing the piston-spring retainer snap ring using a compressor tool.

Transmission Overhaul

4. Remove the spring retainer and the piston return spring(s).
5. Remove the clutch piston by directing air pressure into the piston apply hole in the clutch hub.
6. Remove and discard the piston outer seal from the piston itself and the piston inner seal from the clutch drum.
7. Inspect the drum for wear, damage, and cracks, and the band surface of the drum for severe wear or scoring from metal-to-metal band contact. If the drum is cracked or worn excessively in the band area, replace the drum. If the drum is serviceable, deglaze its band surface by bead blasting it or by sanding around the drum's circumference with 120-180 grit sandpaper or emery cloth (Fig. 9-47). <u>Do not sand this surface in a front-to-back direction.</u> With a cape or bushing chisel remove the drum bushing. Replace the bushing using a driver and driving head as shown in Fig. 9-48.
8. Clean the drum in a suitable cleaning solvent and blow dry with low pressure compressed air.
9. Clean the clutch piston in a suitable cleaning agent and blow it dry with low pressure compressed air.
10. Inspect the piston for wear and cracks. Make sure the residual check ball and retainer are serviceable and the ball operates freely in its bore within the piston.
11. Install a new inner seal over the clutch piston guide inside the drum. Install a new outer seal on the clutch piston itself. Lubricate both the seals with clean transmission fluid and reinstall the piston into the clutch drum, using firm, steady hand pressure. <u>Be careful not to cock the piston in its drum bore; this can tear and ruin the seals.</u>

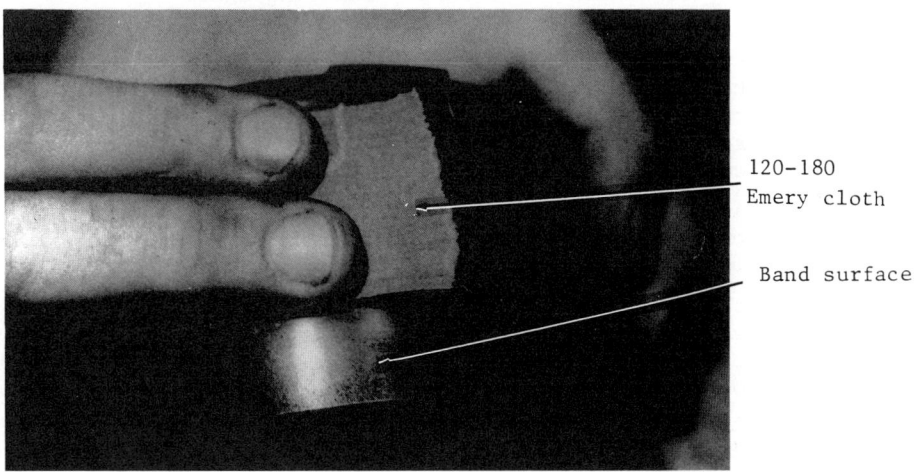

FIG. 9-47 Deglazing the reverse-high drum with sandpaper.

FIG. 9-48 Replacing the reverse-high bushing.

12. Place the clutch piston return spring(s) into position on the clutch piston. Place the spring retainer on top of the springs, and install the snap ring using the tool shown in Fig. 9-46. As you adjust this tool downward, compressing the springs, make sure to keep the spring retainer centered over the snap-ring groove. If the retainer catches in this groove, it will bend.
13. Install the snap ring and remove the compressor tool from the drum.
14. Soak all the new composition (friction) plates in clean transmission fluid for 30 minutes before installing them.
15. It is always wiser to install new steel driven plates. But if the old plates are still serviceable and you are going to reuse them, either bead blast or sand both sides of each plate with 120-320 grit emery or sandpaper until sanding marks are visible over the total bearing surfaces of each plate (Fig. 9-49). Perform the same procedure on the friction plate side of the pressure plate.
16. Clean each plate in a suitable cleaning solvent and blow dry with compressed air. Then lubricate each plate with clean transmission fluid.
17. Install the clutch plates alternately in the drum starting with a steel plate, then a friction (composition) plate (Fig. 9-50). The last plate installed is the pressure plate. <u>Note: For the correct number of reverse high clutch plates required for this transmission model, refer to the manufacturer's specifications in the service manual.</u>
18. Install the pressure plate snap ring with a screw driver (Fig. 9-51). Make sure the snap ring fully seats in its groove in the clutch drum.

Transmission Overhaul

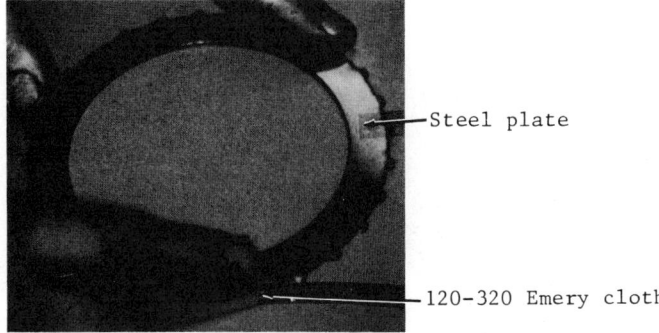

FIG. 9-49 Deglazing the clutch steel plates.

19. With a feeler gauge, measure the clearance between the snap ring and the pressure plate (Fig. 9-52).
20. Hold the pressure plate downward as you make the check. The clearance should be between .050 and .071 inch. If the clearance is not within specifications, selective snap rings are available from your supplier in these thicknesses: .050-.054, .064-.068, .078-.082, and .092-.096 inch. Install the correct size snap ring as necessary and recheck the clearance.

Forward Clutch

1. Remove the forward clutch pressure plate snap ring with a screw driver (Fig. 9-53).
2. Remove the pressure plate, drive, and driven clutch plates from the clutch drum.

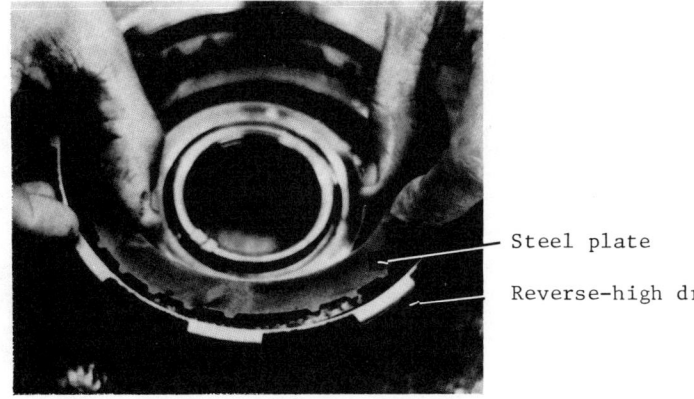

FIG. 9-50 Installation of the plates into the reverse-high drum.

FIG. 9-51 Installation of the reverse-high pressure-plate snap ring.

3. Using the tool and arbor press shown in Fig. 9-54, remove the disc spring snap ring with a screwdriver.
4. To remove the piston from the clutch drum, apply air pressure to the clutch-piston apply-pressure hole in the drum.

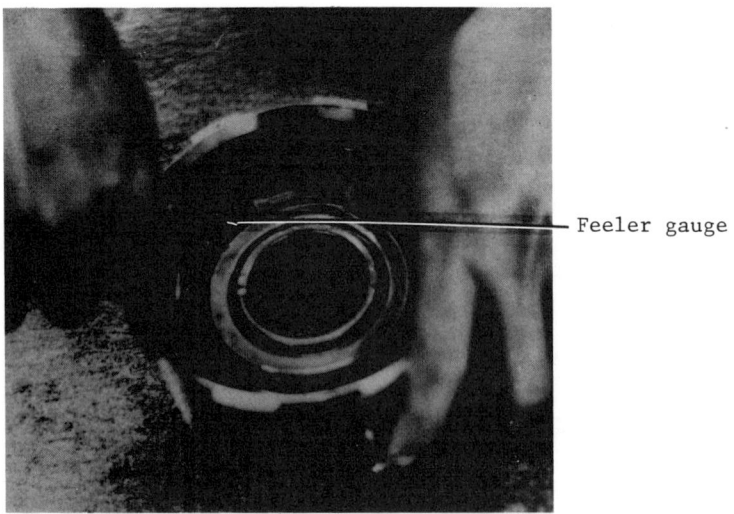

FIG. 9-52 Measuring reverse-high clutch clearance with a feeler gauge.

Transmission Overhaul

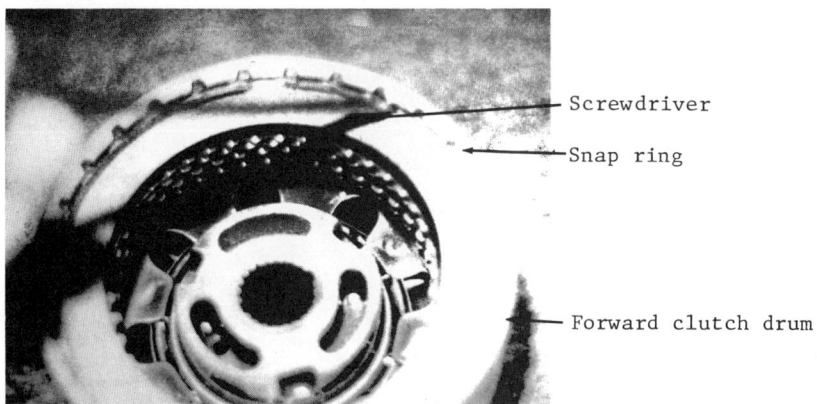

FIG. 9-53 Removing the forward-clutch pressure-plate snap ring.

5. Remove the clutch-piston outer seal and the inner seal from the clutch drum and discard.

6. Wash the piston in a suitable cleaning solvent and blow dry with low pressure compressed air.

7. Inspect the piston for wear and cracks. Make sure the residual check ball and retainer are serviceable, and the ball operates freely in its bore within the piston.

8. Clean the clutch drum in a suitable cleaning solvent and blow dry with low pressure compressed air. Inspect the drum for wear, damage, and cracks; replace it as required.

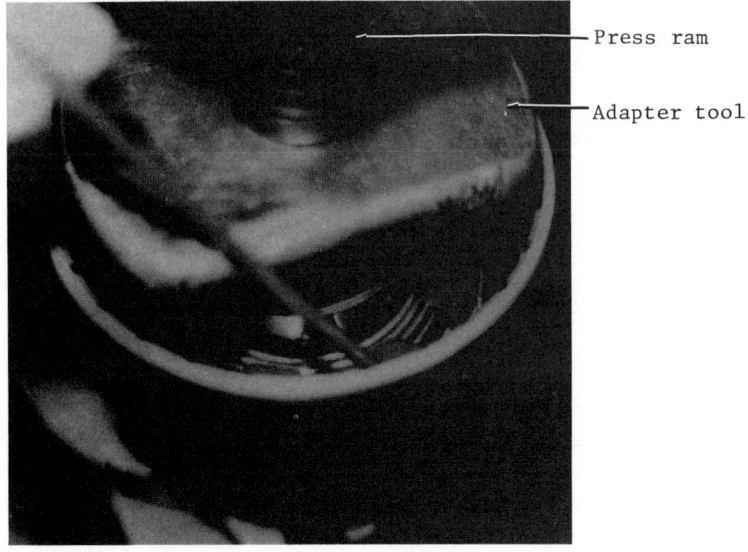

FIG. 9-54 Removing the disc-spring snap ring.

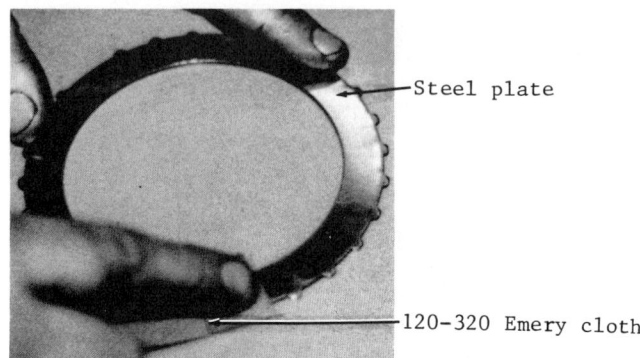

FIG. 9-55 Deglazing the forward clutch steel plates.

9. Install a new clutch piston seal on the piston and one over the piston guide inside the drum. Lubricate these seals with clean transmission fluid.

10. Reinstall the piston into the clutch drum, using firm, steady hand pressure. Be careful not to cock the piston in its drum bore; this can tear and ruin the seals.

11. Install the disc return spring and snap ring using the tools shown in Fig. 9-54.

12. Soak all new composition (friction) plates in clean transmission fluid for 30 minutes before installing them.

13. It is always wiser to install new steel drive plates. But if the old plates are still serviceable and you are going to reuse them, either bead blast or sand both sides of each plate with 120-320 grit emery or sandpaper until sanding marks are visible over the total bearing surfaces of each plate (Fig. 9-55). Perform the same procedure on the friction plate side of each of the pressure plates.

14. Clean each plate in a suitable cleaning solvent and blow dry with compressed air. Lubricate each plate with clean transmission fluid.

15. Install the lower pressure plate with the flat side up and the radius side down.

16. Install one friction (driven) clutch plate and then alternately install the remaining drive and driven plates. The last plate installed will be the upper pressure plate. <u>Note: Refer to the shop manual for the correct number of clutch plates for this particular transmission model.</u>

17. Install the pressure plate snap ring with a screwdriver (Fig. 9-56). Make sure the snap ring fully seats in its groove in the clutch drum.

18. With a feeler gauge, check the clearance between the snap ring and

Transmission Overhaul

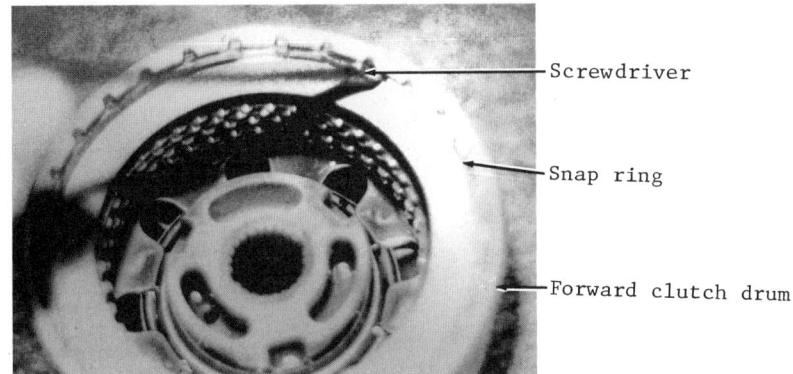

FIG. 9-56 Installation of the forward-clutch pressure-plate snap ring.

the pressure plate (Fig. 9-57). Use downward pressure on the pressure plate when making this check. The clearance should be .025-.050 inch.

19. If the clearance is not within specifications, selective snap rings are available in these thicknesses: .050-.054, .064-.068, .078-.082, and .092-.096 inch. Install the correct size snap ring and recheck the clearance.

Forward Planetary Gear Train

ring gear and clutch hub

1. Inspect the teeth of the ring gear itself for wear and damage (Fig. 9-58).

FIG. 9-57 Measuring the forward clutch clearance with a feeler gauge.

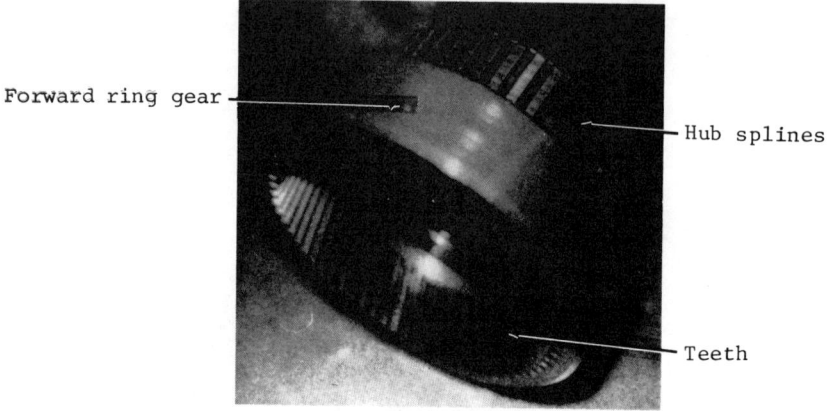

FIG. 9-58 Check the forward ring gear for wear and damage.

2. Inspect the splines of the clutch hub for wear and damage.
3. If either the ring gear or clutch hub require replacement, remove the retaining snap ring and separate the units.
4. Remove and replace the forward clutch hub bushing using the tool shown in Fig. 9-59.
5. Clean the ring gear and clutch hub in a suitable cleaning solvent and blow dry with compressed air.
6. If you have separated the ring gear from the clutch hub to replace

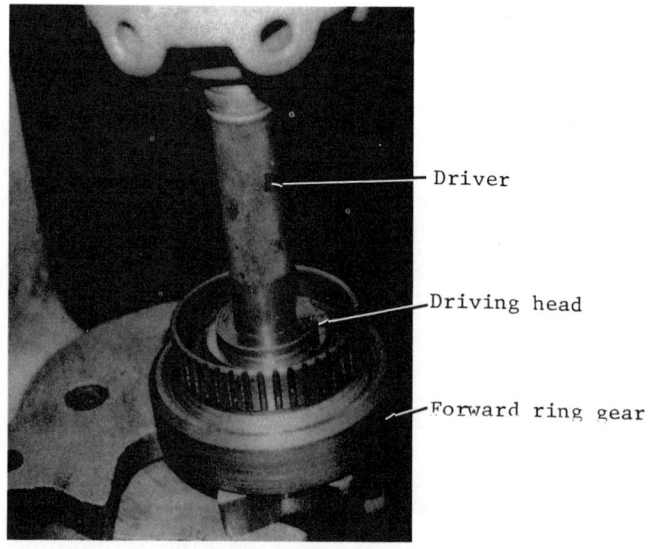

FIG. 9-59 Removing and replacing the forward-clutch hub bushing.

Transmission Overhaul

FIG. 9-60 Inspect the forward carrier and its components for wear and damage.

either part, install the forward clutch hub in the ring gear, making sure the hub bottoms in the groove of the ring gear.

7. Install the forward clutch hub retaining snap ring. Make certain that this snap ring fully seats in the snap ring groove in the ring gear.

forward carrier

1. Clean the forward carrier in a suitable cleaning solvent and blow dry using low pressure compressed air.
2. Inspect the carrier-to-output shaft splines for wear or damage (Fig. 9-60).
3. Inspect the teeth of each pinion gear for wear or damage. Check each pinion for excessive end play and looseness on its carrier support pin. If the carrier splines, pinion teeth, pinion thrust washers, or bearings show damage or excessive wear, replace the carrier assembly.

input shell and sun gear

1. Remove the external snap ring from the sun gear using snap ring pliers (Fig. 9-61).
2. Remove No. 5 thrust washer from the input shell and sun gear.
3. From the inside of the input shell, remove the sun gear and then remove, if necessary, the snap ring from the sun gear.
4. Remove both sun gear bushings using the tools shown in Fig. 9-62. Press both bushings through the gear at the same time.
5. Inspect the input shell driving lugs or slots for wear or damage. Inspect the shell itself for cracks or damage; replace it if necessary.
6. Inspect the sun gear teeth for wear or damage; replace it if

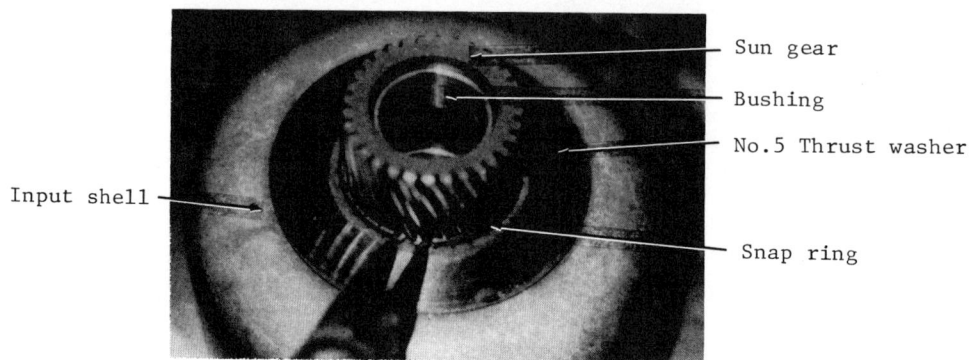

FIG. 9-61 Removal of sun gear snap ring.

necessary. Clean the input shell and sun gear in a suitable cleaning solvent and blow dry with compressed air.

7. Using the tool shown in Fig. 9-62, press a new bushing into each end of the sun gear.
8. If removed, reinstall the internal snap ring on the sun gear, and install the sun gear into the input shell.
9. Reinstall the No. 5 thrust washer on the sun gear and input shell.
10. Reinstall the external snap ring over the sun gear and into its groove (Fig. 9-63).

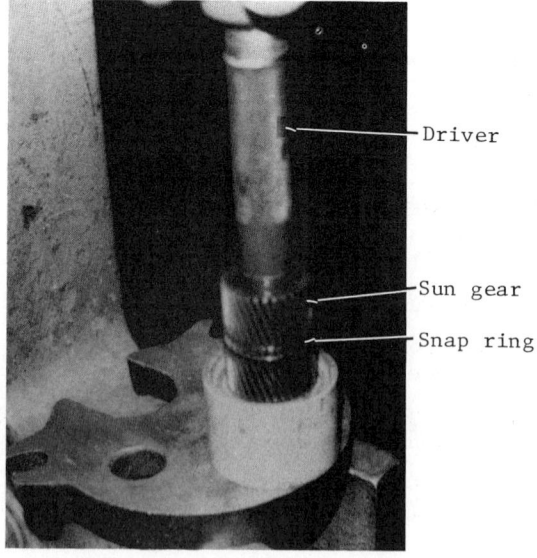

FIG. 9-62 Removal and replacement of sun gear bushings.

Transmission Overhaul

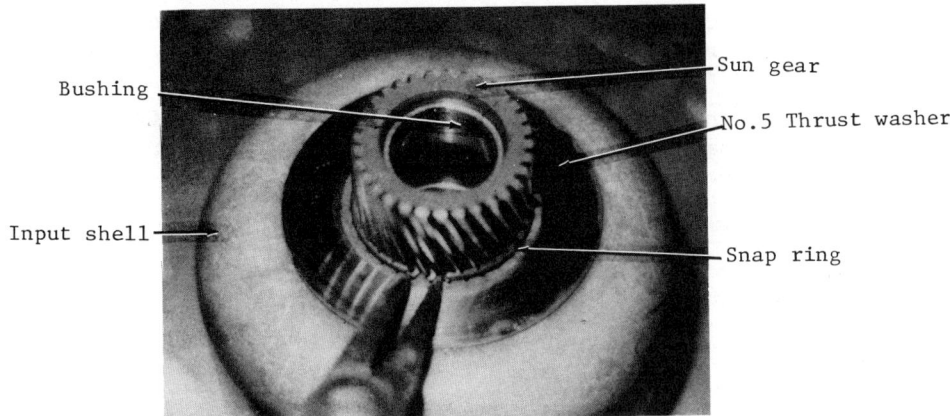

FIG. 9-63 Installation of the sun gear snap ring.

Reverse Planetary Gear Train

ring gear and hub

1. Inspect the teeth of the reverse ring gear for wear or damage (Fig. 9-64).
2. Inspect the output shaft splines on the hub for wear and damage.
3. If either the ring gear or hub require replacement, remove the retaining snap ring and separate the units.
4. Clean the ring gear and hub in a suitable cleaning solvent and blow dry with compressed air.

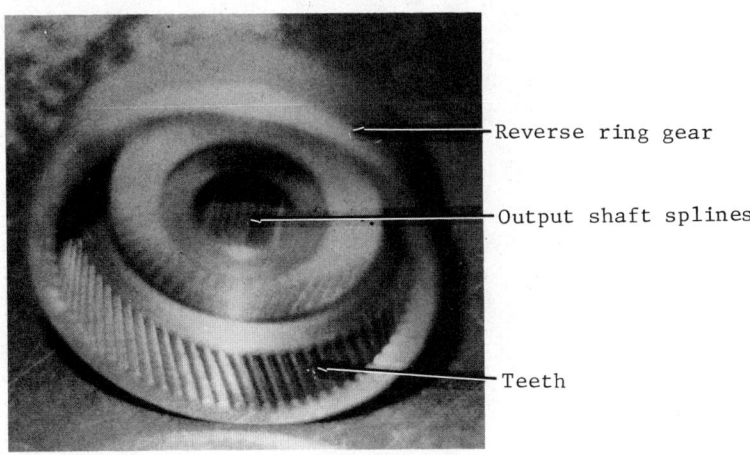

FIG. 9-64 Inspect the reverse ring gear and hub for wear and damage.

5. If separated, to replace either part, install the hub in position within the ring gear. Make sure the hub fully seats in its groove within the ring gear.
6. Reinstall the snap ring in the reverse ring gear. Make certain that the snap ring fully seats in the snap ring groove within the ring gear itself.

low reverse carrier

1. Clean the low reverse carrier in a suitable cleaning solvent and blow dry using low pressure compressed air.
2. Inspect the driving lugs or slots of the carrier for wear or damage (Fig. 9-65).
3. Inspect the teeth of each pinion gear for wear or damage. Check each pinion for excessive end play and looseness on its carrier support pin. If the carrier lugs, pinion teeth, pinion thrust washers, or bearings show damage or excessive wear, replace the carrier assembly.

Low Reverse Brake Drum

1. Inspect the drum along with its driving lugs or slots for wear or damage (Fig. 9-66). Check the splines that support the inner race of the one-way clutch for wear and damage and the band surface of the drum for any severe wear or scoring from metal-to-metal band contact. If the lugs, drum, splines, or band surface are cracked or worn excessively, replace the drum.

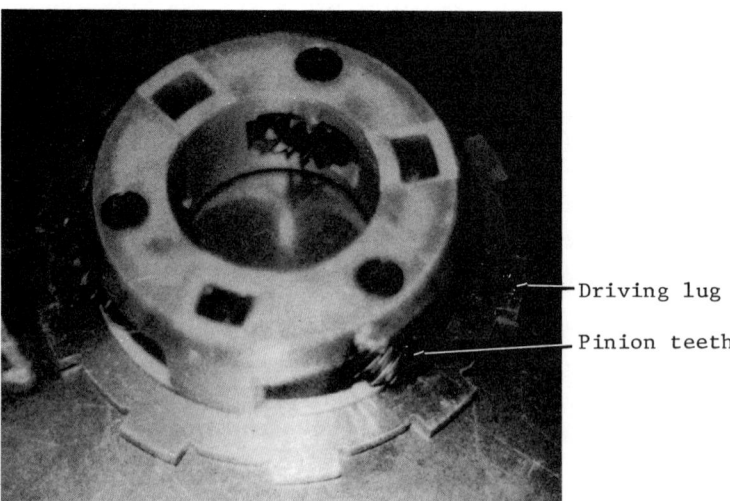

FIG. 9-65 Inspect the low-reverse carrier for wear and damage.

Transmission Overhaul

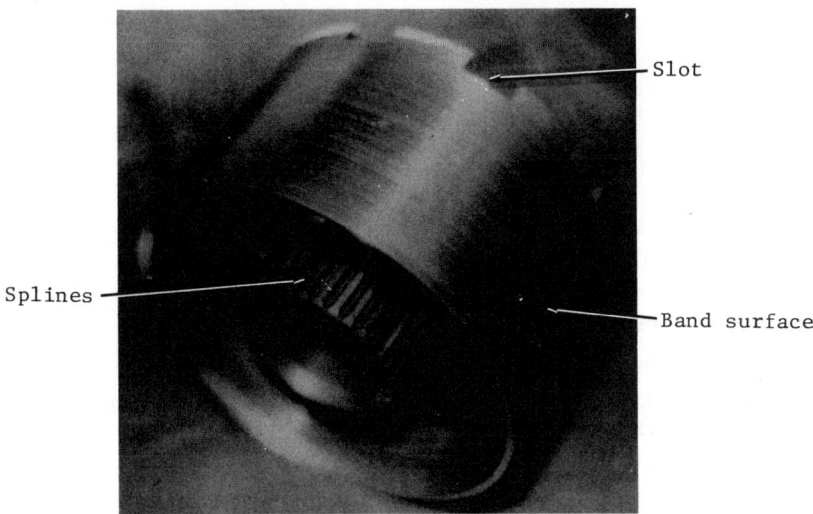

FIG. 9-66 Inspect the low-reverse drum for wear and damage.

2. If the drum is serviceable, deglaze its band surface by bead blasting it or by sanding this surface, front to back, using 40-60 grit sandpaper or emery cloth (Fig. 9-67). <u>Do not sand around the drum.</u>

3. To remove the bushing, use a cape or bushing chisel and cut along the bushing seam until the chisel breaks through the bushing wall. Pry up the loose ends of the bushing with an awl or suitable tool and remove the bushing.

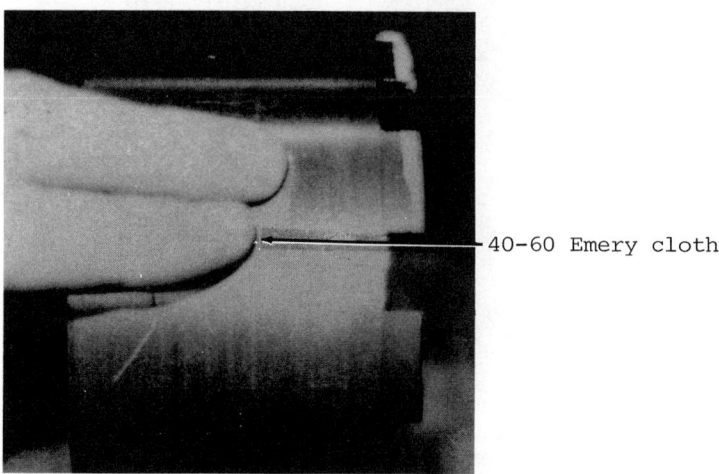

FIG. 9-67 Deglazing the band surface of the low-reverse drum.

4. Install a new bushing with the tool shown in Fig. 9-68.

5. Clean the low reverse drum in a suitable cleaning solvent and blow dry with compressed air.

Output Shaft and Distributor Sleeve

1. Clear the output shaft and distributor in a suitable cleaning solvent and blow them dry with compressed air.

2. Inspect the splines and bearing surfaces of the output shaft for wear or damage; replace the shaft if necessary.

3. Remove the three metal sealing rings from the distributor and discard.

4. Inspect the distributor for visible signs of wear and damage; replace it if necessary.

5. Reinstall the distributor on the output shaft. Install the 1¼-inch distributor retaining ring with snap-ring pliers (Fig. 9-69).

6. Replace the metal sealing rings on the distributor (Fig. 9-70). Make sure the rings fully seat in their respective ring grooves and will rotate freely.

Governor Assembly

1. Remove the primary governor valve retaining ring with snap-ring pliers. Remove the washer, spring, and primary governor valve from the housing (Fig. 9-71).

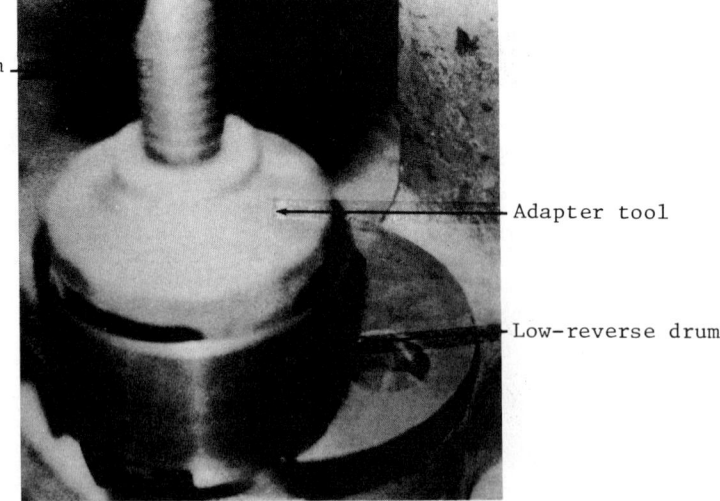

FIG. 9-68 Installation of the low-reverse drum bushing.

Transmission Overhaul

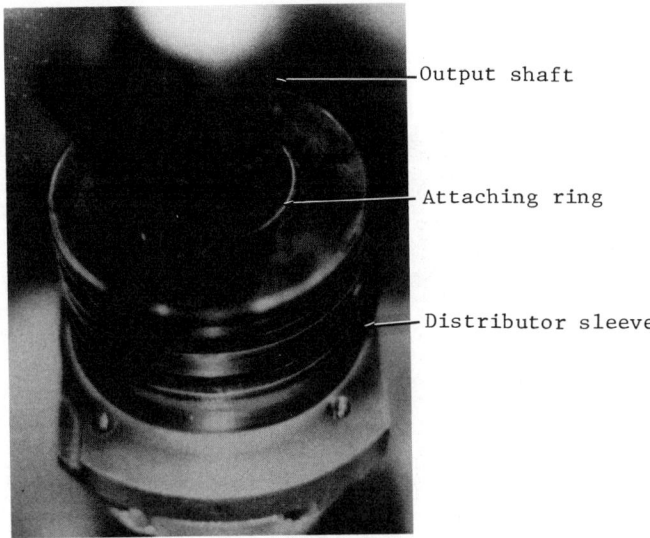

FIG. 9-69 Installation of the distributor-to-output shaft snap ring.

2. Remove the secondary governor valve spring retaining clip, spring, and governor valve from the housing.

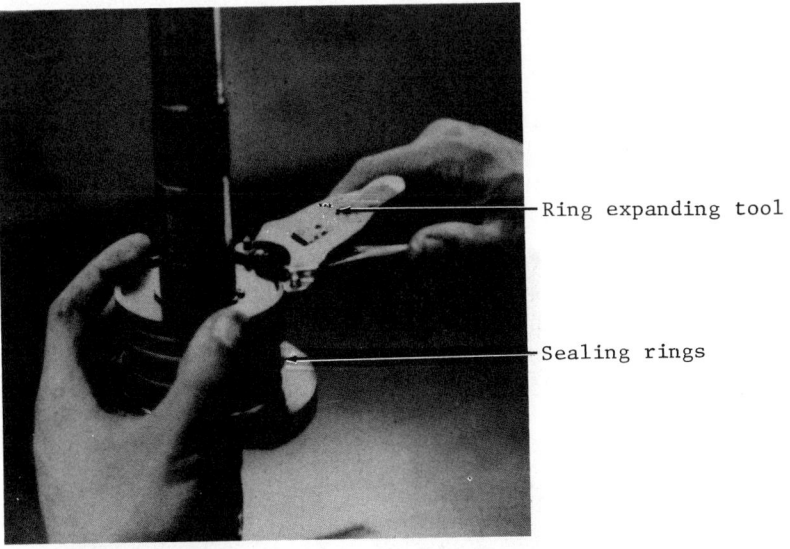

FIG. 9-70 Replacing the metal sealing rings on the distributor, using a special ring expander.

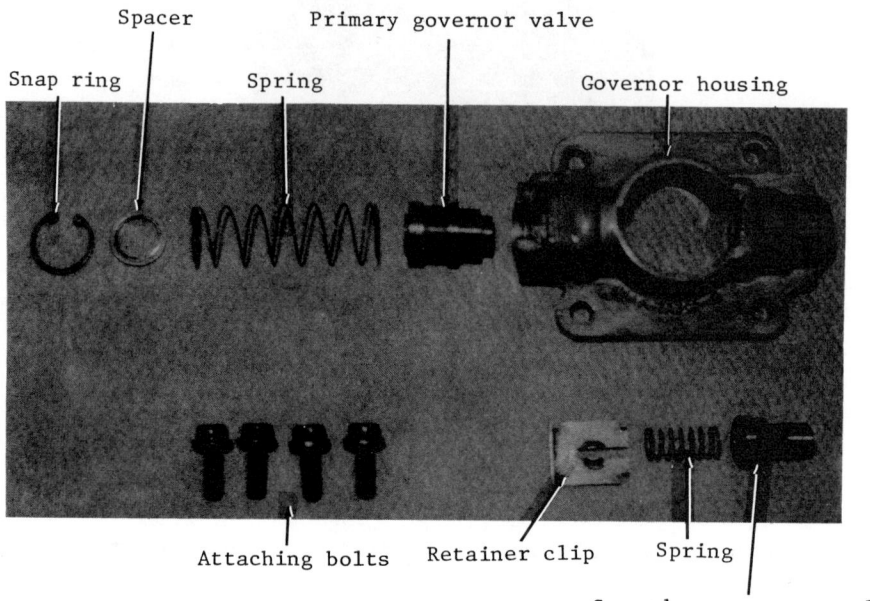

FIG. 9-71 Servicing the governor housing and its components.

3. Clean the governor housing and all its components in a suitable cleaning solvent and blow dry with low pressure compressed air. <u>Be careful not to lose the smaller parts.</u>

4. With a straightedge, check the governor housing-to-distributor mating surface for warpage (Fig. 9-72). If slight warpage exists in this area, resurface it by moving the housing back and forth over a piece of 320-grit emery or sandpaper, resting on an extremely flat surface. Reclean and dry the housing.

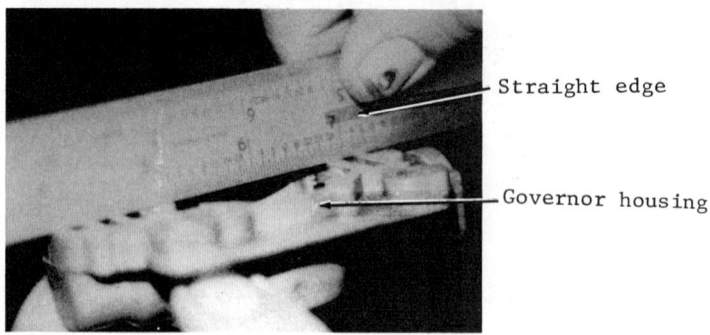

FIG. 9-72 Checking the governor housing for warpage.

Transmission Overhaul

5. Check the primary and secondary governor valves in their respective housing bores. The valves must slide freely back and forth on a dry surface. Remove the valves from the housing and lubricate them in clean transmission fluid.

6. Install the primary governor valve into the housing. Reinstall the spring, washer, and snap ring. <u>Make sure the washer is centered in the housing on top of the spring and the snap ring fully seats in its groove in the housing.</u>

7. Install the secondary governor valve in the housing; reinstall the spring and retaining clip. <u>Make sure to install the clip with the small, concaved area facing the spring; this holds the spring in the correct position.</u>

One-Way Clutch

1. Clean the one-way clutch's outer race, inner race, spring retainer, rollers, and springs in a suitable cleaning solvent and blow dry with low pressure compressed air (Fig. 9-73).

2. Inspect the outer race for worn, pitted, or damaged roller ramps. Check the mount bolts' holes for damaged threads; replace the race if necessary.

3. Inspect the outer circumference of the inner race for wear, pits, or damage. Check the splines for wear and damage; replace the inner race as required.

FIG. 9-73 Inspect the one-way clutch components for wear and damage.

4. Check the spring retainer for warpage, wear, or damage; replace as necessary.
5. Inspect each roller for wear, pits, or other damage.
6. Check each spring for warpage, wear, or damage; replace any roller or spring as required.

Low Reverse Band

1. Inspect the band lining for excessive wear, pits, and signs of overheating.
2. If the band is still serviceable, deglaze the lining by scraping it with a sharp object such as a knife (Fig. 9-74).
3. Clean the band in a suitable cleaning detergent and blow dry with compressed air.
4. Lubricate with clean transmission fluid.

Extension Housing

1. Inspect the extension housing for cracks or other visible damage.
2. With a chisel or other suitable puller, remove the extension housing seal (Fig. 9-75).

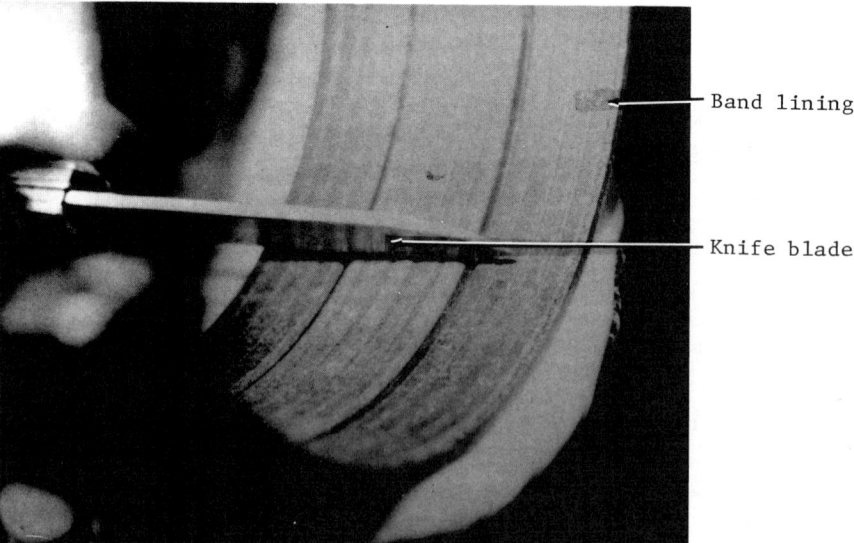

FIG. 9-74 Deglazing the low-reverse band.

Transmission Overhaul

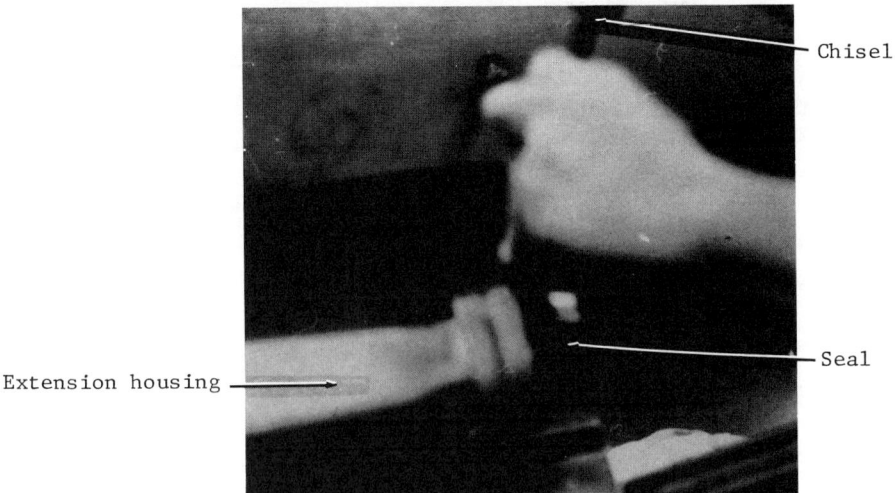

FIG. 9-75 Removing the extension housing seal.

3. With a driver and the proper driving head, remove the extension housing bushing (Fig. 9-76).
4. Wash the housing in a suitable solvent and blow dry with compressed air.
5. With the tools shown in Fig. 9-76, install a new extension housing bushing.
6. Replace the extension housing seal using the driver shown in Fig. 9-77.

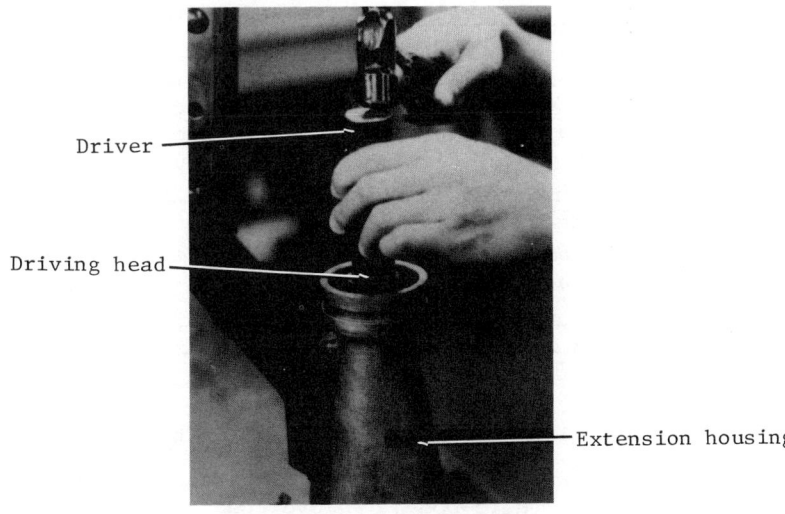

FIG. 9-76 Removing and replacing the extension housing bushing.

FIG. 9-77 Installing the extension housing seal.

Valve Body

Note: Before attempting to disassemble, clean, and reassemble the C-4 body, review the general valve body repair procedures as outlined in Section 8 of this manual.

disassembly

1. Remove the eight 10-24 x 1-3/8-inch machine screws which attach the fluid screen to the valve body and remove the screen and gasket. <u>Be careful not to lose the throttle pressure limit valve and spring when separating the fluid screen from the body.</u>

2. Remove the two 1/4-20 x 1-1/2-inch machine attaching screws from the upper valve body and the nine 10-24 x 7/8-inch attaching screws from the underneath side of the lower valve body. Separate the lower valve body, gasket, and separator plate (Fig. 9-78) from the upper valve body. <u>Be careful not to lose the upper valve-body/shuttle valve and check valve when separating the upper and lower valve bodies.</u>

3. From the upper valve body (Fig. 9-79) remove the manual valve retaining ring and slip the manual valve out of its bore in the body. Note: <u>The manufacturer uses the retaining ring to hold the manual valve in its bore during shipment. Therefore, it is not necessary to reinstall this ring during reassembly.</u>

4. Carefully pry the low servo modulator valve retainer from the body, and remove the retainer plug, spring, and valve from the upper valve body.

5. Using side cutters, carefully pry out the throttle booster-valve plug retaining pin from the valve body. Remove the plug, valve, and spring.

Transmission Overhaul

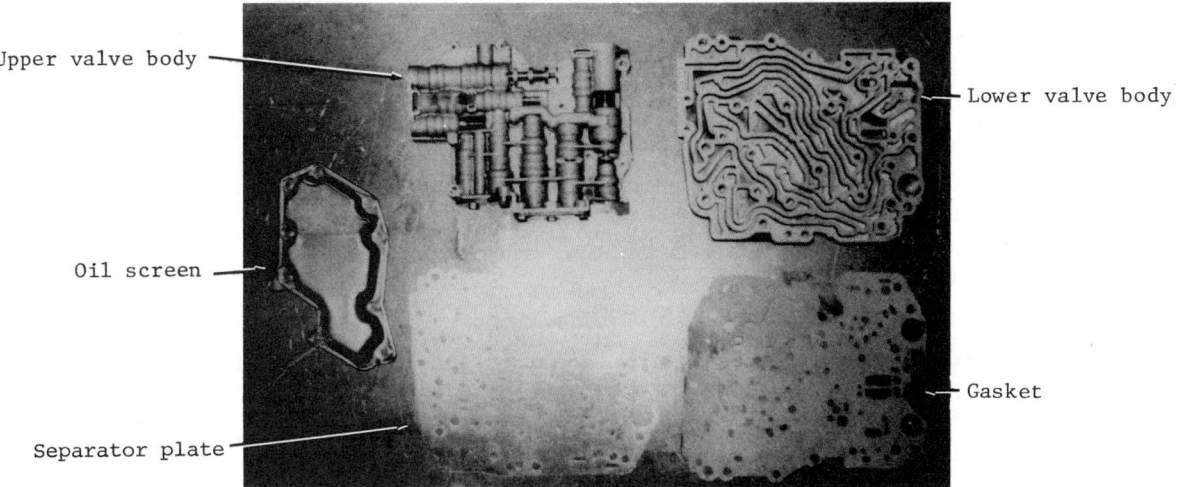

FIG. 9-78 Screen, upper and lower valve bodies, gaskets, and separator plate.

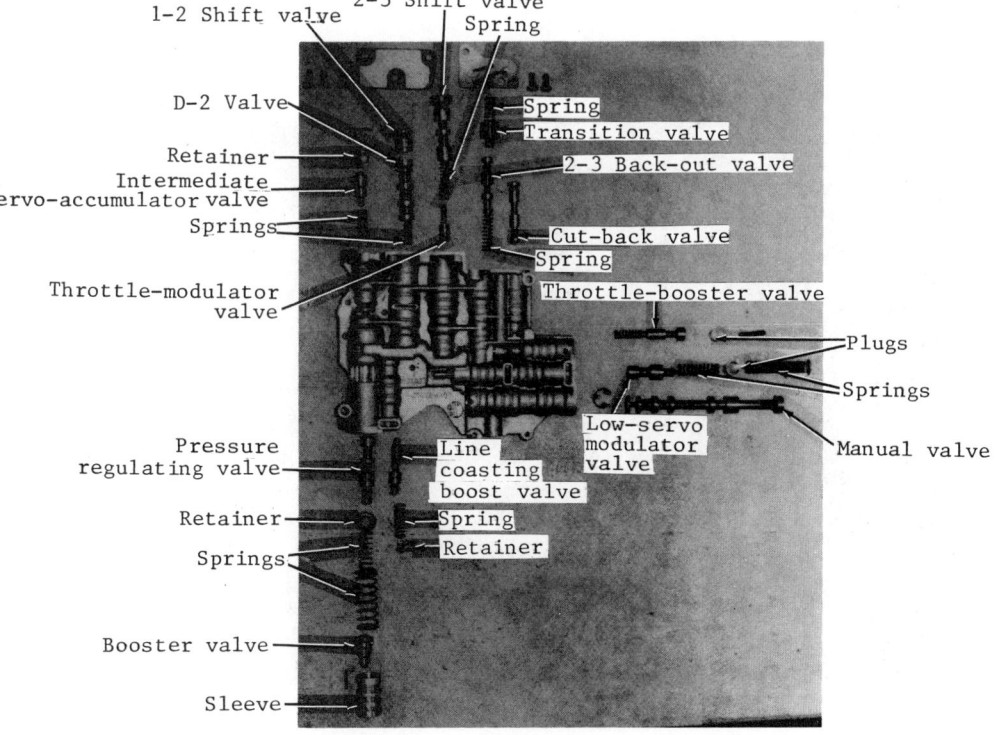

FIG. 9-79 Servicing the components of the upper valve body.

6. Remove the two attaching screws and take off the cut-back valve and transition valve cover plate from the valve body.
7. Remove the cut-back valve from its bore.
8. Remove the transition valve spring, transition valve, 2-3 back-out valve, and spring from the body.
9. Remove the three attaching screws and take off the 1-2 shift and 2-3 shift valve cover plate from the upper valve body.
10. Remove the 2-3 shift valve, spring, and throttle modulator valve from their bore.
11. Remove the 1-2 shift valve, D2 valve, and spring from their bores.
12. Take out the intermediate servo retaining pin and remove the intermediate accumulator retainer, valve, and spring from their bores in the upper valve body.
13. Push the main oil pressure booster valve inward and remove the retaining pin. Next, remove the main oil pressure booster valve, sleeve, springs, retainer, and the main oil pressure regulator valve.
14. Take out the line coasting boost valve retainer from the upper valve body; remove the spring and the line coasting boost valve.
15. Clean and inspect all components as outlined in Section 8.

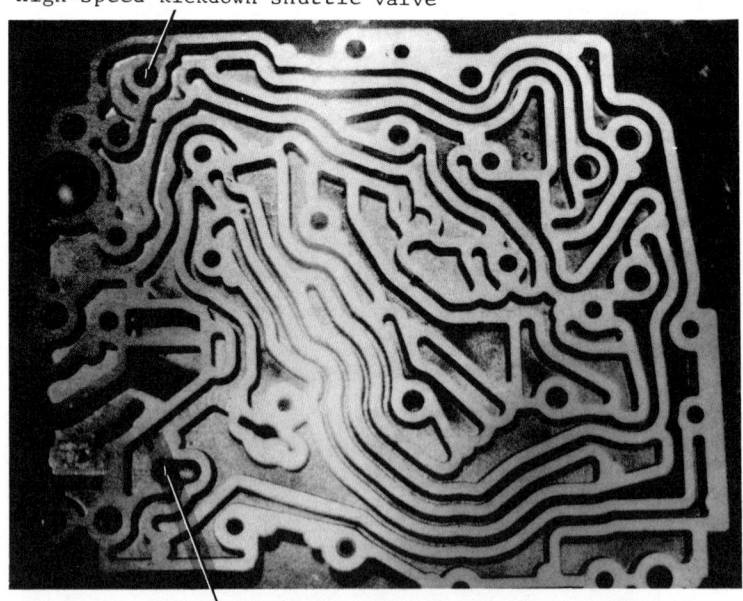

FIG. 9-80 Shuttle ball locations in the lower valve body.

Transmission Overhaul

FIG. 9-81 Shuttle ball and servo-check valve locations within the upper valve body.

reassembly

1. Place all the shuttle valves and servo check valve in the lower and upper valve bodies as shown in Figs. 9-80 and 9-81. Position the new gasket and separator plate on the lower valve body and install <u>but do not tighten</u> the attaching screw.
2. Place the lower valve body and plate assembly on the upper valve body and install the eleven attaching screws finger tight.
3. Install the fluid screen screws loosely, without the screen, to correctly align the upper and lower valve bodies, gasket, and separator plate.
4. Torque the four screws that the screen will cover to 40 to 55 pounds-inch.
5. Reinstall the throttle-valve limit valve and spring in the lower valve body. Remove the screen attaching screws and place the gasket and fluid screen in position on the lower valve body. Reinstall the screen attaching screws.
6. Torque all the (10-24) valve body and screen attaching screws to 40 to 55 pounds-inch; torque the two 1/4-inch upper valve body to lower valve body attaching screws to 80 to 120 pounds-inch.
7. Install the low servo modulator valve, spring, and retainer plug into the valve body; depress the plug and install the retainer.

8. Insert the throttle booster valve spring, valve (small diameter end into spring), and plug into the body. Depress the plug inward and reinstall the retaining pin. <u>Make sure the three grooves are at the top of the pin as you reinstall it.</u>

9. Position the spring, 2-3 back-out valve, and the transition valve and spring into the valve body.

10. Install the cut-back valve into the body.

11. Secure the cut-back and transition valve cover plate to the valve body with the two attaching screws. Torque these screws to 20 to 35 pounds-inch.

12. Place the throttle modulator valve, spring, and 2-3 shift valve into the valve body.

13. Install the spring, D2 valve, and the 1-2 shift valve into the valve body.

14. Secure the 1-2 shift valve and the 2-3 shift valve cover plate to the valve body with its three attaching screws; torque these screws to 20 to 35 pounds-inch.

15. Place the spring, intermediate servo accumulator valve and the retainer into the valve body. Depress the retainer slightly and reinstall the retaining pin.

16. Reinstall the line coasting boost valve and spring into the valve body. Depress the spring and reinstall its retainer.

17. Position the main oil pressure regulator valve and spring retainer into the valve body. Install the two springs, sleeve, and main oil pressure booster valve into the body.

18. While holding the main oil pressure booster valve in place, reinstall the retaining pin.

19. Slide the manual valve into its bore in the valve body. <u>Make sure that the valve end, with the two lands closest together, is inserted first.</u>

TRANSMISSION REASSEMBLY

General Instructions

1. When reassembling the transmission, make sure to install the correct thrust washer between the subassemblies.

2. Use Vaseline to hold the thrust washers in their proper location.

3. Lubricate all bearing surfaces with transmission fluid.

4. If the end play is not within specifications after assembling the transmission, check for a misplaced thrust washer or a thrust washer that came out of position during the assembly operation.

Transmission Overhaul

FIG. 9-82 Location of No. 9 thrust washer and the one-way clutch outer race.

5. Torque all fasteners to factory specifications.

Assembly Procedure

1. Reinstall the No. 9 thrust washer inside and at the rear of the transmission case (Fig. 9-82). Place the one-way clutch outer race inside the case.
2. From the back of the case, reinstall the six outer race-to-case attaching bolts, and with the case in a vertical position, torque these bolts to 13 to 20 pounds-foot (Fig. 9-83).

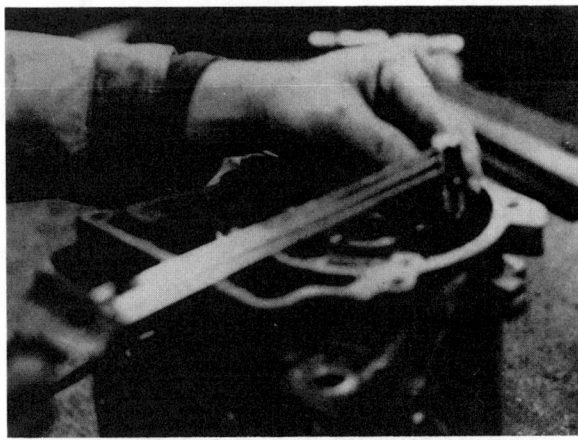

FIG. 9-83 Torquing the one-way clutch outer race-to-case attaching bolts.

FIG. 9-84 Installation of the parking pawl, return-spring, and retaining pin.

3. Install the parking pawl retaining pin in the transmission case (Fig. 9-84).
4. Reinstall the parking pawl on the retaining pin, and install the parking pawl return spring as shown in Fig. 9-84.
5. Install No. 10 thrust washer on the parking gear (Fig. 9-85). Place the gear and thrust washer on the back face of the transmission case.
6. Insert the two distributor tubes in the governor distributor sleeve; reinstall the sleeve onto the case. As you install the sleeve, insert the ends of the two tubes into their openings in the case, and

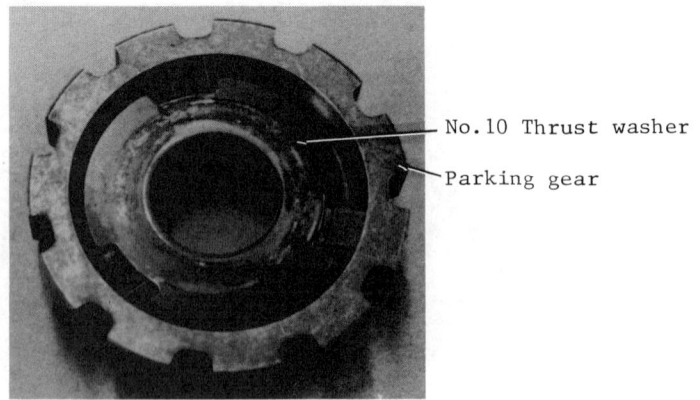

FIG. 9-85 No. 10 thrust washer location.

Transmission Overhaul

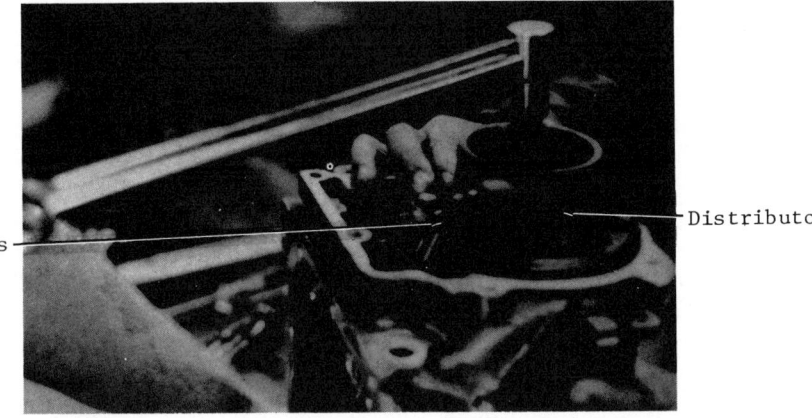

FIG. 9-86 Installing the governor distributor sleeve-to-case attaching bolts.

insert the parking pawl retaining pin into its alignment hole in the distributor sleeve.

7. Reinstall the four governor distributor sleeve-to-case attaching bolts and torque these bolts to 12 to 20 pounds-foot (Fig. 9-86).
8. Install the output shaft and governor distributor assembly into the distributor sleeve and case (Fig. 9-87).
9. Position the case back to a horizontal position. Then reinstall the governor housing on the governor distributor, and install the attaching

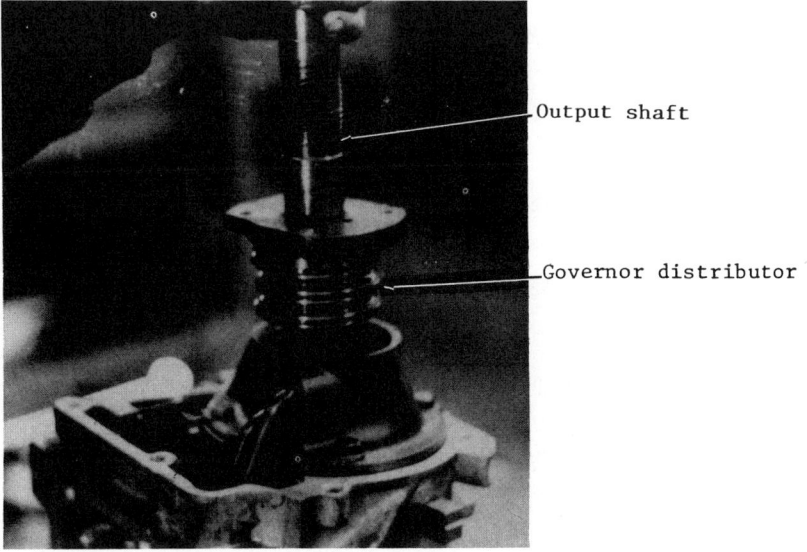

FIG. 9-87 Installing output shaft and governor distributor.

bolts (Fig. 9-88). Place the manual lever into the park position, and torque these bolts to 80 to 120 pounds-inch.

10. Position the transmission case in a vertical position again with the output shaft upward. Place a new extension housing gasket onto the case. Reinstall the extension housing, vacuum-tube clip, and the extension housing-to-case attaching bolts. Torque these bolts to 28 to 40 pounds-foot (Fig. 9-89).

11. Place the transmission case in a holding fixture, if one is available. Rotate the fixture so that the front pump mounting face is up. At this point, <u>make sure that No. 9 thrust washer is still in its position at the bottom of the case (Fig. 9-82).</u>

12. Reinstall the one-way clutch spring retainer into the case.

13. Install the inner race inside of the spring retainer. <u>Be certain that the face with the step faces toward the rear of the case, mating with the thrust washer.</u>

14. Install all the individual springs between the inner and outer races.

15. Starting at the back of the transmission case, reinstall each of the one-way clutch rollers by slightly compressing its spring and positioning the roller between the spring and the spring retainer (Fig. 9-90).

16. After assembling the one-way clutch, rotate the inner race clockwise to center the rollers and springs. Reinstall the low reverse drum (Fig. 9-91). The splines of this drum have to engage with the splines of the one-way clutch inner race. Check the one way clutch operation

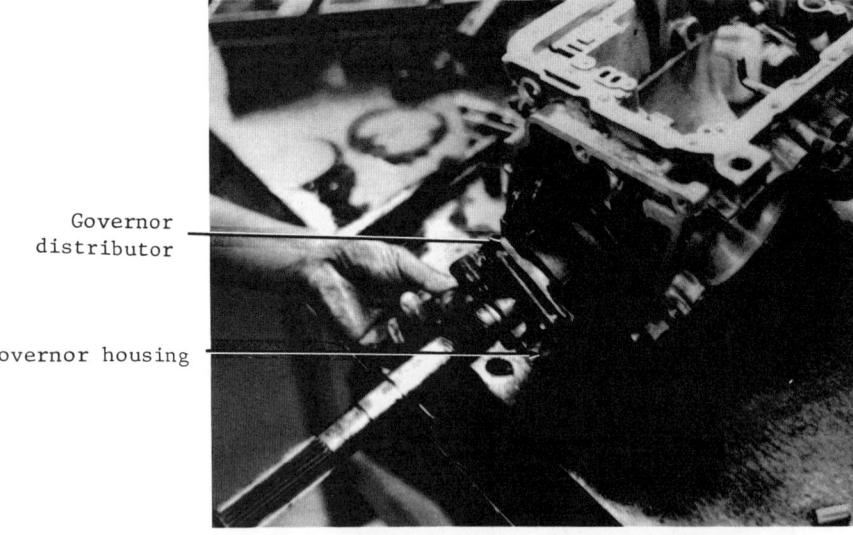

FIG. 9-88 Reinstalling the governor onto the distributor.

Transmission Overhaul

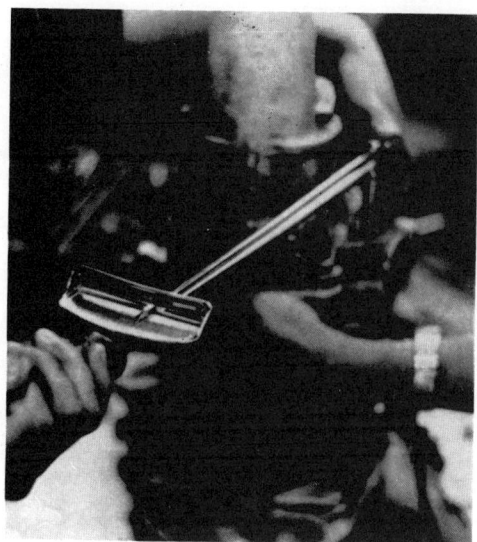

FIG. 9-89 Torquing the extension housing-to-case attaching bolts.

by turning the low reverse drum; the drum should turn clockwise but not counterclockwise.

17. Position No. 8 thrust washer onto the low reverse drum (Fig. 9-92).

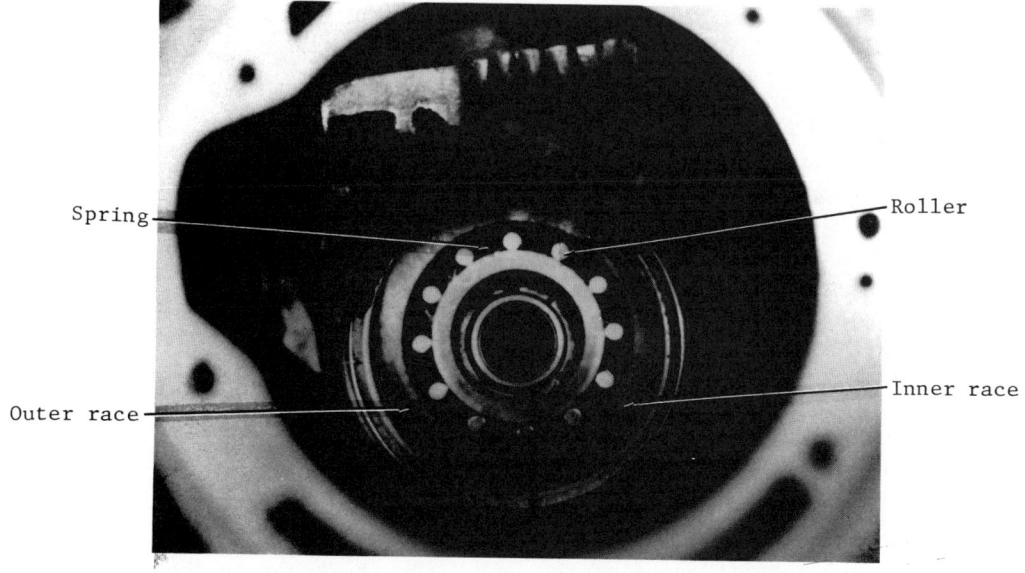

FIG. 9-90 Installing the one-way clutch.

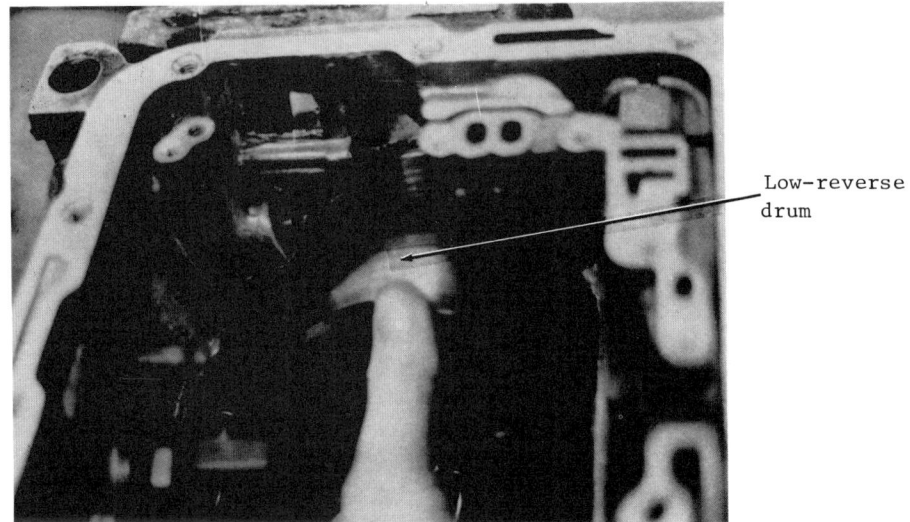

FIG. 9-91 Reinstalling the low-reverse drum.

18. Install the low reverse band into the case with the end of the band, accommodating the small strut toward the low reverse servo (Fig. 9-93).

19. Position the reverse ring gear and hub onto the output shaft. Move the output shaft forward and reinstall the 1-3/16-inch reverse ring gear hub-to-output shaft retaining ring (Fig. 9-94).

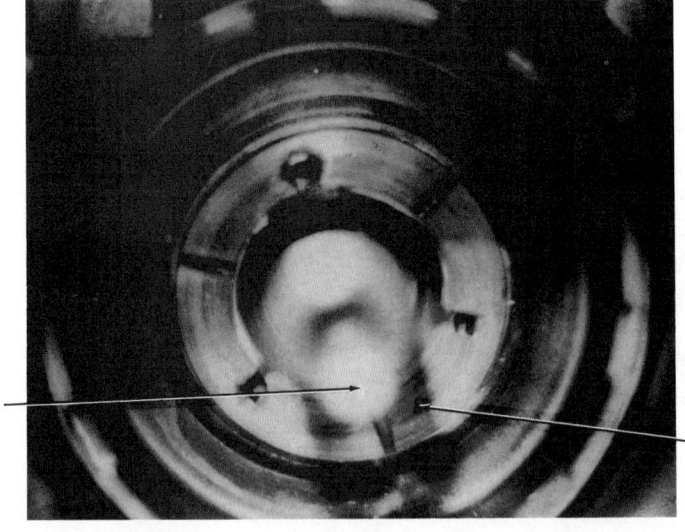

FIG. 9-92 Positioning of the No. 8 thrust washer on the low-reverse drum.

Transmission Overhaul

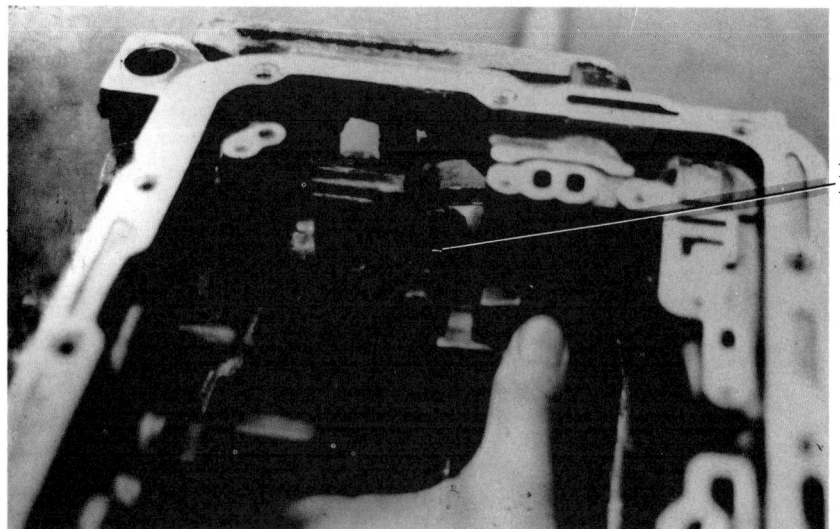

FIG. 9-93 Installing the low-reverse band into the case.

20. Place No. 7 thrust washer into position on the reverse planet carrier (Fig. 9-95).

21. Install the **reverse planet carrier** into the reverse ring gear, and engage the tabs of the carrier with the slots located in the low reverse drum (Fig. 9-96).

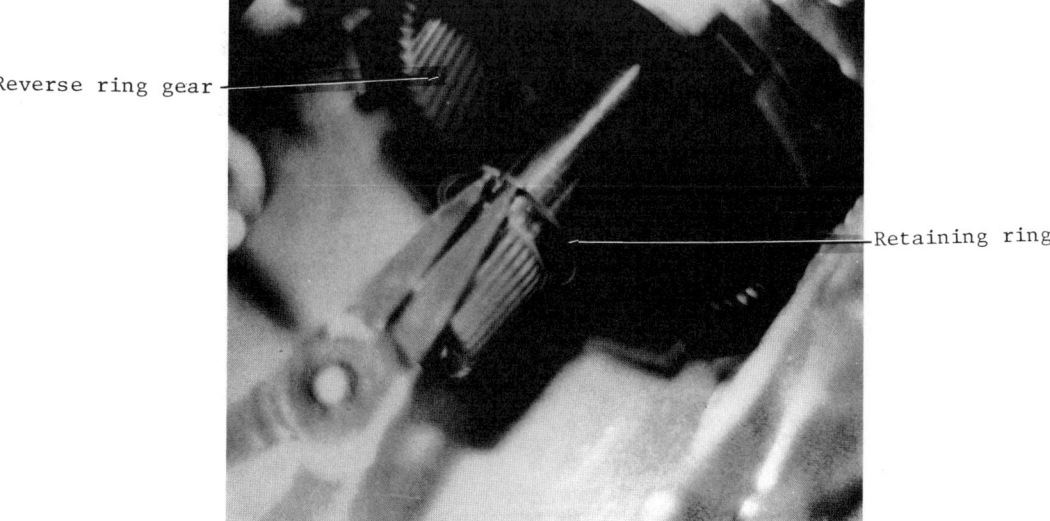

FIG. 9-94 Installing the reverse ring gear hub-to-output shaft retaining ring.

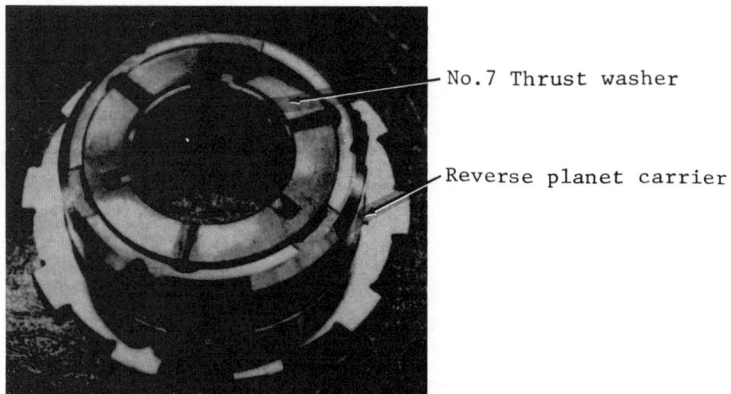

FIG. 9-95 Position of the No. 7 thrust washer.

22. Place No. 6 thrust washer in position in front of the reverse carrier (Fig. 9-97).

23. On a bench, install the forward clutch into the reverse high clutch by rotating the units. This meshes the reverse high clutch plates with the splines of the forward clutch hub (Fig. 9-98).

24. Using as a guide the end play check reading taken during the disassembly process, determine which No. 2 thrust washer is necessary to control end play and then proceed as follows:

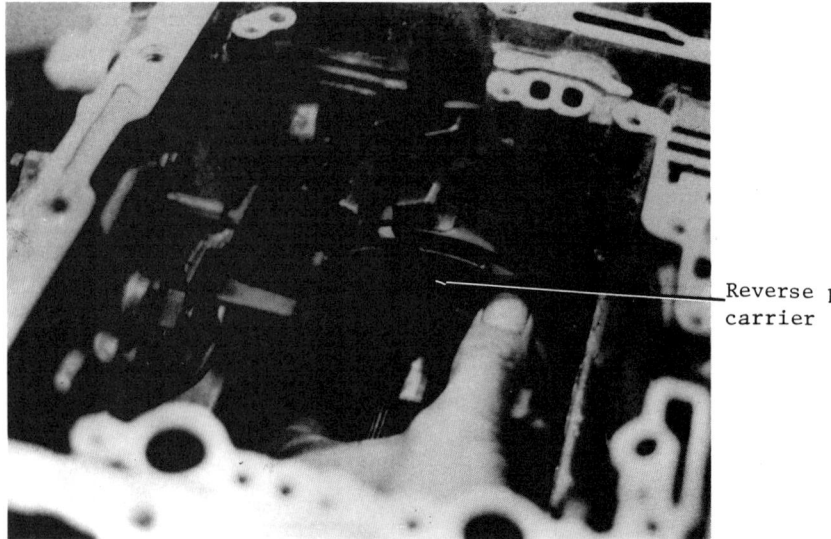

FIG. 9-96 Installing the reverse carrier.

Transmission Overhaul

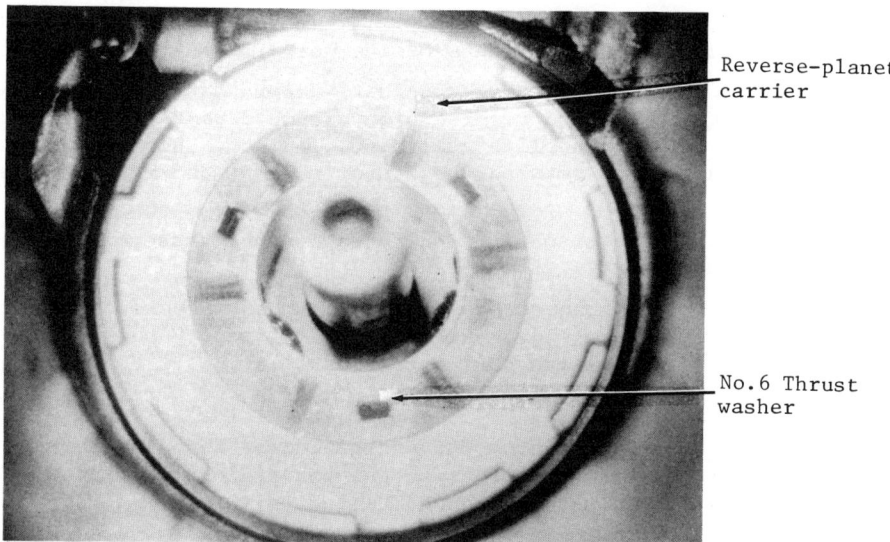

FIG. 9-97 Location of the No. 6 thrust washer.

a. Place the stator support vertically on a workbench and install the correct No. 2 thrust washer or washer with spacer as required to bring the end play up to specifications.

b. Position the reverse high and forward clutch onto the stator support.

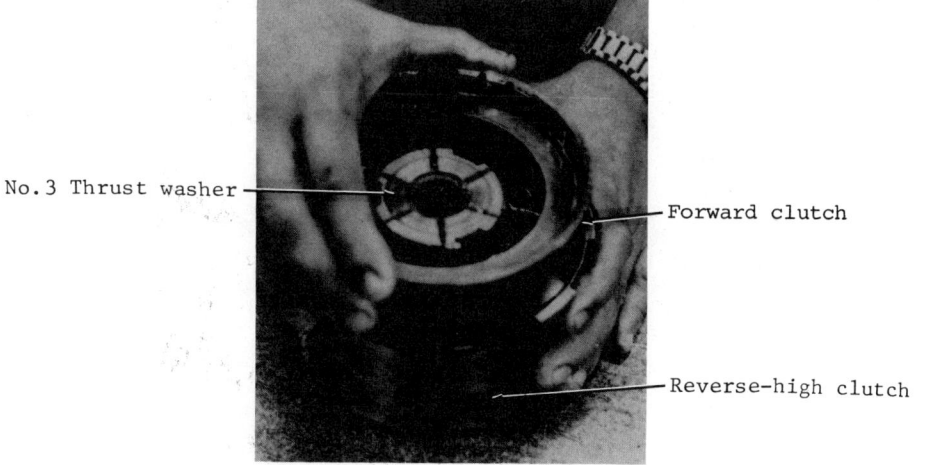

FIG. 9-98 Installing the forward clutch into the reverse-high clutch.

c. Invert the complete assembly, making sure that the reverse high clutch drum bushing seats on its forward clutch mating surface.

d. Select the thickest No. 1 thrust washer that will fit between the stator support and reverse high clutch drum thrust surfaces and still maintain a slight clearance. <u>Do not choose a washer that you must force between the stator support and reverse high clutch drum</u>.

e. Remove the reverse high and forward clutch assemblies from the stator support.

f. Install the selected Nos. 1 and 2 thrust washers on the front pump stator support (Fig. 9-99) using sufficient Vaseline to hold these washers in position during the installation of the front pump.

25. Install No. 3 thrust washer onto the forward clutch (Fig. 9-98).

26. Install the forward clutch hub and ring gear into the forward clutch by turning the ring gear to mesh the forward clutch plates with the splines on the forward clutch hub (Fig. 9-100).

27. Reinstall the No. 4 thrust washer onto the forward planet carrier (Fig. 9-101). Next, install the forward planet carrier into the forward clutch hub and ring gear. <u>Inspect the forward thrust bearing race inside the planet carrier for proper positioning against the thrust bearing. Be certain that the race is centered for alignment with the sun gear, located on the input shell</u>.

28. Install the input shell and sun gear onto the gear train (Fig. 9-102). Turn the input shell to engage its drive lugs with those of the

FIG. 9-99 Correct position of the Nos. 1 and 2 thrust washers on the stator support.

Transmission Overhaul

FIG. 9-100 Installation of the forward ring gear into the forward clutch assembly.

FIG. 9-101 No. 4 thrust washer location.

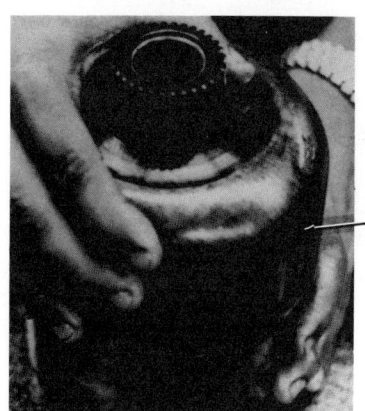

FIG. 9-102 Installation of the input shell and sun gear onto the gear train.

reverse high clutch. If the drive lugs will not engage, then the outer race, inside the forward planet carrier, is not centered and cannot engage the end of the sun gear inside the input shell. Center the thrust bearing race and install the input shell.

29. Hold the gear train together and reinstall the forward part of the gear train assembly into the case (Fig. 9-103). To accomplish this, the front planet carrier internal splines must mesh with the output shaft splines, and the input shell sun gear must mesh with the reverse pinion gears.

30. If you are installing a new intermediate band, soak it in transmission fluid for 30 minutes prior to installation. Then, install the band through the front of the case by aligning the band ends with the clearance holes in the case (Fig. 9-104).

31. Position a new front pump gasket on the case; line up the bolt holes in the gasket with the bolt holes within the case.

32. Install two guide studs into two of the pump bolt holes, and reinstall the front pump stator support into the reverse high clutch. Remove the guide studs and install all but one of the front pump-to-case attaching bolts. Torque these bolts to 38 to 40 pounds-foot (Fig. 9-105). <u>Note</u>: If the pump bolts secure the converter housing to the case, install the housing first before installing and torquing the bolts.

33. Reinstall the input shaft (Fig. 9-106). Make certain you install the short, splined end of the input shaft toward the rear of the transmission.

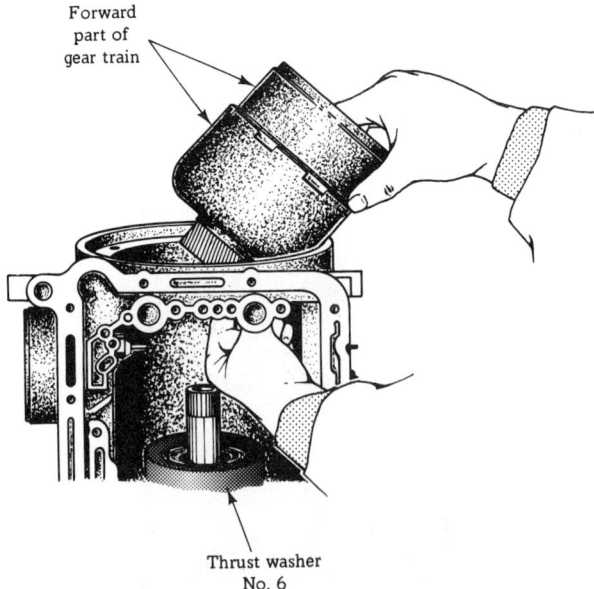

FIG. 9-103 Installation of the forward part of the gear train into the case.

Transmission Overhaul

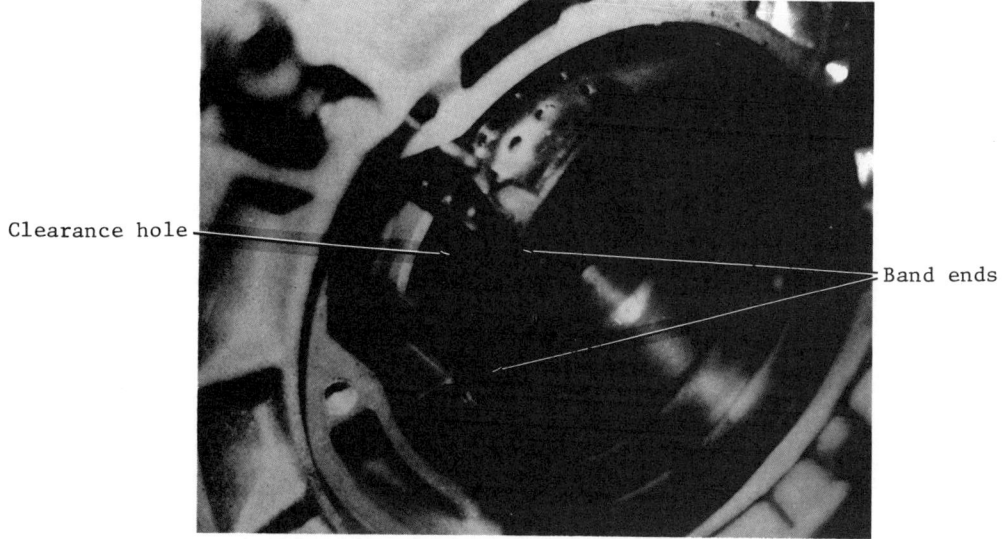

FIG. 9-104 Installation of the intermediate band into the case.

34. Install a dial indicator as shown in Fig. 9-4 and check the transmission end play. If the end play is not within specifications, either you used the wrong selective thrust washers, or one of the ten thrust washers is not in its correct position.

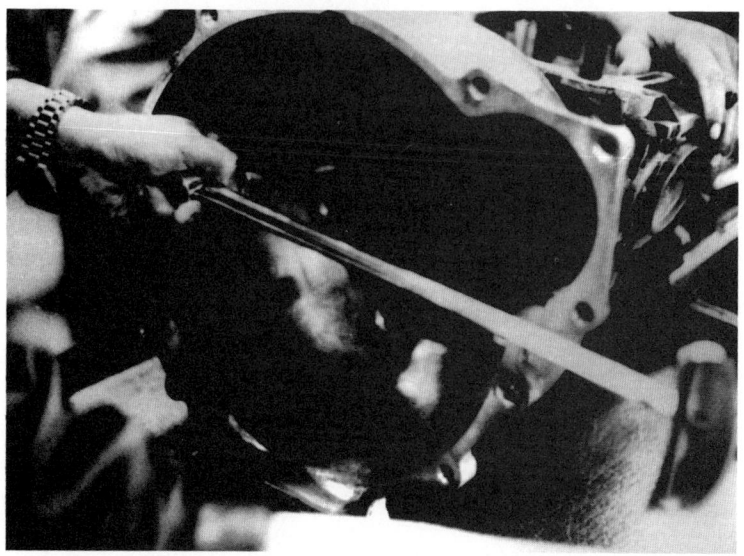

FIG. 9-105 Torquing the front pump-to-case attaching bolts.

FIG. 9-106 Installation of the input shaft.

35. Remove the dial indicator, and install the remaining front pump-to-case attaching bolt. Torque this bolt to 38 to 40 pounds-foot.

36. Position the converter housing onto the transmission case. Then install the five converter housing-to-case attaching bolts; torque these bolts to 28 to 40 pounds-foot.

37. Install the intermediate and low reverse band adjusting screws with new lock nuts into the case. Reinstall the struts for each band (Fig. 9-107).

38. Adjust the intermediate band by torquing the adjustment screw to 10 pounds-foot (Fig. 9-108). Then, back off the adjusting screw 1-3/4

FIG. 9-107 The locations of the band-adjusting screw and band struts after installation.

Transmission Overhaul

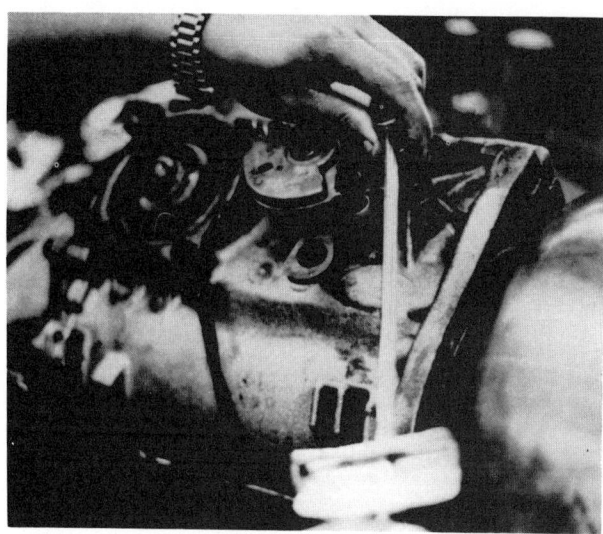

FIG. 9-108 Adjusting the intermediate band.

turns. While holding this adjustment screw from turning, torque the new locknut to 35 to 45 pounds-foot.

39. Adjust the low-reverse band by torquing the adjustment screw to 10 pounds-foot (Fig. 9-109). Then, back off the adjusting screw exactly 3 full turns. While holding this adjustment screw from turning, torque the new locknut to 35 to 45 pounds-foot.

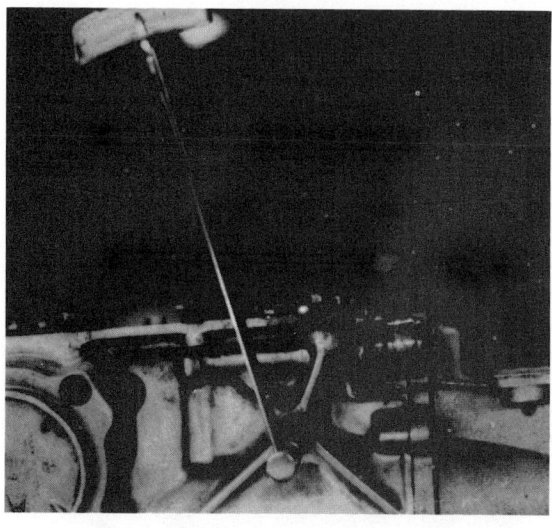

FIG. 9-109 Adjusting the low-reverse band.

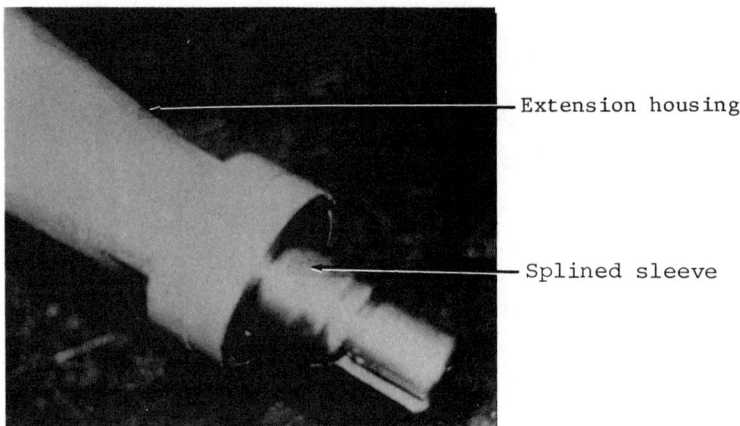

FIG. 9-110 A splined sleeve installed on the output shaft.

40. Install a slip yoke or splined sleeve on the output shaft (Fig. 9-110). Turn the input and output shafts in both directions to check for free rotation of the gear train.
41. Shift the manual lever at the side of the transmission case into the P detent position. Reinstall the valve body onto the case. Position the inner downshift lever between the downshift lever stop and the downshift valve. Make sure the two lands on the ends of the manual valve engage the actuating pin on the manual detent lever. Install seven valve body-to-case bolts, but do not torque these bolts at this time.

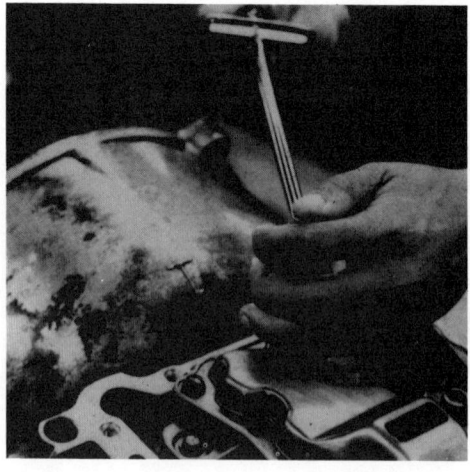

FIG. 9-111 Torquing the valve body-to-case attaching bolts.

Transmission Overhaul

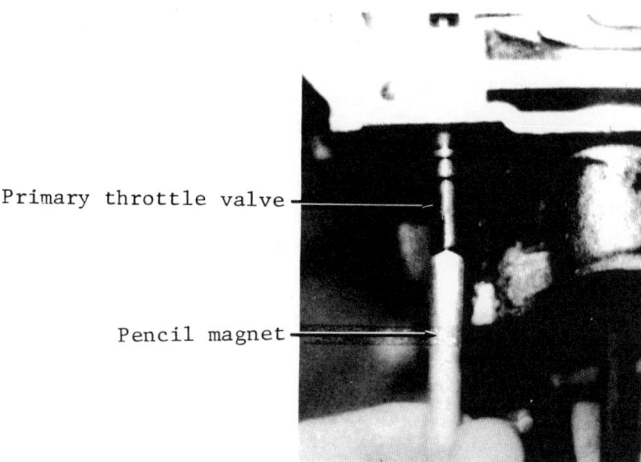

FIG. 9-112 Installation of the primary throttle valve.

42. Position the detent spring on the lower valve body and install the spring-to-case bolt finger tight.
43. While holding the detent spring roller in the center of the manual detent lever, install the detent spring-to-lower valve-body bolt. Torque this bolt to 80 to 120 pounds-inch.
44. Torque all the control valve body-to-case attaching bolts to 80 to 120 pounds-inch (Fig. 9-111).
45. Position a new pan gasket on the transmission case; install the fluid pan and its attaching bolts. Torque these bolts to 12 to 16 pounds-foot.
46. Install the primary throttle valve into the transmission case (Fig. 9-112).
47. Install the vacuum unit, gasket, and control rod into the case. Torque the vacuum unit to 28 to 40 pounds-foot.

CHECK-UP QUESTIONS

    The questions listed below will assist you in determining how well you remember the material contained in this section. Read each question carefully before adding the word or words necessary to complete the sentence. If you can't complete the sentence, review that portion of the section that covers the question.

1. Before disassembling an automatic transmission, the mechanic should _____ _____ the case thoroughly.

2. Always install new _____, _____ _____, and _____ when rebuilding the transmission.
3. Do not dry parts with a rag; instead use _____ _____.
4. End play of the transmission should be checked with a _____ _____.
5. The end play of a C-4 transmission should be _____ inch.
6. Nos. _____ and _____ thrust washers control end play of the C-4 transmission.
7. To remove or replace the intermediate band, align the band ends with the _____ _____ cut into the case.
8. Remove or replace the forward part of the gear train as an _____.
9. The location of the No. 5 thrust washer is on the _____ _____.
10. A _____ _____ secures the reverse ring gear to the output shaft.
11. A _____ _____ secures the governor distributor to the output shaft.
12. The location of the No. 10 thrust washer is on the _____ _____.
13. The low reverse servo piston has a _____ seal.
14. The two front pump gears have a mark on the side of each gear where the gear teeth are _____.
15. The two large sealing rings fit in the oil ring grooves toward the _____ of the stator support.
16. To remove the reverse high piston spring retainer snap ring use a _____ tool.
17. Remove the reverse high clutch drum bushing with a _____ _____.
18. Soak all the new composition plates in clean transmission fluid for _____ _____ before installation.
19. The reverse high clutch plate clearance should be _____ inch.
20. The forward clutch plate clearance should be _____ inch.
21. Deglaze the band surface of the low reverse drum by sanding it from front to back using _____ grit sandpaper.
22. The one-way clutch has _____ rollers and springs.
23. When removing the oil screen from the valve body, be careful not to loose the _____ _____ valve.
24. Use _____ to hold all thrust washers in place.
25. When assembling the transmission, _____ all fasteners to factory specifications.

Transmission Overhaul

26. The governor housing-to-distributor bolt torque is _____ _____.
27. To center the rollers and springs of the one-way clutch, turn the inner race _____.
28. The location of the No. 8 thrust washer is on the _____ _____ _____.
29. Soak a new intermediate band for _____ minutes prior to installation.
30. To adjust the low reverse band, torque the adjusting screw to _____ pounds-foot and back it off _____ turns.

Appendix

SECTION 1

1. recycle
2. vortex
3. drain, plug
4. bench, checks
5. pulleys
6. pressure
7. electrical, kickdown
8. internal, pressures
9. four
10. steam
11. inside
12. air
13. diaphragm
14. incline, plane
15. bearing
16. blind
17. single-post
18. two, four
19. telescoping, ratchet-type
20. wet
21. one
22. Heli-coil, repair, kit
23. smaller, lighter
24. air, ratchet
25. pulling-type, impact-type
26. snap, rings
27. pencil, magnet

APPENDIX

SECTION 2

1. torque, wrench
2. torquing
3. clicker, torque
4. torque, increased
5. micrometer
6. micrometer, 40
7. .001
8. ratchet, micrometer
9. bent, feeler, gauge
10. tapered, gauge
11. dial, indicator
12. dial
13. fastener
14. screw
15. pitch
16. bolt
17. flat
18. cotter, slotted
19. key, ball, pin
20. splines

SECTION 3

1. logical, specific, thorough
2. level, condition
3. suction, gun
4. vacuum-operated, modulator
5. cardboard, paper
6. black, light
7. fluid, cooler
8. manual, valve
9. shift, points and control, pressure
10. vacuum, pump and gauge, assembly
11. wide-open, throttle
12. road, test
13. band, clutch
14. clutch-, band-application
15. pressure
16. 1.5
17. high
18. modulator, selective
19. diagnosis
20. air
21. overhaul
22. rubber, metal
23. stethoscope
24. dynamometer
25. kickdown

SECTION 4

1. oxidized
2. owner's, manual, service, manual
3. pan, converter
4. dipstick
5. rear
6. wear
7. dimple
8. lint
9. paper
10. petroleum, jelly, grease
11. torque, wrench, factory, specifications
12. 6, o'clock
13. pump
14. stop
15. shop, manual

SECTION 5

1. planetary, member
2. tight
3. lining
4. external
5. selective
6. tightens or loads
7. Allen, eight-point
8. specifications
9. fluid, remove, pan
10. control, input
11. manual, valve
12. detent, pawl, shift, gate
13. throttle, kickdown or detent
14. service or transmission, manual
15. cable

SECTION 6

1. seal
2. downward
3. flywheel
4. reassembling
5. short, circuit
6. C-clamp
7. bushing
8. inward
9. pump, rotor, drive, gear
10. bushing
11. wheel, bearing, grease
12. mechanical, linkages

APPENDIX

SECTION 7

1. expensive
2. flush
3. before
4. true
5. two
6. fingernail
7. polish, crocus
8. replace
9. counterclockwise
10. thrust washers, bearings
11. welds
12. flushing, machine
13. drain, plug
14. ring, gear
15. pressure, flow
16. Prussian, blue, indelible, pen
17. passages
18. new, pump
19. straight, edge
20. aligning, tool

SECTION 8

1. housing or framework
2. varnish, shellac
3. low-pressure, compressed, air
4. replace
5. Heli-coil, insert
6. bushing
7. cut
8. shifter-shaft, seal
9. fluid, leak
10. epoxy, cement
11. extension, housing
12. extension, housing, bushing
13. inward
14. input, shaft
15. fine, wire
16. clutch, piston
17. drum, piston
18. feeler, gauge
19. 40-60
20. one, direction
21. accumulator
22. governor
23. square, edges
24. shafts
25. thrust, washers
26. bearing
27. bushing
28. chisel, remover
29. valve body
30. rag
31. 120-180
32. 30
33. scrape
34. Teflon
35. toward

SECTION 9

1. steam, clean
2. gaskets, sealing, rings, seals
3. compressed, air
4. dial, indicator
5. .008-.042
6. 1,2
7. clearance, hole
8. assembly
9. input, shell
10. snap, ring
11. snap, ring
12. parking, gear
13. bonded
14. chamfered
15. front
16. spring-compressor
17. cape or bushing, chisel
18. 30, minutes
19. .050-.071
20. .025-.050
21. 40-60
22. 12
23. throttle-pressure, limit
24. Vaseline
25. torque
26. 12-20, pounds-foot
27. clockwise
28. low-reverse, drum
29. 30
30. 10, 3

INDEX

Abrasive blaster, 20, 26, 27, 29
Accumulator, 241-243
Accumulator piston installer, 59
Air agitation system, 26
Air checks, 118-120
Air manifold, 26
Air-powered tools, 50
Air ratchet, 51
Air-supply gauge, 13
Allen head screw, 76
Arbor press, 34

Ball-peen hammer, 56
Balls, 81
Band adjustment, 150-156
Bands, 258, 259
Barrier filter, 26
Beam-type torque wrenches, 64-66
Bearings, 250, 251
Bell housing bolts, 187
Bent feeler gauge, 71
Black light test, 97
Bolts, 78
Brake drum, 238
Bushing chisel, 36
Bushing equipment, 35-38
Bushing puller, 35, 36

Bushing replacement, 225
Bushings, 251-254

Carriers, 245, 246
Cartridge filter, 26
Case, 223-230, 276, 277
  porosity, 229
Clicker-type wrenches, 64
Clutch drum, 32, 233-238
Clutch plate, 257, 258
Clutch spring compressor, 29-34
Cold tanks, 223
Converter, 195
  cover, 190
  end-play gauge, 5, 6
  fluid, 143, 144, 145, 146, 147
Converter hub, 198
  inspections, 198
Converter mount bolt, 187-188
Cooler leakage test, 98
Cooling systems, 262-269
  air cooled, 262, 264
  water cooled, 262, 263
Cotter pins, 80, 81
Crescent block, 215, 216
Cross member, 187

Delivery hose, 14
Delivery pressure gauge, 14
Delivery volume gauge, 14
Detent pawl, 14
Dial indicators, 73, 74
Dipstick, 88
Disc-brake assembly, 16
Disc-brake control valve, 16
Downshift linkage, 168, 169
Drain plug, 7, 135, 187, 188, 208, 209
Drive-range test, 108, 109
Driveshaft, 17
Drive stud inspection, 196
Drive two test, 109
Drums, 233-238
  brake, 238
  clutch, 233-238
Drying shelf, 26
Dual-port modulator system, 101, 102
Dynamometer, 9-12
  test, 122-131

Electrical kickdown switches, 169
Electrical transmission circuits, 103
End-play checks, 274-276
Extension housing, 230-233, 276, 277, 316, 317, 318
  seal, 185
Exhaust gas recirculation (EGR) system, 101
External snap ring, 83, 84

Fastening devices, 75
  bolts, 78
  cotter pins, 80
  keys, 81
  machine screws, 75-77
  nuts, 79
  snap rings, 82, 84
  splines, 82
  washers, 80
Feeler gauges, 71-73
  bent gauge, 71
  round gauge, 72
  tapered gauge, 71
Feeler gauge strip, 216, 217
Filter, 139, 140
Flange inspection, 196
Flat washers, 80

Fluid condition, 90
Fluid level, 87, 88
Flushing, 205-208
Flushing machine, 2
Forward carrier, 307
Forward clutch, 279-281, 301-305, 331, 332
Forward planetary gear train, 305-307
Friction clutch plates, 258
Front pump, 293
Front seal, 181-184

Gasket, 262, 277
Gear pullers, 53, 54, 55
Gearshift linkage, 161-163
Gearshift rod, 191
Governor, 243-245, 312-315, 326
Governor distributor, 285, 286, 325, 326

Hard parts, 222
  accumulator, 241-243
  bearings, 250, 251
  bushings, 251, 254
  drums, 233-238
  extension housing, 230-233
  governor, 243-245
  one-way clutches, 238-241
  planetary gear trains, 245-247
  servos, 241-243
  thrust washers, 248-250
  transmission case, 223-230
  transmission shaft, 247-248
  valve bodies, 254-256
Heli-coil, 48
  thread repair kit, 48
Hexagonal head screw, 76
Hex nut, 79
Hoist tests, 121, 122
Holding fixture, 46-48
Hot tank, 223
Housing seal, 184, 185
Hub inspection, 198
Hydraulic gauge, 14, 86
  sets, 17, 18
Hydraulic jacks, 40-42
Hydraulic lift, 38
Hydraulic pump, 212-217

Idle speed checks, 98, 99
Impact wrench, 51

INDEX

Impeller, 199
Inner race, 315
Input tachometer, 12
Intermediate servo, 291, 292
Internal snap-ring, 84

Jacks, 40-42, 44, 45
Jet cleaner, 20, 22, 23, 223

Keys, 81
Kickdown linkage, 165, 168, 169
Kickdown valves, 99, 165

Lathe-cut ring, 260
Leakage test, 203, 204
Linkage, 161
  accelerator pedal linkage, 164, 165
  adjustments, 161-163
  checks, 99
  downshift linkage, 161, 168, 169
  gearshift linkage, 161-163
  kickdown valve linkage, 164, 165, 168
  manual valve linkage, 161-163
  throttle valve linkage, 164-166
Lip seal ring, 260, 261
Load application gauge, 13
Locking dog, 39
Lock washer, 80
Low reverse servo, 293
Lug inspection, 196

Machine screws, 75-78
Manual lever, 289, 290
Manual low checks, 109
Manual valve, 161-163
Measuring devices, 63
  dial indicators, 73-75
  feeler gauges, 71-73
  micrometers, 68-71
  torque wrenches, 63-68
Metal clad seals, 261, 262
Metal sealing rings, 259, 260
Micrometer (mike), 68-71
Micrometer wrenches, 64
Mineral spirits, 4
Modulator wrench, 58, 59
Mount assembly, 12
Mount plate, 12

Negative post, 189
Neutral safety switch, 190
Noise detection, 121
Nuts, 79
  hex, 79
  slotted hex, 79

Oil reserve tank, 17
Oil screen, 318, 319
Oil seal, 231-233
One-way clutches, 200, 238-241, 315, 316
Operational tests, 103
  hydraulic pressure test, 110-116
  road test, 106-110
  stall test, 103-106
O-ring, 260, 291
Outer race, 315
Overrunning clutches, 238-241

Parts washers, 20, 23, 25
Pawl, 162
Pencil magnet, 58
Phillips head screw, 76
Pins, 81
Pitch, 78
Planetary gear trains, 245-247
Pneumatic tools, 50
Porosity, 229
Positive post, 189
Powerglide puller, 54
Pressure tests, 110-116
Pullers, 53
  gear pullers, 53, 54
  powerglide puller, 54
  pump puller, 56, 57
  seal puller, 56
Pump puller, 56, 57

Rear mount, 187, 188
Remover plate, 54
Retainer, 315
Retainer press plate, 34
Retaining rings, 82
Reverse band, 316, 328
Reverse brake drum, 310
Reverse carrier, 310, 331
Reverse high clutch, 297-301
Reverse planetary gear train, 309

Reverse planet carrier, 282
Reverse ring gear, 283
Reverse tests, 118
Ring gears, 245, 246
Road tests, 106-110
Roller-type clutch, 240
Rotor tip, 217
Round feeler gauge, 72
Rubber sealing rings, 260, 261

Safety solvent, 4
Safety stands, 42
Safety-type parts washer, 20, 23, 25
Scribe, 58
Sealing washer, 80
Seal protectors, 59
Seal puller, 56
Seal replacement, 227-229
Servo piston, 59, 287
Servos, 241-243, 287, 288, 291
Shifter shaft seal, 227
Shift gate, 162
Single post hoist, 38
Slide-hammer seal puller, 56
Slotted head screw, 76
Slotted hex nut, 79
Snap ring pliers, 57, 58
Snap rings, 57, 82, 83
   external, 83
   internal, 84
   Truarc, 84
Soft parts, 222, 257
   bands, 258
   clutch plates, 257, 258
   gaskets, 262
   metal-clad seals, 261, 262
   metal sealing rings, 259, 260
   rubber sealing rings, 260, 261
   Teflon sealing rings, 259, 260
Solvent pump, 4
Solvent tank, 4
Speedometer gear installer, 59
Splines, 82, 247
Spring-steel washer, 80
Stall tests, 103-106
Satrter assembly, 189
Starter ring-gear, 210
Stator assembly, 198, 238
Stator one-way clutch tests, 200
Stator-to-impeller interference tests, 198-200

Stator-to-turbine interference test, 200
Steam cleaner, 20, 21
Sump, 4
Sun gear, 245, 246, 307, 308

Tachometer, 12, 86
Tapered feeler gauge, 71
Teflon sealing rings, 259, 260
Thread repair, 225
Throttle lever, 190
Throttle rod, 189, 191
Throttle valve, 166, 167
Thrust washer, 248-250, 278, 279, 330
Torque converter, 195
   drain plug, 7
   flushers, 1, 2
Torque shaft, 191
Torque wrenches, 63-68
Transmission case, 223-230
Transmission controlled spark (TCS) system, 103
Transmission installation, 185-193
Transmission removal, 172-181
Transmission shafts, 247, 248
Trip pin, 40
Truarc pliers, 58
Truarc retaining ring, 84
Twin post lift (hoist), 39
Twist drill, 49
Turbine end-play test, 201

Universal drive assembly, 3
Universal emd-play gauge, 6, 7

Vacuum gauge, 14, 86
Vacuum pump, 15, 18, 19, 86, 87
Vacuum system check, 100-102
Valve bodies, 254-256, 318-322
Varsol, 4
Vortex flow, 5
Vortex method, 4

Washers, 80
Wet benches, 45
Wiring plug, 190
Woodruff keys, 80
Work benches, 45, 46